MECHANISMS OF CELL DEATH

THE SECOND ANNUAL CONFERENCE OF THE CELL DEATH SOCIETY

ANNALS OF THE NEW YORK ACADEMY OF SCIENCES

Volume 887

EDITORIAL STAFF

Executive Editor
BARBARA M. GOLDMAN

Managing Editor
JUSTINE CULLINAN

Associate Editor
JOYCE HITCHCOCK

The New York Academy of Sciences
2 East 63rd Street
New York, New York 10021

ANNALS OF THE NEW YORK ACADEMY OF SCIENCES
Volume 887

MECHANISMS OF CELL DEATH
THE SECOND ANNUAL CONFERENCE OF THE CELL DEATH SOCIETY

Edited by Zahra Zakeri, Richard A. Lockshin,
and Luis Benítez-Bribiesca

The New York Academy of Sciences
New York, New York
1999

The cover of the paper-bound edition of this volume shows the logo of the Cell Death Society.

Library of Congress Cataloging-in-Publication Data

Mechanisms of cell death : the second annual conference of the Cell Death Society / editors, Zahra Zakeri, Richard A. Lockshin, Luis Benitez-Bribiesca.

p. cm. — (Annals of the New York Academy of Sciences, v. 887)

Includes bibliographical references and index.

ISBN 1-57331-240-1 (alk. paper). — ISBN 1-57331-241-X (alk. paper)

1. Cell death Congressus. 2. Apotosis Congresses. I. Zakeri, Zahra, 1952– . II. Lockshin, R A. (Richard A.) III. Benitez -Bribiesca, Luis. IV. Series.

Q11 .N5 vol. 887

[QH671]

500 s—dc21

[571.9' 39]

99-42519

CIP

K-M Research/CCP

Printed in the United States of America

ISBN 1-57331-240-1 (cloth)

ISBN 1-57331-241-X (paper)

ISSN 0077-8923

ANNALS OF THE NEW YORK ACADEMY OF SCIENCES

Volume 887

MECHANISMS OF CELL DEATH

THE SECOND ANNUAL CONFERENCE OF THE CELL DEATH SOCIETY[a]

Conference Organizers
ZAHRA ZAKERI AND RICHARD A. LOCKSHIN

Advisory Board Members
RAYMOND BIRGE, MICHAEL HENGARTNER, AND BARBARA OSBORNE

Editors
ZAHRA ZAKERI, RICHARD A. LOCKSHIN, AND LUIS BENÍTEZ-BRIBIESCA

CONTENTS

[a]This volume is the result of a conference entitled **Mechanisms of Cell Death** held by the Cell Death Society at the Queens College Campus of the City University of New York on July 17–18, 1998.

Part II. Programmed (Developmental) Cell Death

Part III. Cell Death in Pathological and Clinical Situations

Financial assistance was received from:

- Calbiochem-Novabiochem Corporation
- National Institutes of Health (National Institutes of Aging)
- Oncogene Science
- Queens College of CUNY
- St. John's University, Graduate School of Arts and Sciences
- Upstate Biotechnology, Inc.

Preface

RICHARD A. LOCKSHIN[a]

Department of Biological Sciences, St. John's University, 8000 Utopia Parkway, Jamaica, New York 11439, USA

This book derives from a meeting of the Cell Death Society held at Queens College, Flushing, New York, in the summer of 1998. Several themes developed at that meeting, most of which hovered around the issue that the context (environment, lineage, and history) in which a cell finds itself is a critical factor in determining whether or not it will undergo apoptosis. Following these arguments, we solicited the manuscripts presented in this volume. The contributing authors responded with a series of articles that expand this theme and take it in new directions. The eighteen chapters that we present subdivide into three basic groups. The first block, **Mechanisms and General Considerations**, begins with an appropriate warning by Huber Warner that control of apoptosis is not a simple issue to be achieved as a generalized process: pathologies result from both excessive and insufficient apoptosis, and the therapeutic goal is to target control to specific tissues and times. This point is amply recapitulated in the final section.

Angermüller *et al.* and Jacotot *et al.* focus on the mitochondrion, with Jacotot *et al.* emphasizing that many different signals may converge on the mitochondrion, and Angermüller *et al.* noting that the mitochondrial changes may originate from physical rupture of the mitochondrial membrane. Kagan *et al.* report that the mitochondrial antioxidant coenzyme Q_{10}, which is considered to be antiapoptotic, both protects mitochondrial membrane potential and prevents apoptosis. However, it is antiapoptotic only in nonproliferating cells, and then only in specific situations. Furthermore, its ability to stabilize mitochondrial membrane potential does not guarantee survival of the cell in all circumstances. These authors, like Warner, feel that in the longer run the mitochondrion may be a target for therapy.

Walker *et al.* and Liu *et al.* take a different tack, noting that many of the preferred criteria for apoptosis, and therefore the targets for research and criteria for evaluating the success of experimental manipulation, are secondary to the initial event, which is a high-level fragmentation of DNA that depends neither on caspases nor on the endonucleases presumed to generate the DNA ladder. Newell *et al.* and Bernassola, Rossi, and Melino raise other issues: Newell *et al.* emphasize that, as incipient apoptosis alters the metabolism of cells, the amount of glucose available to and used by cells affects such primary aspects of apoptosis as the amount of Fas expressed. According to Bernassola, Rossi, and Melino, nitric oxide, generally considered to be a signaling molecule, can directly inactivate (by nitrosylation) transglutaminase, and thereby inhibit either apoptosis or the morphological manifestation of apoptosis. The context in which apoptosis occurs strongly affects the incidence and our interpretation.

[a] 718-990-1854 (voice); 718-990-5958 (fax).
e-mail: Lockshin@stjohns.edu

The second section, **Programmed (Developmental) Cell Death**, is characterized by some surprising and intriguing generalizations and challenges. Liu and Hengartner begin by demonstrating that, like the caspase family and the bcl-2 family, the mechanism of phagocytic engulfment may be conserved across species. Hensey and Gauthier, using early *Xenopus* eggs, highlight the maturation of two separate apoptotic pathways. A maternally prepared pathway rests quiescent, even in badly damaged cells, until after the mid-blastula transition, then activates synchronously. The embryonic pathway appears at gastrulation. The maturation of these pathways provides an excellent opportunity to study the components of apoptosis.

It will be important to understand this sequence as it builds from presumably simple origins. Merion *et al.*, for instance, note that bone morphogenetic proteins can induce either chondrogenesis or cell death, without clear evidence as yet as to how the decision is made. The sequence to apoptosis is complex and dependent on many other communications among the cells.

Finally, in the third section, **Cell Death in Pathological and Clinical Situations**, seven chapters explore the complexity and utility of addressing apoptosis in pathology. Benitez-Bribiesca *et al.* would like an inexpensive, accurate, and rapid assay for apoptosis as a measure of the effectiveness of antineoplastic agents. The comet may be such an assay and, if properly used, could help an oncologist quickly evaluate the effectiveness of a treatment. Hager and Hanahan, using DNA chips, find that the primary markers of apoptosis are typically not affected by p53 mutations. Regulation of the vasculature, IGF-2, or function of bcl-X_L and Loh9, may be more likely to have an impact on the 50% of tumors that are p53-negative than direct attack on apoptosis machinery, since p53-negative cells can undergo apoptosis, but only at a much higher threshold of stress. Next, Kazzaz *et al.* and Mantell *et al.* both address the issue of whether or not hyperoxic lung damage is apoptotic. Though programmed and therefore distinct from necrosis, hyperoxic lung epithelium dies in a manner distinctly different from the image of classical apoptosis. An understanding of the distinctions and common features will be essential for dealing with the damage. For instance, caspase inhibitors may be less useful in this situation than in others. In inflammatory situations, the role of apoptosis is indirect but nevertheless profound: As Huber *et al.* indicate, cytomegalovirus infection generates an autoimmune T cell response to cardiac antigens liberated by virus-infected cells; the autoimmune response generates far more damage than the virus, and the damage can be limited by suppressing T cells. Mevorach postulates a similar mechanism, as massive induced apoptosis can result in apoptotic fragments that can serve as antigens, resulting in autoimmunological diseases such as lupus. Using a different tactic, as Gougeon and Montagnier emphasize, HIV can turn lymphocytes into effectors of apoptosis; anti-retroviral therapies strive to normalize lymphocyte apoptosis.

Thus, the theme that emerges from these several authors is that apoptosis remains a central issue in disease and health but that, to exploit our knowledge, we must be more subtle. It is no longer sufficient to say "apoptosis" as a shibboleth and claim that we have explained anything. We must begin the task of identifying how the signals differ from one cell type to another and, within the same cell types, from one circumstance to another. As Barbara Osborne so eloquently summarized the meeting, "Context is everything."

ACKNOWLEDGMENT

This research was supported in part by NIH Grants IR15GM/AG57614 to R.A.L., R13A615846 to Zahra Zakeri and R.A.L., and 6M568821 to Timothy Carter (St. John's).

Apoptosis: A Two-edged Sword in Aging

HUBER R. WARNER[a]

Biology of Aging Program, Gateway Building, Suite 2C231, National Institute on Aging, National Institutes of Health, Bethesda, Maryland 20892, USA

ABSTRACT: Here we summarize briefly what is known about both the positive and negative impacts of apoptosis during aging in mammalian systems and also update an earlier review. It is important to understand both of these impacts to devise useful interventions. Such interventions include both physiological and molecular approaches, including transgenic interventions. The critical roles of the mitochondria in both generating reactive oxygen species, and in initiating apoptosis are recognized, suggesting that maintaining mitochondrial function could be an important therapeutic goal, especially in post-mitotic tissues. In contrast, the ability to eliminate unwanted, damaged and dysfunctional cells through apoptosis has anti-aging implications in mitotic tissues.

INTRODUCTION

Cell death in biological systems can be separated into two distinct processes: necrotic death and apoptotic death.[1,2] Whereas the former is usually thought to be associated with massive cell damage, such as occurs in stroke and myocardial infarction, the latter is a genetically programmed process often occurring as part of natural physiological processes.[3] It would be easy to assume that all cell death has detrimental effects during aging, but this appears not to be so. For example, apoptosis plays a role in the destruction of unwanted cells not only during development, but also in the negative selection of thymocytes and lymphocytes,[4] the attenuation of autoimmunity,[5] and as a factor in maintaining proliferative homeostasis.[6–8] This review summarizes briefly what is known about both the positive and negative impact of apoptosis during aging in mammalian systems and also serves to update a similar review published earlier.[9]

WHAT IS APOPTOSIS?

Apoptosis refers to a process of rapid, but genetically controlled death of a cell in response to one of a variety of signals, through activation of cell death-inducing proteins. Two genes coding for these proteins, *ced3* and *ced4*, were first identified in the nematode *Caenorhabditis elegans,*[10] but homologs for these genes have now been identified in mammalian systems. However, the mammalian systems involve many additional factors, providing the opportunity for much more exquisite control of the process. For example, instead of the single inhibitor of apoptosis found in *C.*

[a]301-496-6402 (voice); 301-402-0010 (fax).
e-mail: warnerh@exmur.nia.nih.gov

elegans and coded by the *ced9* gene,[10] mammals contain a family of ced9-like proteins including both positive and negative effectors e.g. bax and bcl-2.[11] Also, *C. elegans* contains a single apoptotic protease, the product of the *ced3* gene, whereas at least 11 ced3-like proteases, called caspases, have so far been identified in human cells.[12] These caspases are mainly found in proenzyme forms, and provide a protease cascade upon activation by an appropriate signal. Each caspase has specific targets which include other procaspases[12] and members of the bcl-2 family,[13] as well as unphosphorylated Rb protein,[14] poly (ADP ribose) polymerase,[15] lamin,[16] and actin.[17] These proteolytic reactions apparently prevent DNA replication and repair, and prepare the cell for many of the morphological changes that accompany apoptosis, including nuclear DNA degradation.[18,19]

ROLE OF APOPTOSIS IN REPAIR AND CANCER PREVENTION

Human cells are constantly under assault from a variety of sources, including reactive oxygen species (ROS) produced in the mitochondria during oxidative phosphorylation, and during exposure to radiation and environmental mutagens. These ROS include superoxide anion, hydrogen peroxide and hydroxyl radical. All of these can damage both nuclear and mitochondrial DNA, leading to altered bases, single-strand breaks, and double-strand breaks in the DNA. Although cells contain robust systems to both neutralize ROS and to repair DNA, when damage is excessive cell replacement may be a preferred alternative to DNA repair. This requires that the damaged cell first die by apoptosis, and then be replaced by division of a nearby functional cell. The checkpoint protein p53 is activated by DNA damage, and is presumably involved in the decision whether to stop replication and repair the DNA, or die by apoptosis.[20] The details of how this decision is made are not known.

Caloric restriction retards both aging and tumor induction;[21] the latter could be due to up-regulation of apoptosis to eliminate pre-neoplastic cells.[22,23] Holt *et al.*[24] have shown that caloric restriction also stimulates apoptosis in aging rat intestine. Warner *et al.*[25] have hypothesized that retardation of aging could also be due to up-regulation of apoptosis. One cause of aging may be slow accumulation of damaged and/or senescent cells, especially if functional changes in these cells compromise overall tissue function. Senescent cells have been detected in human tissue by Dimri *et al.*[26] Wang *et al.*[27,28] have shown that senescent fibroblasts do not die readily in response to serum withdrawal, suggesting that they are inherently resistant to apoptosis. Senescent fibroblasts secrete matrix-degrading enzymes,[29] so if they do actually accumulate with increasing age, this could induce matrix degradation and loss of proliferative homeostasis in nearby cells,[30] possibly leading to cancer. Masson *et al.*[31] have shown that stromelysin-3 metalloproteinase, which degrades matrix proteins, does promote epithelial cell malignancy.

APOPTOSIS AND AUTOIMMUNITY

Apoptosis plays a critical role in the immune system, and it has been estimated that as many as 95% of the T-lymphocytes produced are eliminated by apoptosis,[32] presumably because they recognize self-antigens. Thus, apoptosis is a critical element in defense against autoimmune disease, and the importance of this process during aging has recently been elucidated. Two mutations in mice known as *lpr* (lymphoproliferation) and *gld* (generalized lymphoproliferative disease) result in autoimmune disease; when present in an MRL background, *lpr* mice are also short-lived. The *lpr* and *gld* genes code for Fas and Fas ligand proteins, respectively.[33,34] The interaction of Fas ligand with the Fas receptor protein normally induces apoptosis of T-lymphocytes; thus both *lpr* and *gld* mice appear to be deficient in eliminating lymphocytes recognizing self antigens. Caloric restriction increases the rate of apoptosis of T-lymphocytes in the MRL/*lpr* mouse, and also extends its life span,[35] suggesting an association among autoimmune disease, apoptosis, and life span in this mouse strain.

A Fas gene homolog has been identified in humans, and mutations in this gene lead to abnormal lymphoproliferation and autoimmunity.[36,37] It is not known whether these mutations affect human longevity, but it has been shown that bcl-2 levels are elevated in circulating T-lymphocytes of patients with systemic lupus erythematosus, suggesting that apoptosis in these lymphocytes is suppressed.[38]

Total thymocyte count and Fas expression on T-lymphocytes decrease with increasing age in mice, and transgenic overexpression of the Fas gene prevents these age-related losses.[39] Maintenance of Fas-induced apoptosis in these mice suggests that T-cell senescence with increasing age is associated with defective apoptosis.

APOPTOSIS AND OSTEOPOROSIS

Osteoporosis occurs when the appropriate balance between bone deposition by osteoblasts and bone resorption by osteoclasts is not maintained. This balance could be maintained and regulated not only by differential production of these cells by differentiation of stem cells, but also by selective apoptosis of each kind of cell. Hughes *et al.*[40] have reported that estrogen induces apoptosis of osteoclasts, but not osteoblasts, providing one explanation for the association between menopause and osteoporosis. It would be useful to understand how apoptosis of osteoclasts is regulated in order to develop interventions to prevent or delay age-related bone loss by stimulating osteoclast apoptosis. It is known that caspase inhibitors extend the life span of osteoclasts in culture upon withdrawal of serum,[41] indicating that caspases are involved in the regulation of osteoclast survival. One possible strategy to reduce bone loss could be induction of apoptosis of osteoclasts by bisphosphonates.[42]

APOPTOSIS IN POST-MITOTIC CELLS

In post-mitotic tissues cell replacement is not possible in response to damage, so the possible outcomes are either complete repair, continued survival of a dysfunctional cell, or apoptosis. The final outcome may eventually be development of significant age-related pathology.

The major source of most cell damage is assumed to be oxidative stress due to mitochondrial production of ROS. These ROS presumably subject the mitochondrion itself to substantial oxidative damage of not only DNA, but also proteins and lipids. Particularly relevant in understanding the relationships among damage, aging, and apoptosis may be oxidation of enzymes critical in ATP production, and maintenance of mitochondrial membrane integrity. Two mitochondrial proteins that have been shown to be particularly sensitive to oxidative stress are aconitase and adenine nucleotide translocase.[43,44] Their level of oxidative alteration increases with age in house flies, suggesting that ATP production could become limiting with increasing age owing to limiting activity of either one or both of these enzymes. The essential role of adenine nucleotide translocase has been examined in mice in which this gene has been knocked out.[45] These mice exhibit muscle mitochondrial myopathy, ragged red muscle fibers, defects in respiration and premature death. Presumably this pathology is the result of a deficiency in the rate of supply of ATP to the cytoplasm in muscle cells.

Maintenance of mitochondrial membrane integrity is also compromised by oxidative stress. Oxidative stress can induce the mitochondrial permeability transition, mitochondrial depolarization, decrease in ATP concentration and eventually cell death.[46] Because mitochondria lack catalase activity, hydrogen peroxide may build up in mitochondria as the result of superoxide dismutase activity, and this hydrogen peroxide can react with unsaturated lipids in the mitochondrial membrane. Combined with the possible limiting production of ATP as discussed above, the membrane may become permeable to cytochrome c, which can leak out of the mitochondria and initiate apoptosis.[3] One way that the bcl-2 protein inhibits apoptosis is to block the release of cytochrome c from the mitochondria in response to various apoptotic signals.[47] Because of the critical role of mitochondria in energy production, it is particularly intriguing that mitochondrial damage may be one trigger for apoptosis. Green and Reed[48] have suggested that the permeability transition appears to be the mastermind that orchestrates apoptosis.

Neurodegenerative Disease

That cell death is important in neurodegenerative disease is unquestioned. How and why neurons die, particularly why specific neurons die, are important unanswered questions. Evidence is now accumulating that apoptosis is associated with loss of neurons in Alzheimer's disease,[49–51] Huntington's disease,[51,52] and amyotrophic lateral sclerosis.[52]

The lines of evidence that apoptosis is involved in the above neurodegenerative diseases include demonstrations that β-amyloid induces apoptosis in cultured neurons,[50] and either caspase inhibition or overexpression of the *bcl-2* gene slows onset of ALS in transgenic mouse models.[53,54] DNA degradation and other morphological features characteristic of apoptosis are also observed.[51,52] Similar observations suggest that at least some neurons die by apoptosis following ischemia as well.[55–57] Two recent short reviews by Barinaga[58,59] discuss the evidence for the conclusion that apoptosis plays a role in neurodegenerative disease. These findings provide optimism that strategies can be developed to prevent, or at least slow, the loss of neurons through the use of peptide inhibitors of caspases.[56]

That oxidative stress is a causal factor in neuron death is supported by several lines of evidence. Over-expression of the human gene for the cytoplasmic form of superoxide dismutase (*SOD1*) in mice reduces the size of the damaged brain area following cerebral ischemia.[60] Also, over-expression of this gene in motor neurons in *Drosophila* extends the maximum life span, suggesting that death of motor neurons induced by oxidative damage could be a causative factor in aging in this organism.[61]

Retinal Degeneration

Evidence that inherited retinal degeneration may also occur by apoptosis of photoreceptor cells has been recently reviewed by Travis.[62] Apoptosis appears to be routinely, but gradually, induced by the synthesis of mutant proteins, particularly rhodopsin.[63] Other mutant proteins unique to photoreceptor cells also induce apoptosis, which may explain the identification of so many genetic risk factors for retinitis pigmentosa. In every case, mouse models carrying the defective genes provide morphological and biochemical evidence that the deaths occur by apoptosis.[62] Over-expression of the baculovirus inhibitor of caspases, p35, prevents blindness in *Drosophila* retinal degeneration mutants carrying an altered rhodopsin gene.[64] However, it is not clear why the presence of these mutant proteins induces apoptosis, although it may be relevant that photoreceptors are highly aerobic cells.

Cardiovascular Disease and Frailty

Cardiac and skeletal muscle are two other tissue types which have, at best, a limited potential for replacement of damaged cells. They are also highly aerobic, suggesting they may be particularly vulnerable to oxidative stress. For example, mice born lacking manganese superoxide dismutase, the mitochondrial form of the enzyme, die very quickly after birth with cardiomyopathy.[65] These mice also had a variety of other pathologies, including reductions in skeletal muscle, and central nervous system degeneration.[66] Even heteozygous mice carrying one good gene for this enzyme show definite signs of oxidative stress,[67] although these mice are quite viable, no overt pathology has yet been reported, and it is not known whether their life span is normal.

The major age-related disease leading to mortality in humans is still cardiovascular disease, and death of cardiac myocytes is a critical factor following a myocardial infarction. Apoptotic cardiomyocytes are localized to the borders of infarcted myocardium in human autopsy material, in contrast to necrosis, which occurs primarily within the infarcted area.[68] Several laboratories have reported similar evidence that apoptosis is associated with cardiac pathology in both humans and rats,[9] and the induction of Fas may be one of the apoptotic initiators in rats.[69] The general consensus is that both necrosis and apoptosis occur after ischemia reperfusion, and that antioxidant intervention early after the ischemic event may attenuate cell death and limit the amount of damage.[68] A more recent view is that strategies to prevent apoptosis by inhibition of caspases might be useful by limiting cell loss to the initial necrosis-inducing damage following a myocardial infarction.[12]

The major cause of frailty may simply be the loss of muscle mass, presumably by apoptosis. A largely unexplored area is whether this loss is due to the slow accumulation of damage within muscle cells, including mitochondrial damage, leading eventually to gradual apoptosis of these damaged cells. It is known that mitochondrial deletions do accumulate with increasing age in skeletal muscle of rhesus monkeys,[70] but it is not clear whether this has any causal relationship to aging, especially because the percent of mitochondrial genomes mutated remains quite small, even in older individuals. What seems more likely is that increasing oxidation of both the DNA and mitochondrial proteins could eventually trigger the permeability transition and apoptosis as discussed above, leading to a gradual loss of myocytes and muscle mass. What is not known is how much cytochrome c must be leaked from damaged mitochondria into the cytoplasm to induce apoptosis of that cell.

SUMMARY

TABLE 1 summarizes the known relationships among apoptosis, aging and age-related disease. Apoptosis has both positive and negative impacts during aging, and it is important to understand both of these in order to devise useful interventions. Such interventions include both physiological and molecular approaches, including transgenic interventions. The critical roles of the mitochondria in both generating reactive oxygen species, and in initiating apoptosis are recognized, suggesting that maintaining mitochondrial function could be an important therapeutic goal, especially in post-mitotic tissues. In contrast, the ability to eliminate unwanted, damaged

TABLE 1. Roles of apoptosis in aging

Process	Physiological impact	Comments
Death of lymphocytes reactive to self.	Prevents autoimmune responses.	Up-regulated by caloric restriction?
Estrogen stimulation of apoptosis of osteoclasts.	Maintain proper balance of osteoblasts and osteoclasts, thus preventing osteoporosis?	
Death of damaged and potentially neoplastic cells.	Prevent/delay tumorigensis.	Up-regulated by caloric restriction?
Death of neurons following stroke.	Loss of general brain functions.	Death occurs by both necrosis and apoptosis.
Neurodegenerative diseases such as AD, PD, ALS, etc.	Loss of specific brain functions.	Death occurs by apoptosis in response to localized oxidative stress?
Retinal degeneration.	Loss of retinal cells.	Induced by mutations in rhodopsin and other retinal proteins.
Death of cardiomyocytes following myocardial infarction.	Loss of cardiac function.	Death occurs by both necrosis and apoptosis.
Death of skeletal myocytes.	Loss of muscle mass.	Contributes to frailty in late life?

and dysfunctional cells through apoptosis has anti-aging implications in mitotic tissues. Thus, understanding tissue specific differences in initiation and mechanisms of apoptosis is necessary before appropriate interventions will become possible.

REFERENCES

1. COLUMBANO, A. 1995. Cell death: Current difficulties in discriminating apoptosis from necrosis in the context of pathological processes in vivo. J. Cell. Biochem. **58:** 181–190.
2. KERR, J.F.R., A.H. WYLLIE & A.R. CURRIE. 1972. Apoptosis: A basic biological phenomenon with wide-ranging implications in tissue kinetics. Br. J. Cancer **26:** 239–257.
3. RAFF, M. 1998. Cell suicide for beginners. Nature **396:** 119–122.
4. NAGATA, S. & P. GOLSTEIN. 1995. The Fas death factor. Science **267:** 1449–1456.
5. OGAWA, N., H. DANG & N. TALAL. 1995. Apoptosis and autoimmunity. J. Autoimmun. **8:** 1–19.
6. THOMPSON, C.B. 1995. Apoptosis in the pathogenesis and treatment of disease. Science ***267:*** 1456–1462.
7. VAUX, D.L., S. CORY & J.H. ADAMS. 1988. Bcl-2 gene promotes haemopoietic cell survival and cooperates with c-myc to immortalize pre-B cells. Nature **335:** 440–442.
8. WILLIAMS, G.T. 1991. Programmed cell death: Apoptosis and oncogenesis. Cell **65:** 1097–1098.
9. WARNER, H.R., R.J. HODES & K. POCINKI. 1997. What does cell death have to do with aging? J. Am. Ger. Soc. **45:** 1140–1146.
10. ELLIS, R.E., J. YUAN & H.R. HORVITZ. 1991. Mechanisms and functions of cell death. Ann. Rev. Cell Biol. **7:** 663–698.
11. ADAMS, J.M. & S. COREY. 1998. The Bcl-2 protein family: Arbiters of cell survival. Science **281:** 1322–1326.
12. THORNBERRY, N.A. & Y. LAZEBNIK. 1998. Caspases: enemies within. Science **281:** 1312–1316.
13. CLEM, R.J., E.H.-Y. CHENG, C.L. KARP, D.G. KIRSCH, K. UENO, A. TAKAHASHI, M.B. KASTAN, D.E. GRIFFIN, W.C. EARNSHAW, M.A. VELIUONA & J.M. HARDWICK. 1998. Modulation of cell death by Bcl-XL through caspase interaction. Proc. Natl. Acad. Sci. USA **95:** 554–559.
14. FATTMAN, C.L., B. AN & Q.P. DOU. 1997. Characterization of interior cleavage of retinoblastoma protein in apoptosis. J. Cell. Biochem. **67:** 399–408.
15. TEWARI, M., L.T. QUAN, K. O'ROURKE, S. DESNOYERS, Z. ZENG, D.R. BEIDLER, G.G. POIRIER, G.S. SALVESAN & V.M. DIXIT. 1995. Yama/CPP32β, a mammalian homolog of CED-3, is a CrmA-inhibitable protease that cleaves the death substrate poly (ADP-ribose) polymerase. Cell **81:** 801–809.
16. TAKAHASHI, H., E.S. ALNEMRI, Y.A. LAZEBNIK, T. FERNANDES-ALNEMRI, G. LITWACK, R.D. MOIR, R.D. GOLDMAN, G.G. POIRIER, S.H. KAUFMANN & W.C. EARNSHAW. 1996. Cleavage of lamin A by Mch2α but not CPP32: Multiple interleukin 1β-converting enzyme-related proteases with distinct substrate recognition properties are active in apoptosis. Proc. Natl. Acad. Sci. USA **93:** 8395–8400.
17. KAYALAR, C., T. ÖRD, M.P. TESTA, L.-T. ZHONG & D.E. BREDESEN. 1996. Cleavage of actin by interleukin 1β-converting enzyme to reverse DNase I inhibition. Proc. Natl. Acad. Sci. USA **93:** 2234–2238.
18. LIU, X., H. ZOU, C. SLAUGHTER & X. WANG. 1997. DFF, a heterodimeric protein that functions downstream of caspase-3 to trigger DNA fragmentation during apoptosis. Cell **89:** 175–184.

19. ENARI, M., H. SAKAHIRA, H. YOKOYAMA, K. OKAWA, A. IWAMATSU & S. NAGATA. 1998. A caspase-activated DNase that degrades DNA during apoptosis, and its inhibitor ICAD. Nature **391:** 43–50.

20. LEVINE, A.J. 1997. p53, the cellular gatekeeper for growth and division. Cell **88:** 323–331.

21. WEINDRUCH, R. 1989. Dietary restriction, tumors, and aging in rodents. J. Gerontol. **44:** 67–71.

22. GRASL-KRAUPP, B., W. BURSCH, B. RUTTKAY-NEDECKY, A. WAGNER, B. LAUER & R. SCHULTE-HERMANN. 1994. Food restriction eliminates preneoplastic cells through apoptosis and antagonizes carcinogenesis in rat liver. Proc. Natl. Acad. Sci. USA **91:** 9995–9999.

23. JAMES, S.J. & L. MUSKHELISHVILI. 1994. Rates of apoptosis and proliferation vary with caloric intake and may influence incidence of spontaneous hepatoma in C57BL/6×C3HF1 mice. Cancer Res. **54:** 5508–5510.

24. HOLT, P.R., S.F. MOSS, A.R. HEYDARI & A. RICHARDSON. 1998. Diet restriction increases apoptosis in the gut of aging rats. J. Gerontol. **53A:** B168–B172.

25. WARNER, H.R., G. FERNANDES & E. WANG. 1995. A unifying hypothesis to explain the retardation of aging and tumorigenesis by caloric restriction. J. Gerontol. **50A:** B107–B109.

26. DIMRI, G.P., X. LEE, G. BASILE, M. ACOSTA, G. SCOTT, C. ROSKELLEY, E.E. MEDRANO, M. LINSKENS, I. RUBELJ, O. PEREIRA-SMITH, M. PEACOCKE & J. CAMPISI. 1995. A biomarker that identifies senescent human cells in culture and in aging skin in vivo. Proc. Natl. Acad. Sci. USA **92:** 9363–9367.

27. WANG, E. 1995. Senescent human fibroblasts resist programmed cell death, and failure to suppress bcl-2 is involved. Cancer Res. **55:** 2284–2292.

28. WANG, E., M.J. LEE & S. PANDEY. 1994. Control of fibroblast senescence and activation of programmed cell death. J. Cell. Biochem. **54:** 432–439.

29. MILLIS, A.J., M. HOYLE, H.M. MCCUE & H. MARTINI. 1992. Differential expression of metalloproteinase and tissue inhibitor of metalloproteinase genes in aged human fibroblasts. Exp. Cell Res. **201:** 373–379.

30. CAMPISI, J. 1997. Aging and cancer: the double edged sword of replicative senescence. J. Am. Ger. Soc. **45:** 482–488.

31. MASSON, R., O. LEFEBVRE, A. NOËL, M. EL FAHIME, M.P. CHENARD, C. WENDLING, F. KEBERS, M. LEMEUR, A. DIERICH, J.M. FOIDART, P. BASSET & M.C. RIO. 1998. In vivo evidence that stromelysin-3 metalloproteinase contributes in a paracrine manner to epithelial cell malignancy. J. Cell Biol. **140:** 1535–1541.

32. JACOBSON, M.D., M. WEIL & M.C. RAFF. 1997. Programmed cell death in animal development. Cell **88:** 347–354.

33. WATANABE-FUKUNAGA, R., C.I. BRANNAN, N.G. COPELAND, N.A. JENKINS & S. NAGATA. 1992. Lymphoproliferation disorder in mice explained by defects in Fas antigen that mediates apoptosis. Nature **356:** 314–317.

34. TAKAHASKI, T., M. TANAKA, C.I. BRANNAN, N.A. JENKINS, N.G. COPELAND, T. SUDA & S. NAGATA. 1994. Generalized lymphoproliferative disease in mice, caused by a point mutation in the Fas ligand. Cell **76:** 969–976.

35. LUAN, X., W. ZHAO, B. CHANDRASEKAR & G. FERNANDES. 1995. Calorie restriction modulates lymphocyte subset phenotype and increases apoptosis in MRL/lpr mice. Immunol. Lett. **47:** 181–186.

36. RIEUX-LAUCAT, F., F. LE DEIST, C. HIVROZ, I.A.G. ROBERTS, K.M. DEBATIN, A. FISCHER & J.P. DEVILLARTAY. 1995. Mutations in Fas associated with human lymphoproliferative syndrome and autoimmunity. Science **268:** 1347–1349.

37. Fisher, G.H., F.J. Rosenberg, S.E. Straus, J.K. Dale, L.A. Middeton, A.Y. Lin, W. Strober, M.J. Lenardo & J.M. Puck. 1995. Dominant interfering Fas gene mutations impair apoptosis in a human autoimmune lymphoproliferative syndrome. Cell **81:** 935–946.
38. Aringer, M., W. Wintersberger, C.W. Steiner, H. Kiener, E. Presterl, U. Jaeger, J.S. Smolen & W.B. Graninger. 1994. High levels of bcl-2 protein in circulating T-lymphocytes, but not B-lymphocytes, of patients with systemic lupus erythematosus. Arthritis Rheum. **37:** 1423–1430.
39. Zhou, T., C.K. Edwards & J.D. Mountz. 1995. Prevention of age-related T cell apoptosis defect in CD2-fas-transgenic mice. J. Exp. Med. **182:** 129–137.
40. Hughes, D.E., A. Dai, J.C. Tiffee, H.H. Li, G.R. Mundy & B.F. Boyce. 1996. Estrogen promotes apoptosis of murine osteoclasts mediated by TGF-β. Nature Med. **2:** 1132–1136.
41. Okahashi, N., M. Koide, E. Jimi, T. Suda & T. Nishihara. 1998. Caspases (interleukin-1β-converting enzyme family proteases) are involved in the regulation of the survival of osteoclasts. Bone **23:** 33–41.
42. Rogers, M.J., K.M. Chilton, F.P. Coxom, J. Lawry, M.O. Smith, S. Suri & R.G. Russell. 1996. Bisphosphonates induce apoptosis in mouse macrophage-like cells in vitro by a nitric oxide-independent mechanism. J. Bone Mineral Res. **11:** 1482–1491.
43. Yan, L.-J. R.L. Levine & R.S. Sohal. 1997. Oxidative damage during aging targets mitochondrial aconitase. Proc. Natl. Acad. Sci. USA **94:** 11168–11172.
44. Yan, L.-J. & R.S. Sohal. 1998. Mitochondrial adenine nucleotide translocase is modified oxidatively during aging. Proc. Natl. Acad. Sci. USA **95:** 12896–12901.
45. Graham, B.H., K.G. Waymire, B. Cottrell, I.A. Trounce, G.R. MacGregor & D.C. Wallace. 1997. A mouse model for mitochondrial myopathy and cardiomyopathy resulting from a deficiency in the heart/muscle isoform of the adenine nucleotide translocator. Nature Genetics **16:** 226–234.
46. Nieminen, A.-L, A.K. Saylor, S.A. Tesfai, B. Herman & J.J. Lemasters. 1995. Contribution of the mitochondrial permeability transition to lethal injury after exposure of hepatocytes to t-butylhydroperoxide. Biochem. J. **307:** 99–106.
47. Yang, J., X. Liu, K. Bhalla, C.N. Kim, A.M. Ibrado, J. Cai, T.I. Peng, D.P. Jones & X. Wang. 1997. Prevention of apoptosis by bcl-2: Release of cytochrome c from mitochondria blocked. Science **275:** 1129–1132.
48. Green, D.R. & J.C. Reed. 1998. Mitochondria and apoptosis. Science **281:** 1309–1312.
49. Smale, G., N.R. Nicholas, D.R. Brady, C.E. Finch & W.E. Horton, Jr. 1995. Evidence for apoptotic cell death in Alzheimer's disease. Exp. Neurol. **133:** 225–230.
50. Cotman, C.W. & A.J. Anderson. 1995. A potential role for apoptosis in neurodegeneration and Alzheimer's disease. Mol. Neurobiol. **10:** 19–45.
51. Dragunow, M., R.L. Faull, P. Lawlor, E.J. Beilharz, K. Singleton, E.B. Walker & E. Mee. 1995. In situ evidence for DNA fragmentation in Huntington's disease striatum and Alzheimer's disease temporal lobes. Neuroreport **9:** 1053–1057.
52. Thomas, L.B., D.J. Gates, E.K. Richfield, T.F. O'Brien, J.B. Schweitzer & D.A. Steinder. 1995. DNA end labelling (TUNEL) in Huntington's disease and other neuropathological conditions. Exp. Neurol. **133:** 265–272.
53. Kostic, V., V. Jackson-Lewis, F. de Bilbao, M. Dubois-Dauphin & S. Przedborski. 1997. Bcl-2: prolonging life in a transgenic mouse model of familial amyotrophic lateral sclerosis. Science **277:** 559–562.
54. Friedlander, R.M., R.H. Brown, V. Gagliardini, J. Wang & J. Yuan. 1997. Inhibition of ICE slows ALS in mice. Nature **388:** 31.

55. HARA, H., R.M. FRIEDLANDER, V. GAGLIARDINI, C. AYATA, K. FINK, Z. HUANG, M. SHIMIZU-SASAMATA, J. YUAN & M.A. MOSKOWITZ. 1997. Inhibition of interleukin 1β converting enzyme family proteases reduces ischemic and excitotoxic neuronal damage. Proc. Natl. Acad. Sci. USA **94:** 2007–2012.

56. MILLIGAN, C.E., D. PREVETTE, H. YAGINUMA, S. HOMMA, C. CARDWELL, L.C. FRITZ, K.J. TOMASELLI, R.W. OPPENHEIM & L.M. SCHWARTZ. 1995. Peptide inhibitors of the ICE protease family arrest programmed cell death of neurons in vivo and in vitro. Neuron **15:** 385–393.

57. FRIEDLANDER, R.M., V. GAGLIARDINI, V., H. HARA, K.B. FINK, W. LI, G. MACDONALD, M.C. FISHMAN, A.H. GREENBERG, M.A. MOSKOWITZ & J. YUAN. 1997. Expression of a dominant negative mutant of interleukin-1β converting enzyme in transgenic mice prevents neuronal cell death induced by trophic factor withdrawal and ischemic brain injury. J. Exp. Med. **185:** 933–940.

58. BARINAGA, M. 1998. Stroke-damaged neurons may commit cellular suicide. Science **281:** 1302–1303.

59. BARINAGA, M. 1998. Is apoptosis key in Alzheimer's disease? Science **281:** 1303–1304.

60. YANG, G., P.H. CHAN, J. CHEN, E. CARLSON, S.F. CHEN, P. WEINSTEIN, C.J. EPSTEIN & H. KAMII. 1994. Human copper-zinc superoxide dismutase transgenic mice are highly resistant to reperfusion injury after focal cerebral ischemia. Stroke **25:** 165–170.

61. PARKES, T.L., A.J. ELIA, D. DICKINSON, A.J. HILLIKER, J.P. PHILLIPS & G.L. BOULIANNE. 1998. Extension of Drosophila lifespan by overepression of human SOD1 in motorneurons. Nature Genetics **19:** 171–174.

62. TRAVIS, G.H. 1998. Mechanisms of cell death in inherited retinal degeneration. Am. J. Hum. Genet. **62:** 503–508.

63. HUMPHRIES, M.M., D. RANCOURT, G.J. FARRAR, P. KENNA, M. HAZEL, R.A. BUSH, P.A. SIEVING, D.M. SHIELS, N. MCNALLY, P.CREIGHTON, A. ERVEN, A. BOROS, K. GULYA, M.R. CAPECCHI & P. HUMPHRIES. 1997. Retinopathy induced in mice by targeted disruption of the rhodopsin gene. Nature Genet. **15:** 216–219.

64. DAVIDSON, F.F. & H. STELLER. 1998. Blocking apoptosis prevents blindness in Drosophila retinal degeneration mutants. Nature **391:** 587–591.

65. LI, Y., T.-T. HUANG, E.J. CARLSON, S. MELOV, P.C, URSELL, J.L. OLSON, L.J. NOBLE, M.P. YOSHIMURA, C. BERGER, P.H. CHAN, D.C. WALLACE & C.J. EPSTEIN. 1995. Dilated cardiomyopathy and neonatal lethality in mutant mice lacking manganese superoxide dismutase. Nat. Genet. **11:** 376–381.

66. LEBOVITZ, R.M., H. ZHANG, H. VOGEL, J. CARTWRIGHT, L. DIONNE, N. LU, S. HUANG & M.M. MATZUK. 1996. Neurodegeneration, myocardial injury, and perinatal death in mitochondrial superoxide dismutase-deficient mice. Proc. Natl. Acad. Sci. USA **93:** 9782–9787.

67. WILLIAMS, M.D., H. VAN REMMEN, C.C. CONRAD, T.-T. HUANG, C.J. EPSTEIN & A. RICHARDSON. 1998. Increased oxidative damage is correlated to altered mitochondrial function in heterozygous manganese superoxide dismutase knockout mice. J. Biol. Chem. **273:** 28510–28515.

68. SARASTE, A., K. PULKKI, M. KALLAJOKI, K. HENDRIKSEN, M. PARVINEN & L.M. VOIPIO-PULKKI. 1997. Apoptosis in human acute myocardial infarction. Circulation **95:** 320–323.

69. KAJSTURA, J., W. CHENG, K. REISS, W.A. CLARK, E.H. SONNENBLICK, S. KRAJEWSKI, J.C. REED, G. OLIVETTI & P. ANVERSA. 1996. Apoptotic and necrotic myocyte cell deaths are independent contributing variables of infarct size in rats. Lab. Invest. **74:** 86–107.

70. LEE, C.M., S.S. CHUNG, J.M. KACZKOWSKI, R. WEINDRUCH & J.M. AIKEN. 1993. Multiple mitochondrial DNA deletions associated with age in skeletal muscle of rhesus monkeys. J. Gerontology **48:** B201–B205.

Ultrastructural Alterations of Mitochondria in Pre-apoptotic and Apoptotic Hepatocytes of TNFα-treated Galactosamine-sensitized Mice[a]

S. ANGERMÜLLER,[b,d] J. SCHÜMANN,[c] H.D. FAHIMI,[b] AND G. TIEGS[c]

[b]*Department of Anatomy and Cell Biology II, University of Heidelberg, Germany*
[c]*Institute of Experimental and Clinical Pharmacology and Toxicology, University of Erlangen-Nürnberg, Germany*

ABSTRACT: The electron microscopical studies presented here show that characteristic morphological alterations in mitochondria are a very early hallmark of the hepatocellular apoptotic program. Before chromatin condensation occurs, the outer mitochondrial membrane is focally disrupted and the inner membrane protrudes through this gap forming a hernia. The demonstration of cytochrome oxidase in mitochondria revealed a very strong activity in pre-apoptotic and apoptotic cells as well as in apoptotic bodies.

In the past, electron microscopical observations of apoptotic cells have mainly focused on nuclear alterations such as chromatin condensation, cellular shrinkage, and fragmentation.[1–4] Other ultrastructural changes such as those of mitochondria were rarely discussed and mitochondria were usually interpreted either as normal[3, 5–7] or as slightly swollen.[8]

In contrast, functional alterations of mitochondria in apoptotic cells have been clearly established by many authors. The reduction of the mitochondrial transmembrane potential ($\Delta\Psi_m$) causing the opening of mitochondrial permeability transition pores is an early event in the apoptotic process.[6, 9–12] Those "apoptotic" mitochondria contain a protein of about 50 kD in the intermembranous space which is released when the outer membrane is disrupted.[13] Enari *et al.* described similar results using a cell lysate containing a substance inducing apoptosis.[14] The authors combined this substance with the purified nuclei of intact cells and observed chromatin condensation and DNA fragmentation of the nuclei. The release of this protein was significantly inhibited by Bcl-2 protein.[15]

Bcl-2 is a protooncogene that blocks programmed cell death.[16,17] Recently, a functional association between Bcl-2 protein and cytochrome c was postulated by several authors.[18,19] Overexpression of Bcl-2 protein prevents the outflow of cytochrome c from mitochondria, thereby blocking caspase-3 activation.[18,19] The apoptotic protease cascade leading to caspase-3 activation is initiated by the formation of

[a]This study was supported by the grants An 192/1-3 and Ti 169/4-1 by the Deutsche Forschungsgemeinschaft, Bonn Bad-Godesberg.

[d]Address for correspondence: Department of Anatomy and Cell Biology II, University of Heidelberg Im Neuenheimer Feld 307 D-69120 Heidelberg, Germany. 06221/548665 (voice); 06221/544952 (fax).
e-mail: Sabine.Angermueller@urz.uni-Heidelberg.de

the Apaf-1 (apoptotic protease activating factor-1) caspase-9 complex in the presence of cytochrome c and dATP. This event leads to caspase-9 activation which in turn cleaves and activates caspase-3.[20,21]

Activated caspases play a central role in the biochemical apoptotic pathway.[22] In this study we have investigated the ultrastructural alterations of mitochondria in the tumor necrosis factor (TNF)-induced apoptotic cell death of hepatocytes in galactosamine (GalN)-sensitized mouse liver. GalN 700mg/kg (Roth Chemicals, Karlsruhe, Germany) was given intraperitoneally in a volume of 200 μl saline. Recombinant murine TNF (10 μg/kg) was injected into the tail vein in a volume of 300 μl saline containing 0.1% human serum albumin 20 minutes after GalN administration.[23] The administration of only TNF or only GalN served as negative controls for the induction of apoptosis. Five hours after treatment, animals were anesthetized by Nembutal (150 mg/kg i.p.) (Sanofi-Ceva; Hannover, Germany). The livers were fixed via the portal vein with a fixative containing 0.25% glutaraldehyde and 2% sucrose in 100mM Piperazine-N,N'-bis (2-ethanesulfonic acid) (Pipes) buffer at a pH value of 7.4 for 5 minutes. For the ultrastructural demonstration of cytochrome oxidase activity we applied a modified 3,3-diaminobenzidine-tetrahydrochloride (DAB) technique[24]: Sections were incubated for 60 min at 37°C in a medium containing 15mg DAB and 0.05% cytochrome c in 100mM Pipes buffer at a pH value of 7.2. Sections were postfixed with aqueous osmium tetroxide, dehydrated in graded ethanol and embedded in Epon 812. Ultrathin sections were counterstained with lead citrate for 1min and examined in a Philips EM 301 electron microscope.

Our electron microscopic studies revealed clearly that the mitochondrial alterations appeared prior to the nuclear alterations, particularly the chromatin condensation at the nuclear envelope (FIG.1). The outer membrane of the mitochondria was disrupted while parts of the inner membrane appeared herniated through this gap. Whereas the cristae were tightly folded in that part of the mitochondrion in which the outer membrane appeared intact, the cristae were reduced or absent in the defective herniated part of the inner membrane (FIG. 2). In apoptotic cells with typical margination of condensed chromatin at the nuclear envelope similar alterations were also found in mitochondria (FIG. 3). Using the DAB-technique for the demonstration of cytochrome oxidase activity in mitochondria,[24] we found a strong staining in the altered mitochondria (FIG. 3). Even when an apoptotic body was engulfed by a neighboring hepatocyte, the heavy reaction product could be recognized in the altered mitochondria (FIG. 4).

It appears that mitochondrial apoptotic factors and all structures involved in the pre-apoptotic ruptures of the $\Delta\Psi_m$ are present in mitochondria lacking mitochondrial DNA. Thus, apoptosis is controlled by the nuclear rather than by the mitochondrial genome.[25,26] The enzymes of the respiratory chain do not seem to be involved in the apoptotic process.[8,27] The data presented in our study indicate that mitochondrial cytochrome oxidase in apoptotic cells can still function when cytochrome c is offered as an electron donor. Krippner *et al.* noted that the activity of cytochrome oxidase is apparently normal in Fas-mediated apoptotic cells but that the inhibition and inactivation of cytochrome c may be the potentially fatal component of the apoptotic program.[8] This suggests that the function of the respiratory chain can be disrupted due to the lack of cytochrome c. Indeed, even in cytochrome c-containing mitochondria

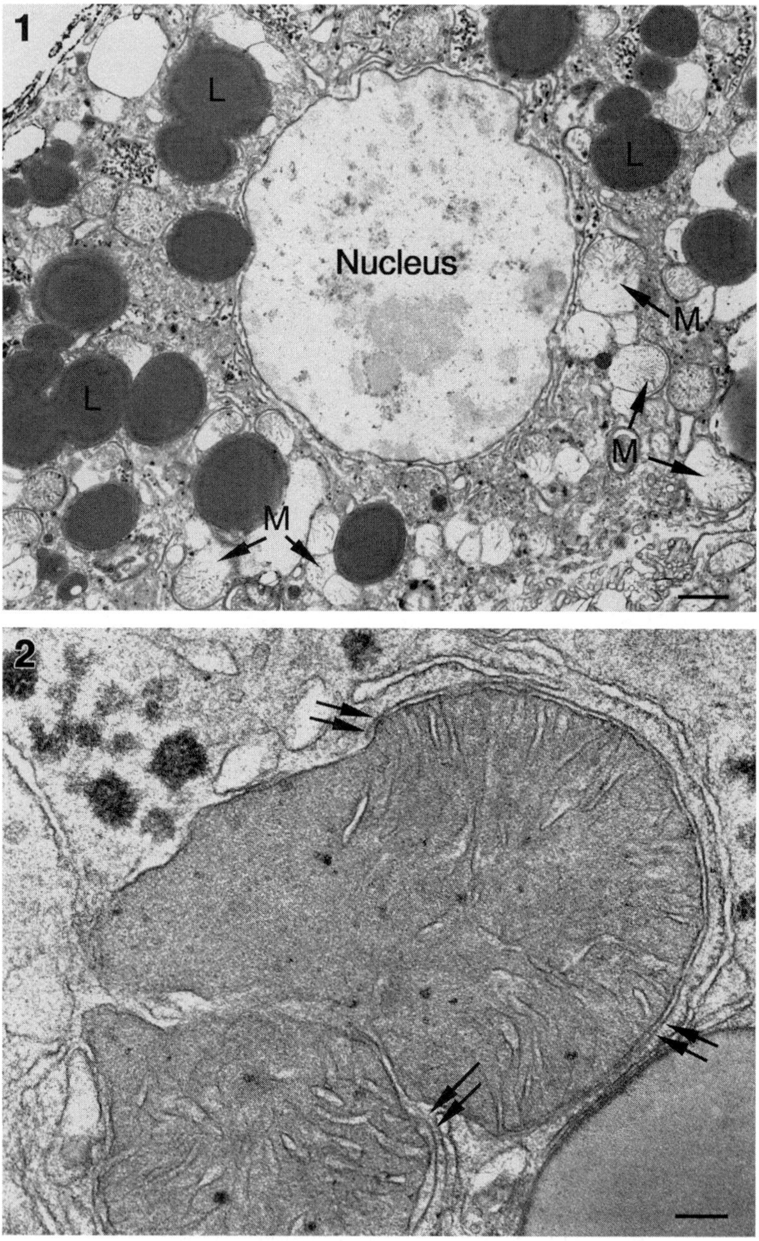

FIGURE 1–2. Electron micrographs of hepatocytes of GalN/TNF-treated mouse liver. **1**. Pre-apoptotic hepatocyte with normal feature of the nucleus. Note the mitochondrial alterations showing tightly folded cristae beside a pale part without cristae. L: lipid, M: mitochondria. *Bar* = 1 µm. **2**. Higher magnification of a mitochondrion with disruption of the outer membrane and herniated part of the inner mitochondrial membrane through this gap (arrows). *Bar* = 0.125 µm.

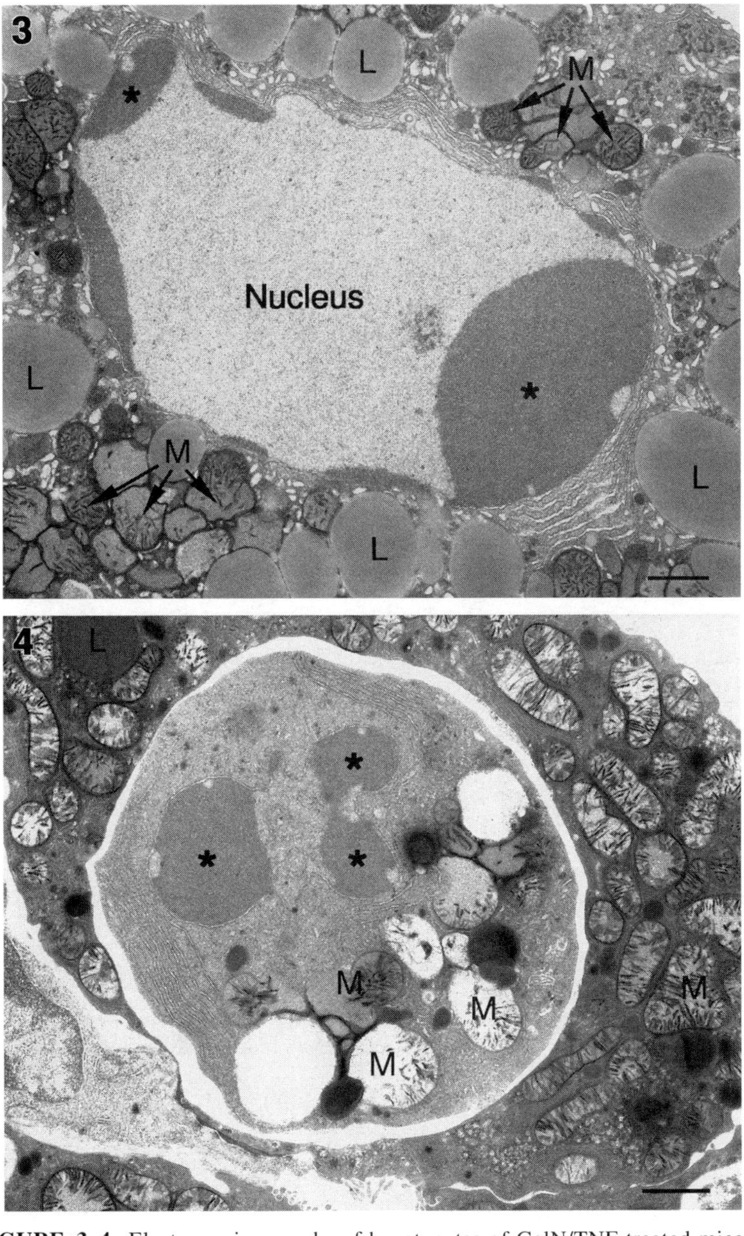

FIGURE 3–4. Electron micrographs of hepatocytes of GalN/TNF-treated mice incubated for the demonstration of cytochrome oxidase activity. **3**. Note the typical apoptotic feature of the nucleus with margination of condensed chromatin at the nuclear envelope (asterix) and heavy reaction product demonstrating strong cytochrome oxidase activity in the altered mitochondria. *Bar* = 0.5 μm. **4**. Hepatocyte having engulfed an apoptotic body which contains a nucleus with condensed chromatin (asterisk) and typical apoptotic mitochondria with disrupted outer membrane and herniated inner membrane. Note again the strong staining of the apoptotic mitochondria. L: lipid, M: mitochondria. *Bar* = 1μm.

of apoptotic cells, cytochrome oxidase is unable to function when cytochrome c is inactivated by an inhibitor.[8]

In summary, our electron microscopical studies demonstrate characteristic morphological alterations in mitochondria as a very early hallmark of the hepatocellular apoptotic program. Before chromatin condensation occurs, the outer mitochondrial membrane is focally disrupted and the inner membrane protrudes through this gap forming a hernia. The demonstration of cytochrome oxidase in mitochondria revealed a very strong activity in pre-apoptotic and apoptotic cells as well as in apoptotic bodies.

REFERENCES

1. KERR, J.F.R., A.H. WYLLIE & A.R.CURRIE.1972. Apoptosis: A basic biological phenomenon with wide-ranging implications in tissue kinetics. Br. J. Cancer **26:** 239–257.
2. WYLLIE, A.H. & E. DUVALL. 1992. Cell injury and death. *In* Oxford Textbook of Pathology, Vol. 1, Principles of Pathology. J. O'D. McGee, P.G. Isaacson & N.A. Whright, Eds. :141-157. Oxford University Press.
3. MAJNO, G. & I. JORIS. 1995. Apoptosis, oncosis, and necrosis. An overview of cell death. Am. J. Pathol. **146:** 3–15.
4. PATEL, T. & G.J. GORES. 1995. Apoptosis in hepatobiliary disease. Hepatology **21:** 1725–1741.
5. SEARLE, J., B.V. HARMON, C.J. BISHOP & J.F.R. KERR. 1987. The significance of cell death by apoptosis in hepatobiliary disease. J. Gastroenterol. Hepatol. **2:** 77–96.
6. COSSARIZZA, A., C. FRANCESCHI, D. MONTI, S. SALVIOLI, E. BELLESIA, R. RIVABENE, L. BIONDO, G. RAINALDI, A. TINARI & W. MALORNI. 1995. Protective effect of N-acetylcysteine in tumor necrosis factor-α-induced apoptosis in U937 cells: the role of mitochondria. Exp. Cell. Res. **220:** 232–240.
7. KIMURA, K., H. SASANO, T. SHIMOSEGAWA, K.KATO, T. NOGUCHI, S. MOCHIZUKI, T. SAWAI, M. KOIZUMI, T. TOYOTA & H. NAGURA. 1997. Ultrastructural and confocal laser scanning microscopic examination of TUNEL-positive cells. J. Pathol. **181:** 235–242.
8. KRIPPNER, A., A. MATSUNO-YAGI, R.A. GOTTLIEB & B.M. BABIOR. 1996. Loss of function of cytochrome c in Jurkat cells undergoing Fas-mediated apoptosis. J. Biol. Chem. **271:** 21629–21636.
9. CASTEDO, M., A. MACHO, N. ZAMZAMI, T. HIRSCH, P. MARCHETTI, J. URIEL & G. KROEMER. 1995. Mitochondrial perturbations define lymphocytes undergoing apoptotic depletion in vivo. Eur. J. Immunol. **25:** 3277–3284
10. PETIT, P.X., H.LECOEUR, E. ZORN, C. DAUGUET, B. MIGNOTTE, M.-L. GOUGEON. 1995. Alterations in mitochondrial structure and function are early events of dexamethasone-induced thymocyte apoptosis. J. Cell Biol. **130:** 157–167.
11. ZAMZAMI, N., P. MARCHETTI, M. CASTEDO, C. ZANIN, J.-L. VAYSSIÈRE, P.X. PETIT & G. KROEMER. 1995. Reduction in mitochondrial potential constitutes an early irreversible step of programmed lymphocyte death in vivo. J. Exp. Med. **181:** 1661–1672.
12. ZAMZAMI, N., S.A. SUSIN, P. MARCHETTI, T. HIRSCH, I. GÓMES-MONTERREY, M. CASTEDO & G. KROEMER. 1996. Mitochondrial control of nuclear apoptosis. J. Exp. Med. **183:** 1533–1544.
13. SUSIN, S.A., N. ZAMZAMI, M. CASTEDO, T. HIRSCH, P. MARCHETTI, A. MACHO, E. DAUGAS, M. GEUSKENS & G. KROEMER. 1996. Bcl-2 inhibits the mitochondrial release of an apoptogenic protease. J. Exp. Med. **184:** 1331–1341.
14. ENARI, M., A. HASE & S. NAGATA. 1995. Apoptosis by a cytosolic extract from Fas-activated cells. EMBO J. **14:** 5201–5208.

15. KROEMER, G., P. PETIT, N. ZAMZAMI, J.-L. VAYSSIÈRE & B. MIGNOTTE. 1995. The biochemistry of programmed cell death. FASEB J. **9:** 1277–1287.
16. HOCKENBERY, D., G. NUÑEZ, C. MILLIMAN, R.D. SCHREIBER & S.J. KORSMEYER. 1990. Bcl-2 is an inner mitochondrial membrane protein that blocks programmed cell death. Nature **348:** 334–336.
17. HOCKENBERY, D.M., M. ZUTTER, W.HICKEY, M. NAHM & S.J. KORSMEYER. 1991. BCL2 protein is topographically restricted in tissues characterized by apoptotic cell death. Proc. Natl. Acad. Sci. USA **88:** 6961–6965.
18. KLUCK, R.M., E. BOSSY-WETZEL, D.R. GREEN & D.D. NEWMEYER. 1997. The release of cytochrome c from mitochondria: A primary site for Bcl-2 regulation of apoptosis. Science **275:** 1132–1136
19. YANG, J., X. LIU, K. BHALLA, C.N. KIM, A.M. IBRADO, J.CAI, T.-I. PENG, D.P. JONES & X. WANG. 1997. Prevention of apoptosis by Bcl-2: Release of cytochrome c from mitochondria blocked. Science **275:** 1129–1132.
20. LI, P., D. NIJHAWAN, I. BUDIHARDJO, S.M. SRINIVASA, M. AHMAD, E.S. ALNEMRI & X. WANG. 1997. Cytochrome c and dATP-dependent formation of Apaf-1/caspase-9 complex initiates an apoptotic protease cascade. Cell **91:** 479–489.
21. ZOU, H., W.J. HENZEL, X. LIU, A. LUTSCHG & X. WANG. 1997. Apaf-1, a human protein homologous to *C. elegans* CED-4, participates in cytochrome c–dependent activation of caspase-3. Cell **90:** 405–413.
22. NICHOLSON, D.W. & N.A. THORNBERRY. 1997. Caspases: killer proteases. Trends Biochem. Sci. **22:** 299–306.
23. ANGERMÜLLER, S. G. KÜNSTLE & G. TIEGS. 1998. Pre-apoptotic alterations in hepatocytes of TNFα-treated galactosamine-sensitized mice. J. Histochem. Cytochem. **46:** 1175–1183.
24. ANGERMÜLLER, S. & H.D. FAHIMI. 1981. Selective cytochemical localization of peroxidase, cytochrome oxidase and catalase in rat liver with 3,3′-Diaminobenzidine. Histochemistry **71:** 33–44.
25. JACOBSON, M.D., J.F. BURNE, M.P. KING, T. MIYASHITA, J.C. REED & M.C. RAFF. 1993. Bcl-2 blocks apoptosis in cells lacking mitochondrial DNA. Nature **361:** 365–369.
26. MARCHETTI, P., S.A. SUSIN, D. DECAUDIN, S. GAMEN, M. CASTEDO, T. HIRSCH, N. ZAMZAMI, J. NAVAL, A. SENIK & G. KROEMER. 1996. Apoptosis-associated derangement of mitochondrial function in cells lacking mitochondrial DNA. Cancer Res. **56:** 2033–2038.
27. O'DONNELL, V.B., S. SPYCHER & A. AZZI. 1995. Involvement of oxidants and oxidant-generating enzyme(s) in tumor-necrosis-factor-α-mediated apoptosis: role for lipoxygenase pathway but not mitochondrial respiratory chain. Biochem. J. **310:** 133–141.

Mitochondrial Membrane Permeabilization during the Apoptotic Process

ETIENNE JACOTOT, PAOLA COSTANTINI, ERIC LABOUREAU, NAOUFAL ZAMZAMI, SANTOS A. SUSIN, AND GUIDO KROEMER[a]

Centre National de la Recherche Scientifique, Unité Propre de Recherche 420, 19 rue Guy Môquet, F-94801 Villejuif, France

ABSTRACT: Apoptosis may be viewed as a triphasic process. During the pre-mitochondrial initiation phase, very different pro-apoptotic signal transduction or damage pathways can be activated. These pathways then converge on the mitochondrion, where they cause the permeabilization of the inner and/or outer membranes with consequent release of soluble intermembrane proteins into the cytosol. The process of mitochondrial membrane permeabilization would constitute the decision/effector phase of the apoptotic process. During the post-mitochondrial degradation phase downstream caspases and nucleases are activated and the cell acquires an apoptotic morphology. Recently, a number of different second messengers (calcium, ceramide derivatives, nitric oxide, reactive oxygen species) and pro-apoptotic proteins (Bax, Bak, Bid, and caspases) have been found to directly compromise the barrier function of mitochondrial membranes, when added to isolated mitochondria. The effects of several among these agents are mediated at least in part via the permeability transition pore complex (PTPC), a composite channel in which members of the Bcl-2 family interact with sessile transmembrane proteins such as the adenine nucleotide translocator. These findings suggest that the PTPC may constitute a pharmacological target for chemotherapy and cytoprotection.

INTRODUCTION: A TRIPHASIC MODEL OF APOPTOSIS

In 1995 we first proposed a working hypothesis[1] according to which alterations in mitochondrial membrane permeability would constitute a decisive step in the cascade of events leading to apoptotic cell death. Our triphasic model[2–4] predicts a pre-mitochondrial phase (initiation phase) of the apoptotic process which would be marked by the heterogeneity of pro-apoptotic signal transduction or damage pathways; a common mitochondrial phase (decision/effector phase) during which the banier function of the inner and/or outer mitochondrial membranes is lost; and a common post-mitochondrial phase (degradation phase) during which downstream caspases and nucleases are activated and produce the typical biochemical features of apoptotic cell death. In the meantime, the major principles of this model have been adopted by some investigators[5–9] and criticized by others.[10,11] The "mitochondrial" hypothesis has been particularly well accepted among those groups concerned with the mode of action of pro- and anti-apoptotic proteins from the Bcl-2 family,[6–9]

[a]Address for correspondence: Dr. Guido Kroemer 19 rue Guy Môquet, B.P. 8, F-94801 Villejuif, France; Phone: 33-1-49-58-35-13; fax 33-1-49-58 -35-09.
e-mail: kroemer@infobiogen.fr

which indeed act on mitochondria to disturb or stabilize membrane barrier function, respectively.[12,13] In contrast, scientists who conceive caspase activation cascades as the motor of apoptosis have been reluctant to adopt the mitochondrial hypothesis and rather suggest that mitochondrial membrane permeabilization may be just one among several pathways leading to caspase activation, dispensable for most cases of apoptosis.[10,14,15] However, the debate on the relative contribution of mitochondria and caspases may be futile owing to the multiple bidirectional links existing between caspases and mitochondria.[16] This paper will concentrate on the mechanisms linking upstream (pre-mitochondrial) signal transduction pathways to mitochondrial membrane permeabilization. In particular, we will concentrate on two questions. First, which are the final effectors acting on mitochondria and causing membrane permeabilization? Second, which molecules located within mitochondria participate in membrane permeabilization? After having treated these problems, we will discuss to what degree mitochondria participate in apoptosis regulation.

EFFECTORS CONNECTING PRO-APOPTOTIC SIGNALS TO MITOCHONDRIA

During the last years, it has become clear that a number of proapoptotitc signal transduction pathways can act to increase mitochondrial membrane permeability, in both cells and isolated mitochondria (TABLE 1). Such signs of mitochondrial membrane permeabilization include a dissipation of the inner mitochondrial transmembrane potential ($\Delta\Psi_m$)[b] and/or the release of soluble intermembrane proteins such as cytochrome c, via the outer mitochondrial membrane, to the cytosol. The sequence of these alterations may depend on the cell type and on the apoptosis inducer. Endogenous effectors that can permeabilize mitochondrial membranes include several pro-apoptotic members of the Bcl-2 family, in particular Bax. Upon induction of apoptosis, Bax distributes from the cytosol to mitochondria in a reaction that requires Bax homodimerization as well as a conformational change in Bax.[17–19] The transfection-enforced overexpression of Bax or the micro-injection of recombinant Bax provoke a $\Delta\Psi_m$ collapse and the mitochondrial release of cytochrome c.[6,20,21] These effects can also be obtained by adding recombinant Bax protein to purified mitochondria.[21–23] Similarly, other pro-apoptotic proteins from the Bcl-2 family can act on isolated mitochondria. This applies to recombinant Bak,[24] as well to recombinant Bid. Cleavage of Bid by caspase 8 generates a truncated Bid molecule which acts much more effectively on mitochondria than does the intact Bid molecule.[25,26] This latter observation allows us to delineate a scenario according to which activation of caspase 8 due to cross-linking of the CD95/Fas receptor can cause mitochondrial changes via Bid. Peptides derived from the functionally important Bcl-2 homology region 3 (BH3) have also been shown to cause cytochrome c release from isolated mitochondria *in vitro.*[27] Importantly enough, it appears that the effects of Bax, Bak, Bid, and BH3 peptides are at least in part neutralized by the local presence of anti-

[b]ABBREVIATIONS: $\Delta\Psi_m$, mitochondrial transmembrane potential; mClCCP, carbonylcyanide *m*-chlorophenylhydrazone; $DiOC_6(3)$, 3,3′dihexyloxacarbocyanine iodide; PT, permeability transition; Z-VAD.fmk, N-benzyloxycarbonyl-Val-Ala-Asp-fluoromethylketone.

TABLE 1. Endstage effectors directly acting on mitochondria to increase membrane permeability[a]

Substance	Reported effects	Reference
Endogenous		
bax	causes cytochrome *c* release, AIF release, dissipation of $\Delta\Psi_m$	(21,22)
bak	dissipation of $\Delta\Psi_m$	(24)
caspase-8-digested Bid	cytochrome *c* release	(25,26)
BH-3-derived peptides	cytochrome *c* release	(27)
caspases 1 and 3	all hallmarks of PT pore opening[b]	(28,29)
ganglioside GD3	all hallmarks of PT pore opening	(32,33)
nitric oxide	all hallmarks of PT pore opening	(34–36)
ROS	all hallmarks of PT pore opening	(31)
calcium	all hallmarks of PT pore opening	(31)
Xenobiotics (examples)		
diamide	all hallmarks of PT pore opening	(51)
PKI1195	all hallmarks of PT pore opening	(56)
betulinic acid	all hallmarks of PT pore opening	(57)
lonidamine	all hallmarks of PT pore opening	(52)
heavy metals (Cd, Hg, As)	all hallmarks of PT pore opening	(31,58)

[a]Effects obtained with isolated mitochondria *in vitro*.
[b]Dissipation of $\Delta\Psi_m$, large amplitude swelling, release of soluble intermembrane proteins.

apoptotic members of the Bcl-2 family such Bcl-2 itself or Bcl-XL.[25–27] Other proteins that can induce mitochondrial membrane permeabilization are recombinant caspases.[28,29] At present, it is not known which mitochondrial proteins are cleaved by such caspases. It is tempting to speculate that they could act on the anti-apoptotic proteins Bcl- 2 or Bcl-XL, which become pro-apoptotic once they have been cleaved by caspases 1 or 3, respectively.[30] In addition to proteins, a number of small effector molecules can directly affect mitochondria.[31] This applies to endogenous effectors such as the ceramide derivative ganglioside GD3 (refs. 32 and 33 and unpublished results), nitric oxide and peroxynitrite,[34–36] reactive oxygen species, and calcium.[31] Similarly, a number of cytotoxic xenobiotics with potent pro-apoptotic effects can permeabilize the membranes of isolated mitochondria (TABLE 1).

Is the list of endogenous effectors regulating mitochondrial membrane permeability complete? The answer is no. Thus, post-translational modifications of Bcl-2-like proteins are likely to affect the mitochondrial effects of these proteins. The anti-apoptotic protein Bcl-2 is phosphorylated and activated by mitochondrial protein kinase C α,[37] whereas the proapoptotic protein Bad is subject to (de)phosphorylation reactions determining its subcellular localization.[38] It appears that similar post-translational regulatory modifications affect Bcl-2 interacting proteins such as Bag-1.[39] These examples illustrate that multiple apoptosis- or survival-regulatory pathways can directly or indirectly affect the barrier function of mitochondrial membranes.

THE PERMEABILITY TRANSITION PORE COMPLEX (PTPC): A POSSIBLE INTEGRATOR OF DIFFERENT PRO-APOPTOTIC SIGNALS AT THE MITOCHONDRIAL LEVEL

When studying the loss of mitochondrial function occurring during apoptosis, two possibilities might be considered. The disruption of the $\Delta\Psi_m$ could be attributed to non-specific damage of membrane or, alternatively, to a more specific process involving pore-forming proteins from the Bcl-2 family sessile mitochondrial proteins. The so-called permeability transition (PT) is due to the opening of a regulated proteaceous pore, the mitochondrial PT pore, also called megachannel.[31,40,41] The PT pore is a dynamic multiprotein complex probably located at the contact site between the inner and the outer mitochonchial membranes. Although the exact composition of the PT pore complex is elusive, it is thought to involve proteins from the cytosol (hexokinase), the outer membrane (voltage-dependent anion channel, VDAC), the inner membrane (the adenine nucleotide translocator, ANT), and the matrix (cyclophilin D).[21,29,41–43] The PT pore participates in the regulation of matrix Ca^{2+}, pH, $\Delta\Psi_m$, and volume, functioning as a Ca^{2+}-, voltage-, pH-, and redox-gated channels with several levels of conductance and little if any ion selectivity. At its high level of conductance, it can provoke irreversible $\Delta\Psi_m$ dissipation and matrix swelling.[31,40] The PT pore functions as a sensor of multiple physiological parameters (TABLE 2) and thus processes information on the general metabolic condition of the cell.

A ligand of the ANT, bongkrekic acid, and ligands of cyclophilin D such as cyclosporin A and *N*-methyl-4-Val-cyclosporin, which inhibit PT, prevent the loss of the $\Delta\Psi_m$ in several models of apoptosis.[44,45] Since cyclosporin A, for unknown reasons, is a transient inhibitor of PT,[31] its $\Delta\Psi_m$-stabilizing effect is revealed only in experiments designed to last a few hours or less[21,44,45] or when it is combined with phospholipase A2 inhibitors such as aristolochic acid.[20,46] This latter manipulation is thought to prolong the PT pore-inhibitory effect of cyclosporin A. In contrast, bongkrekic acid alone has a long-term protective effect on mitochondrial function.[45,47,48] These findings establish the implication of the PT pore in the apoptosis-associated mitochondrial dysfunction. In addition, inhibitors of the PT pore such as bongkrekic acid or cyclosporin A plus aristolochic acid prevent the mitochondrial and postmitochondrial signs of apoptosis.[20,21,44–48] As a consequence, opening of the PT pore appears necessary for the induction of cell death in at least some models, e.g. apoptosis induced by Bax,[20,21] p53,[48] tumor necrosis factor,[46,49] or glucocorticoids.[44,45]

Recent attempts to purify and to reconstitute the functional activity of the PTPC have yielded conflicting results in different laboratories. Brdiczka and co-workers have proposed that the ANT could be the molecule responsible for PT pore formation.[50] In contrast, Halestrap and co-workers suggest that two proteins, namely ANT plus cyclophilin D, may form a Ca^{2+}-responsive, CsA-inhibitable channel whose

TABLE 2. Functions of the permeability transition pore[a]

Function	Principles of modulation	Example
voltage sensor	Low $\Delta\Psi_m$ facilitates PT. High $\Delta\Psi_m$ inhibits PT.	Anoxia, respiratory inhibitors induce PT. Hyperpolarization (nigericin) inhibits PT.
thiol sensor	Oxidation of a critical matrix dithiol (in the ANT?, regulated by GSH) induces PT.	Prooxidants and thiol crosslinking induce PT. Prevention of thiol crosslinking prevents PT.
sensor of pyridine oxidation	Oxidation of $NAD(P)H_2$ favors PT (in equilibrium with GSH oxidation).	$NAD(P)H_2$ prevents PT. Antioxidants prevent PT. Prooxidants favor PT.
matrix pH sensor	Reversible histidine protonation (of cyclophilin D ?) prevents PT.	Alkalinization (pH = 7.3) of matrix favors PT. Neutral or acidic matrix pH inhibits PT.
cation sensor	Ca^{2+} induces PT. Other divalent cations inhibit PT.	Increase in matrix Ca^{2+} induce PT. Mg^{2+} and Zn^{2+} prevent PT.
ADP/ATP sensor	ADP (and ATP) inhibit PT.	Extra ATP (glycolytic substrates) prevents PT. Oligomycin (F_1 ATPase inhibitor) favors PT.
Bcl-2/Bax sensor	Senses the ratio of Bax and Bcl-2-like proteins.	Bax induces PT. Bcl-2 and Bcl-XL prevent PT.
protease sensor	Direct action of proteases on outer membrane proteins.	Caspases induce PT. Calpain-like enzyme may favor PT.
lipid sensor ?	Long chain lipid acids induce PT.	Palmitate and stearate induce PT. Carnitine prevents PT.
peptide sensor ?	Amphipathic peptides induce PT.	Mastoparan induces PT.
PBR sensor	Ligands of the peripheral benzodiazepin receptor induce or facilitate PT.	Protoporphyrin IX induces PT. PKI1195 facilitates PT induction.

[a]For details consult main text, ref. 33, and TABLE 3.

characteristics correspond to the PT pore.[41] In contrast to these authors, but in accord with the initial work performed by the Brdiczka group,[42] our data suggest that the PT pore contains additional proteins,[21,29] including porin and hexokinase-I (TABLE 3). In our hands, the PTPC from rat brain co-purifies with apoptosis-regulatory proteins from the Bcl-2 family such as Bax and the Bcl-2-interacting protein Bag-I.[29] This PTPC can be reconstituted into artificial membranes and causes the permeabilization of liposomes, provided that inducers of PT pore opening are added.[29] Based on the characterization of the PTPC, we have recently shown that Bax requires a functional and physical interaction with ANT to exert its pro-apoptotic effects.[21] Thus, immunodepletion of Bax from PTPC or purification of PTPC from Bax knock-out mice yields a complex that fails to permeabilize membranes in response to the ANT ligand Atr. Experiments involving ANT-deficient yeast cells demonstrate that ANT is required for Bax to become cytotoxic. Purified ANT and recombinant ANT form an atractyloside-responsive channel when reconstituted together into liposomal membranes.[21] These data suggest that Bax acts on PTPC, via the ANT, to induce apoptosis. We have also demonstrated that recombinant Bcl-2 or Bcl-XL inhibit the permeabilization of liposomal membranes containing the purified PTPC. Thus while PTPC liposomes are permeabilized when treated with prooxidants, calcium, the ANT ligand atractyloside, or lonidamine, PTPC liposomes containing recombinant Bcl-2 or Bcl-X_L resist these stimuli.[29,51,52] In contrast, Bcl-2 or BCl-X_L fail to confer protection against PT pore opening induced by caspases or diamide in three different experimental systems: intact cells, isolated mitochondria, and PTPC liposomes.[28,29,51] Moreover, mutant Bcl-2 proteins that have lost their cytoprotective function and mutant Bax proteins which have lost their cytocidal potential have no effects on purified PTPC reconstituted into liposomes.[21,28,29,51] Altogether, these data suggest that PTPC am closely cooperating with Bcl-2-like proteins in the control of apoptosis.

Although some proteins within the PTPC have been identified, the overall composition of the PTPC remains an open question. At present, it appears possible that different proteins from the outer mitochondrial membrane (e.g. porin/VDAC, Bax) interact with proteins from the inner membrane (e.g. ANT and perhaps other mitochondrial carriers) in different combinations to regulate the permeability of both mitochondrial membranes in a coordinate or independent fashion. This issue is currently under intense investigation in our laboratory.

MITOCHONDRIA AND PTPC: FACULTATIVE OR OBLIGATE APOPTOSIS REGULATORS?

Although the triphasic scheme of apoptosis (see INTRODUCTION) which places mitochondria in the center of the decision/effector phase provides an attractive theoretical framework for the dissection of apoptotic pathways, it is unclear whether this model applies to all cases of apoptosis. Thus, in several rare examples of apoptosis, cytochrome c release does not occur, at least according to immunoblot analyses of subcellular fractions.[53,54] However, careful the quantitation of cytochrome c using in situ detection methods will be required to exclude the possibility that part of the mitochondrial cytochrome c has been released. A further critique of this model comes from the observation that Bcl-2 is not a universal inhibitor of apoptosis and

TABLE 3. Molecules in the PT pore complex as targets for pharmacological apoptosis modulation

Molecule (topology)	Normal function	Ligands, modulators, and role in PT	Reference
Adenine nucleotide translocator (ANT) (inner membrane, three tissue-specific isoforms in man, two in mouse)	ATP/ADP antiport.	Bongkrekic acid: favors *m*-state and inhibits PT, like ADP.	(59,60)
		Atractyloside: favors *c*-state and induces PT.	(59,60)
		ATP, ADP: inhibit PT.	(43)
		Fatty acids: favor PT.	(31)
		$\Delta\Psi_m$ regulates affinity of ANT to ADP, correlating with its PT-modulatory effect (high $\Delta\Psi_m$ prevents PT).	(61)
		Pro-oxidants and thiol reagents reduce affinity of ANT for ADP (and induce PT).	(61)
Cyclophilin D (matrix)	Peptidyl prolyl isomerase (chaper-one function).	Cyclosporin A: inhibits interaction with ANT (inner membrane) and inhibits PT.	(62)
		N-methyl-Val-4-cyclosporine A (non-immunosuppressive) acts like cyclosporin A.	(62)
		Protons (matrix pH) prevent interaction with inner membrane and inhibit PT.	(62)
		Prooxidants and thiol reagents favor binding to the inner membrane and induce PT.	(63) (41)
Bax	Proapoptotic protein tumor suppressor.	Interacts with ANT to facilitate ATR-induced PT.	(21)
		Induces mitochondrial membrane permeabilization when present in excess.	(22)
		Interacts with Bcl-2, which neutralizes its function	(64)

TABLE 3/continued.

Molecule (topology)	Normal function	Ligands, modulators, and role in PT	Reference
Peripheral benzodiazepin receptor (PBR) (outer membrane)	Receptor for endozepin (=CoA-binding protein). Normal function controversial.	Interacts with ANT and porin.	(65)
		Protoporphyrin IX: induces PT.	(66,67)
		PK1195: facilitates PT.	(46,56)
Hexokinase isoenzymes (cytosolic, binds to porin)	Phosphorylates hexasaccharides (mainly glucose).	Glucose: inhibits PT. Glucose-6-phosphate inhibits the glucose effect.	(42, 43)
Porin (outer membrane) (two isoforms)	Voltage-dependent anion channel (VDAC) transport of metabolites through the outer membrane.	Voltage determines opening or closing; relationship with PT is unclear.	(42)
		Interacts with ANT.	(68)
Mitoch. creatine kinase (intermembrane)	Transfers phosphate from creatinephosphate to ADP or from ATP to creatine.	Inhibits PT in its octamer form.	(42,69)
Ca^{2+} sensitive sites	Sensor for divalent cations.	Ca^{2+} favors PT. pH and Mg^{2+} influence Ca^{2+} binding and modulate PT.	(31,70)

actually fails to prevent apoptosis induced by CD95 crosslinking in certain cell lines in which CD95 ligation results in the rapid activation of caspase-8 within the CD95 receptor complex.[10] Since Bcl-2 is supposed to interrupt apoptosis at the mitochondrial level, this observation has been employed as an argument against the involvement of mitochondria in apoptosis.[10] However, in such cells Bcl-2 also fails to protect mitochondria against the $\Delta\Psi_m$ loss,[10,28] which closely correlates with the later advent of nuclear apoptosis. Moreover, Bcl-2 fails to prevent caspase-induced PT pore opening in isolated mitochondria or in PTPC liposomes.[28,29] Thus, the absence of Bcl-2-mediated cytoprotection does not rule out the possibility that mitochondria are involved in CD95-induced apoptosis. *In vitro* experiments suggest that mitochondria accelerate caspase-8-mediated caspase-3 activation,[55] although this activation can also occur in the absence of mitochondria.[11] The mitochondrial amplification of caspase-8-triggered caspase-3 activation can be inhibited by Bcl-2, but this inhibitory effect is overcome by increasing the amount of caspase-8, suggesting that the extent of caspase-8 activation may determine whether Bcl-2 can preserve the integrity of mitochondrial membranes.[55] Thus, the question whether mitochondria play a major decisive role in this particular apoptosis pathway remains open.

Another issue that remains elusive concerns the involvement of PTPC in the regulation of mitochondrial membrane permeability. Indeed, when added to isolated mitochondria, Bax permeabilizes the outer mitochondrial membrane and causes a $\Delta\Psi_m$ dissipation which is not accompanied by large amplitude swelling, as would be induced by most agents acting on PTPC.[22] However, according to several reports,[20–22] the effects of Bax on mitochondria, *in cellula* or *in vitro*, are inhibited by cyclosporin A and bongkrekic acid, suggesting that Bax functionally interacts with the PTPC. According to one paper, the mitochondrial effects of Bax are not inhibited by cyclosporin A or bongkrekic acid.[23] Again, the doubt resides in the question whether the inhibitor has been simply inefficient in one particularly experimental setting or whether Bax (and by extension other members of the Bcl-2 family) may selectively affect the outer mitochondrial membrane permeability via a direct, PT pore-independent process. The future will tell.

ACKNOWLEDGMENTS

Supported by Agence Nationale pour la Recherche sur le SIDA, Association pour la Recherche contre le Cancer, Centre Nationale de la Recherche Scientifique, Fondation pour la Recherche Médicale, Ligue Franqaise contre le Cancer, Institut National de la Santé et de la Recherche Médicale, Programme de Recherche Fondamentale en Microbiologie et Maladies Infectieuses et Parasitaires du Ministère de la Recherche, Sidaction (to G.K.).

REFERENCES

1. KROEMER, G. *et al*. 1995. The biochemistry of apoptosis. FASEB J. **9:** 1277–1287.
2. KROEMER, G., N. ZAMZAMI & S.A. SUSIN. 1997. Mitochondrial control of apoptosis. Immunol Today. **18:** 44–51.

3. KROEMER, G. 1997. Mitochondrial implication in apoptosis. Towards an endosymbiotic hypothesis of apoptosis evolution. Cell Death Diff. **4:** 443–456.
4. KROEMER, G., B. DALLAPORTA & M. RESCHE-RIGON. 1998. The mitochondrial death/life regulator in apoptosis and necrosis. Annu. Rev. Physiol. **60:** 619–642.
5. LIU, X. S. *et al.* 1996. Induction of apoptotic program in cell-free extracts: requirement for DATP and cytochrome C. Cell. **86:** 147–157.
6. XIANG, J., D.T. CHAO & S.J. KORSMEYER. 1996. Bax-induced cell death may not require interleukin 16-converting enzyme-like proteases. Proc. Natl. Acad. Sci. USA. **93:** 14559–14563.
7. KLUCK, R.M. *et al.* 1997. The release of cytochrome c from mitochondria: a primary site for Bcl-2 regulation of apoptosis. Science. **275:** 1132–1136.
8. VANDER HEIDEN, M.G. *et al.* 1997. Bcl-XL regulates the membrane potential and volume homeostasis of mitochondria. Cell. **91:** 627–637.
9. GREEN, D.R. & J.C. REED. 1998. Mitochondria and apoptosis. Science **281:** 1309–1312.
10. SCAFFIDI, C. *et al.* 1998. Two CD95 (APO-I/Fas) signaling pathways. EMBO J. **17:** 1675–1687.
11. STENNICKE, H.R. *et al.* 1998. Pro-caspase-3 is a major physiological target of caspase-8. J Biol. Chem. **273:** 27084–27090.
12. KROEMER, G. 1997. The proto-oncogene Bcl-2 and its role in regulating apoptosis. Nat. Med. **3:** 614–620.
13. ZAMZAMI, N. *et al.* 1998. Subcellular and submitochondrial mechanisms of apoptosis inhibition by Bcl-2-related proteins. Oncogene **16:** 2265–2282.
14. NICHOLSON, D.W. & N.A. THORNBERRY. 1997. Caspases: killer proteases. Trends Biochem. Sci. **22:** 299–306.
15. ASHKENAZI, A. & V. M. DIXIT. 1998. Death receptors: signaling and modulation. Science **276:** 111–113.
16. GREEN, D.R. & G. KROEMER. 1998. The central execution of apoptosis: mitochondria or caspases? Trends Cell Biol. **8:** 267–271.
17. GROSS, A. *et al.* 1998. Enforced dimerization of Bax results in its translocation, mitochondrial dysfunction and apoptosis. EMBO J. **17:** 3878–3885.
18. WOLTER, K.G. *et al.* 1997. Movement of Bax from the cytosol to mitochondria during apoptosis. J. Cell Biol. **139:** 1281–1292.
19. GOPING, I.S. *et al.* 1998. Regulated targeting of Bax to mitochondria. J. Cell Biol. **143:** 207–215.
20. PASTORINO, J.G. *et al.* 1998. The overexpression of Bax produces cell death upon induction of the mitochondrial permeability transition. J. Biol. Chem. **273:** 71770–7777.
21. MARZO, I. *et al.* 1998. Bax and adenine nucleotide translocator cooperate in the mitochondrial control of apoptosis. Science **281:** 2027–2031.
22. JÜRGENSMEIER, J.M. *et al.* 1998. Bax directly induces release of cytochrome c from isolated mitochondria. Proc. Natl. Acad Sci. USA. **95:** 4997–5002.
23. ESKES, R. *et al.* 1998. Bax-induced cytochrome c release from mitochondria is independent of the permeability transition pore but highly dependent on Mg^{2+} ions. J. Cell Biol. **143:** 217–224.
24. SHIMIZU, S. *et al.* 1998. Bcl-2 prevents apoptotic mitochondrial dysfunction by regulating proton flux. Proc. Natl. Acad Sci. USA. **95**: 1455–1459.
25. LUO, X., I. BUDIHARJO, H. ZOU, C. SLAUGHTER & X. WANG. 1998. Bid, a Bcl-2 interacting protein, mediates cytochrome c release from mitochondria in response to activation of cell surface death receptors. Cell. 481–490.
26. LI, H. *et al.* 1998. Cleavage of BID by caspase 8 mediates the mitochondrial damage in the Fas pathway of apoptosis. Cell. **94:** 491–501.
27. COSULICH, S.C. *et al.* 1997. Regulation of apoptosis by BH3 domains in a cell-free system. Curr. Biol. **12:** 913–920.

28. SUSIN, S.A. *et al.* 1997. The central executioner of apoptosis. Multiple links between protease activation and mitochondria in Fas/Apo-I/CD95- and ceramide-induced apoptosis. J. Exp. Med. **186:** 25–37.
29. MARZO, I. *et al.* 1998. The permeability transition pore complex: a target for apoptosis regulation by caspases and Bcl-2 related proteins. J. Exp. Med. **187:** 1261–1271.
30. CHENG, E.H. Y. *et al.* 1997. Conversion of Bcl-2 to a Bax-like death effector by caspases. Science **278:** 1966–1968.
31. ZORATTI, M. & I. SZABÒ. 1995. The mitochondrial permeability transition. Biochem. Biophys. Acta-Rev. Biomembranes **1241:** 139–176.
32. DEMARIA, R. *et al.* 1997. Requirement for GD3 ganglioside in CD95- and ceramide-induced apoptosis. Science **277:** 1652–1655.
33. SUSIN, S.A., N. ZAMZAMI & G. KROEMER. 1998. Mitochondrial regulation of apoptosis. Doubt no more. Biochim. Biophys. Acta (Bioenergetics). **1366:** 151–165.
34. PACKER, M.A., R. MIESEL & M.P. MURPHY. 1996. Exposure to the parkinsonian neurotoxin 1-methyl-4-phenylpyridinium (MPP^+) and nitric oxide simultaneously causes Cyclosporin A-sensitive mitochondrial calcium efflux and depolarisation. Biochem. Pharmacol. **51:** 267–273.
35. SCARLETT, J.L., M.A. PACKER, C.M. PORTEOUS & M.P. MURPHY. 1996. Alterations to glutathione and nicotinamide nucleotides during the mitochondrial permeability transition induced by peroxynitrite. Biochem. Pharmacol. **52:** 1047–1055.
36. HORTELANO, S. *et al.* 1997. Nitric oxide induces apoptosis via triggering mitochondrial permeability transition. FEBS Lett. **410:** 373–377.
37. RUVOLO, P.P. *et al.* 1998. A functional role for mitochondrial protein kinase C alpha in Bcl2 phosphorylation and suppression of apoptosis. J. Biol. Chem. **273:** 25436–25442.
38. ZHA, J.P. *et al.* 1996. Serine phosphorylation of death agonist BAD in response to survival factor results in binding to 14-3-3 not BCL-X(L). Cell **87:** 619–628.
39. BARDELLI, A., *et al.* 1996. HGF receptor associates with the anti-apoptotic protein BAG-1 and prevents cell death. EMBO J. **15:** 6205–6212.
40. BERNARDI, P. 1996. The permeability transition pore. Control points of a cyclosporin A-sensitive mitochondrial channel involved in cell death. Biochim. Biophys. Acta (Bioenergetics) **1275:** 5–9.
41. HALESTRAP, A.P. *et al.* 1998. Elucidating the molecular mechanism of the permeability transition and its role in reperfusion injury of the heart. Biochim. Biophys. Acta **1366:** 79–94.
42. BEUTNER, G. *et al.* 1996. Complexes between kinases, mitochondrial porin, and adenylate translocator in rat brain resemble the permeability transition pore. FEBS Lett. **396:** 189–195.
43. BEUTNER, G. *et al.* 1998. Complexes between porin, hexokinase, mitochondrial creatine kinase and adenylate translocator display properties of the permeability transition pore. Implication of regulation of permeability transition by the kinases. Biochim. Biophys. Acta-Biomembranes **1368:** 7–18.
44. ZAMZAMI, N. *et al.* 1995. Sequential reduction of mitochondrial transmembrane potential and generation of reactive oxygen species in early programmed cell death. J. Exp. Med. **182:** 367–377.
45. ZAMZAMI, N. *et al.* 1996. Mitochondrial control of nuclear apoptosis. J. Exp. Med. **183:** 1533–1544.
46. PASTORINO, J. G. *et al.* 1996. The cytotoxicity of tumor necrosis factor depends on induction of the mitochondrial permeability transition. J. Biol. Chem. **271:** 29792–29799.
47. MARCHETTI, P. *et al.* 1996. Mitochondrial permeability transition is a central coordinating event of apoptosis. J. Exp. Med. **184:** 1155–1160.

48. POLYAK, K. *et al.* 1997. A model for p53-induced apoptosis. Nature **389:** 300–305.
49. BRADHAM, C.A. *et al.* 1998. The mitochondrial permeability transition is required for tumor necrosis factor alpha-mediated apoptosis and cytochrome c release. Mol. Cell. Biol. **18:** 6353–6364.
50. RÜCK, A., *et al.* 1998. Reconstituted adenine nucleotide translocator forms a channel for small molecules comparable to the mitochondrial permeability transition pore. FEBS Lett. **426:** 97–101.
51. ZAMZAMI, N. *et al.* 1998. The thiol-crosslinking agent diamide overcomes the apoptosis-inhibitory effect of Bcl-2 by enforcing mitochondrial permeability transition. Oncogene **16:** 1055–1063.
52. RAVAGNAN, L. *et al.* 1998. Lonidamine triggers apoptosis via a direct, Bcl-2-inhibited effect on the mitochondrial permeability transition pore. Oncogene. In press.
53. TANG, D. G. *et al.* 1998. Apoptosis in the absence of cytochrome c accumulation in the cytosol. Biochem. Biophys. Res. Commun. **242:** 380–384.
54. ADACHI, S.,R.A. GOTTLIEB & B.M. BABIOR. 1998. Lack of release of cytochrome c from mitochondria into cytosol early in the course of fas-mediated apoptosis of Jurkat cells. J. Biol. Chem. **273:** 19892–19894.
55. KUWANA, T. *et al.* 1998. Apoptosis induction by caspase-8 is amplified through the mitochondrial release of cytochrome c. J. Biol. Chem. **273:** 16589–16594.
56. HIRSCH, T. *et al.* 1998. PK11195, a ligand of the mitochondrial benzodiazepin receptor, facilitates the induction of apoptosis and reverses Bcl-2-mediated cytoprotection. Exp. Cell Res. **241:** 426–434.
57. FULDA, S. *et al.* 1998. Molecular ordering of apoptosis induced by anti-cancer drugs in neuroblastoma cells. Cancer Res. **58:** 4453–4460.
58. COSTANTINI, P. *et al.* 1996. Modulation of the mitochondrial permeability transition pore by pyridine nucleotides and dithiol oxidation at two separate sites. J. Biol. Chem. **271:** 6746–6751.
59. KLINGENBERG, M. 1980. The ADP-ATP translocation in mitochondria, a membrane potential controlled transport. J. Membrane Biol. **56:** 97–105.
60. BRUSTOVETSKY, N. & M. KLINGENBERG. 1996. Mitochondrial ADP/ATP carrier can be reversibly converted into a large channel by Ca^{2+}. Biochemistry **35:** 8483–8488.
61. HALESTRAP, A.P., K.Y. WOODFIELD & C.P. CONNERN. 1997. Oxidative stress, thiol reagents, and membrane potential modulate the mitochondrial permeability transition by affecting nucleotide binding to the adenine nucleotide translocase. J. Biol. Chem. **272:** 3346–3354.
62. NICOLLI, A. *et al.* 1996. Interactions of cyclophilin with mitochondrial inner membrane and regulation of the permeability transition pore, a cyclosporin A-sensitive channel. J. Biol. Chem. **271:** 2185–2192.
63. CONNERN, C.P. & A.P. HALESTRAP. 1994. Recruitment of mitochondrial cyclophilin to the mitochondrial inner membrane under conditions of oxidative stress that enhance the opening of a calcium-sensitive non-specific channel. Biochem. J. **302:** 321–324.
64. OLTVAI, Z.N., C.L. MILLIMAN & S.J. KORSMEYER. 1993. Bcl-2 heterodimerizes in vivo with a conserved homolog, Bax, that accelerates programed cell death. Cell. **74:** 609–619.
65. MCENERY, M.W. *et al.* 1992. Isolation of the mitochondrial benzodiazepine receptor: Association with the voltage-dependent anion channel and the adenine nucleotide carrier. Proc. Natl. Acad. Sci. USA. **89:** 3170–3174.
66. PASTORINO, J.G. *et al.* 1994. Protoporphyrin IX, an endogenous ligand of the peripheral benzodiazepin receptor, potentiates induction of the mitochondrial permeability transition and the killing of culture hepatocytes by rotenone. J. Biol. Chem. **269:** 31041–31046.

67. Marchetti, P. *et al.* 1996. Mitochondrial permeability transition triggers lymphocyte apoptosis. J. Immunol. **157:** 4830–4836.
68. Hodge, T. & M. Colombini. 1997. Regulation of metabolite flux through voltage-gating of VDAC channels. J. Membr. Biol. **157:** 271–279.
69. O'Gorman, E. *et al.* 1997. The role of creatine kinase in inhibition of mitochondrial permeability transition. FEBS Lett. **414:** 253–257.
70. Bernardi, P. & V. Petronilli. 1996. The permeability transition pore as a mitochondrial calcium release channel; a Critical appraisal. J. Bioenerg. Biomembr. **28:** 129–136.

Coenzyme Q_{10} Can in Some Circumstances Block Apoptosis, and This Effect Is Mediated through Mitochondria

TERRI KAGAN, CLAUDETTE DAVIS, LIN LIN, AND ZAHRA ZAKERI[a]

Department of Biology, Queens College and The Graduate School and University Center of CUNY, Flushing, New York 11367, USA

ABSTRACT: The mitochondrial component coenzyme Q_{10} (CoQ_{10}[b]) has been used for many years as a dietary supplement intended to promote good health by trapping free radicals, thus preventing lipid peroxidation and DNA damage. We have tested its use as a generic anti-apoptotic compound and have found that its ability to protect against apoptosis varies depending on both cell type and mode of cell death induction. We have further established that this protection may be mediated by its effect on mitochondrial function and viability. We provide additional evidence that CoQ_{10}'s protective effect on mitochondrial membrane potential does not always result in altered mitochondrial enzyme activity and neither does it guarantee survival. These observations open the way for further investigations into the mechanisms involved in mitochondrial control of apoptosis.

INTRODUCTION

Substantial experimental data implicate both the generation of free oxygen radicals and the failure of mitochondrial function in programmed cell death.[1] For example, reactive oxygen species (ROS), which include free radicals, cause DNA strand breaks and base modifications that result in cell death.[2] DNA can be attacked by lipid peroxyl and alkoxyl radicals, resulting in base oxidation and strand breaks.[3–5] This DNA damage may be mediated by the mitochondria as a result of their mishandling of ROS.[6] Lipid peroxidation may also damage proteins, especially those associated with membranes.[7] These reports on mitochondrial activation during cell death suggest a means for the involvement of a mitochondrial component that can trap free radicals, such as CoQ_{10}.

Although its molecular and cellular mode of action is not fully understood, CoQ_{10} has been used clinically for several decades to treat patients suffering from ischemic heart disease,[8] congestive heart failure,[9] hypertension,[10] HIV infection,[11]

[a]Address for correspondence: Department of Biology, Queens College, 65-30 Kissena Blvd., Flushing, NY 11367, USA. 718-997-3450 (voice) 718-997-3429 (fax).
e-mail: Zahra_Zakeri@qc.edu

[b]ABBREVIATIONS: CoQ_{10}, coenzyme Q_{10}; ROS, reactive oxygen species; PC12, Pheochromocytoma; CHX, cycloheximide; NGF, nerve growth factor; PBS, phosphate buffered saline; TUNEL, TDT (terminal deoxynucleotidyltransferase) mediated dUTP nick end labeling; MTP, mitochondrial permeability transition; MTT, 3-(4,5-dimethylthiazol-2-yl)-2,5-diphenyltetrazolium; Rh123, Rhodamine 123.

and mitochondrial encephalomyopathies.[12] CoQ_{10} has also been shown to have potential for treatment of human cancer. The marked fatality of pancreatic cancer has been linked to exceptionally low levels of CoQ_{10}.[13–14] Furthermore, patients suffering from breast and other cancers are significantly deficient in CoQ_{10}.[15]

CoQ_{10} is found in a wide variety of cell types and is likely present in all cells. It is synthesized in the endoplasmic reticulum and Golgi apparatus from where it is translocated to other cytosolic compartments.[16] A vital component of the respiratory chain, CoQ_{10} is localized in the inner mitochondrial membrane where it serves as a highly mobile carrier of electrons and protons between the flavoproteins and the cytochromes. In this process, one molecule of CoQ_{10} can transfer two electrons and two protons. The reduced form of CoQ_{10} is known as ubiquinone; in this form it picks up a proton and changes to its free radical form, known as ubisemiquinone. This intermediate product can pick up another proton to give ubiquinol, which is the oxidized form of CoQ_{10}.

Ubiquinol is a mobile carrier and is used to transport protons to the b-c1 complex of the mitochondria[17]; however ubisemiquinone has been identified as the critical site where mitochondrial ROS are generated.[18] Ubiquinol also functions as an antioxidant, and ubiquinol has been shown to protect fatty acid emulsions, mitochondria and low-density lipoproteins against lipid peroxidation.[19–22] Antioxidants that scavenge free radicals, like ascorbic acid, α-tocopherol, or β-carotene, have also been shown to protect cells against cell death.[23] These reports indicate that ubiquinol may prevent both the initiation and propagation of lipid peroxidation, owing to its location in the hydrophobic region of the membrane phospholipid bilayer where the lipid peroxidation takes place.

Ubiquinone has been implicated as an important factor in cell death. First, in mouse proximal tubules HMG-CoA reductase inhibitors induce apoptosis via inhibition of isoprenoid production, a major step in the biosynthesis of ubiquinone.[24] Ubiquinone may also initiate downstream signaling pathways in response to oxidative damage *in vitro*. Second, the incidence of apoptosis induced by serum withdrawal appears to be dependent on the plasma membrane content of ubiquinone. In HL-60 cells, supplementation with ubiquinone during serum withdrawal-stimulated apoptosis inhibits cellular ceramide accumulation resulting in net decrease in the induction of apoptosis.[25]

CoQ_{10} has also been shown to stimulate cell proliferation in serum free conditions *in vitro*,[26–27] and thus may also play a role in cell growth. Nevertheless, the mechanisms by which CoQ_{10} elicits its responses are unknown. Therefore we have asked whether CoQ_{10} can interfere with the cell's apoptotic and proliferative machinery and if so, which pathways are involved. We examined the effect of lipid soluble CoQ_{10} on several cell lines chosen to represent tissues from different origins and states of differentiation, malignancy, and life span. We chose human histiocytic lymphoma (U937) and mouse melanoma (B16F10) to represent freely proliferative cells. We used differentiated rat adrenal pheochromocytoma (PC12) cells to represent a differentiated non-proliferative phenotype, in this instance neuronal.

Since distinct alterations in mitochondrial mass and function characterize different models of apoptosis,[28] we chose to look at several specific mediators of cell death and cell growth. For instance, ceramide is a cytosolic second messenger that has been implicated in actions as diverse as the induction of apoptosis, proliferation, senescence, and cell survival in a wide variety of cell types.[29–31] Cycloheximide inhibits apoptosis in a manner consistent with cell death mechanisms requiring protein synthesis,[32] but in other instances, cycloheximide has been shown to kill cells.[33] Ethanol is a consistent cell killer that both enhances intracellular free radical formation and alters membrane structure and fluidity.[34–35] Finally, the presence or absence of nerve growth factor (NGF) has also been shown to affect both cell death and cell proliferation.[36]

Our hypothesis is that CoQ_{10}, as an antioxidant and through its role in the electron transport chain, can suppress the accumulation of ROS and stabilize mitochondria, thus rendering cells more resistant to stress. This may then lead to a longer life span as well as a better quality of life for the individual. In this chapter we present initial data obtained on the protective effect offered by the antioxidant CoQ_{10} against various cellular insults. Our results suggest a selective protective role for CoQ_{10} that may be mediated by alterations in mitochondrial membrane potential.

MATERIALS AND METHODS

Cell Culture

PC12, B16F10 and U937 cells were obtained from ATCC (Rockville, MD). U937 and B16F10 cells were grown in RPMI-1640 (Biofluids). PC12 cells were supplemented with 5% fetal bovine serum (FBS) and 10% horse serum (Biofluids), while U937 and B16F10 cells were supplemented with 10% FBS. All media were further supplemented with 50 U/ml penicillin (Biofluids) and 50 mg/ml streptomycin (Biofluids). All cell lines were maintained in a humid atmosphere at 37°C in 5% CO_2 and fed two times/week. Naïve PC12 cells were induced to differentiate to a neuronal state by transfer to DMEM supplemented with 2% FBS, 1% horse serum and 0.5g/ml NGF (Harlan Bioproducts).

Cell Treatment

Cell death was induced by the addition of various agents. Inducers were diluted in appropriate solvents (ethanol or DMSO) before addition to the cultures, and final solvent concentrations did not exceed 0.1% or have detectable effects. Specific concentrations used were established from dose-response curves (data not shown) and previous studies.[38] C_8-ceramide (Biomol) was used at 10 μM concentration for 18–24 hours. All other agents were obtained from Sigma-Aldrich (St. Louis, MO). Cycloheximide (CHX) was used at 50 μg/ml. 25–200 μM CoQ_{10} was given to cell culture simultaneous with cell death stimuli. Incubation times for all treatments did not exceed 24 hours.

Analysis of Cell Death and Characterization of Apoptotic Cells

Cell death was assessed by various methods.

1. Phase contrast microscopy identified cell shrinkage, blebbing and neurite retraction.

2. Cell viability and membrane integrity were identified by two methods: trypan blue exclusion[37] and LIVE/DEAD® Assay (Molecular Probes, Eugene, OR). Viability was determined by direct counting of cells with a haemocytometer after exposure to trypan blue. The percentage of cells excluding the dye was calculated with respect to the corresponding control and confirmed by 3–4 independent trials. Viability was further confirmed by LIVE/DEAD assay. Briefly, treated cells are incubated in assay solution consisting of calcein acetoxymethyl ester (AM) and ethidium homodimer. Living cells contain calcein esterase and are able to exclude ethidium homodimer. Calcein esterase activity then hydrolyzes the ester compound and allows the AM group to fluoresce green. Dead cells show no esterase activity and have also lost membrane integrity, allowing entry of the ethidium homodimer; as a result they fluoresce red.
3. The integrity of the nucleus was determined by *in situ* staining[38] with Hoechst 33258 (*bis* benzimide) which binds to DNA and allows the structure of the nuclei or nuclear fragments to be seen. Briefly, cells either in suspension or on coverslips were washed in 1×PBS, fixed in 3% paraformaldehyde, washed again in 1×PBS and stained with bis benzimide (16 mg/ml) for 25 minutes. Following this staining process, cells were washed again in 1×PBS and examined by fluorescence microscopy with a DAPI filter (excitation 375–380 nm; emission 450–460 nm).

DNA Fragmentation Assessment by Agarose Gel Electrophoresis

DNA isolation was carried out as described previously.[38] Briefly, cells were harvested and washed in 1× PBS then lysed and allowed to incubate on ice for 15 minutes. After centrifugation, the supernatant was transferred to new Eppendorf tubes and RNase A was added (final concentration of 100 μg/ml); the tubes were then incubated in a 37°C water bath for 1 hour. Lipid and protein were removed from low molecular weight DNA with two phenol:chloroform:isoamyl alcohol (24:24:1) extractions and one chloroform:isoamyl alcohol (24:1) extraction. NaCl (300 mM) and 100% ethanol (2.5 vol.) were added to the tubes and the preparations were allowed to precipitate overnight at –20°C. DNA was pelleted at 12,000×*g* for 30 minutes at 4°C, washed in 75% ethanol and dried in a Speed-Vac (ATR, Laurel, MD). Finally, the pellets were resuspended in 1×TE buffer, loaded onto a 2% agarose gel and allowed to run at 80 volts for approximately 1 hour. Fragmented DNA was then visualized by ethidium bromide staining.

DNA Fragmentation Assessed by **in Situ** *Labeling of DNA Nicked Ends (TUNEL)*

In situ detection of DNA fragmentation was performed using the ApopTag kit (Intergen, Purchase, NY) essentially as described by Zakeri *et al.*[39] Briefly, cells were grown on coverslips or slides. Following induction of cell death, slides were pretreated with terminal deoxynucleotidyltransferase (TDT) buffer for 10 min at RT and then incubated with TDT for 1 h at 37°C in a humidified chamber. After the slides were washed in 1×PBS, they were incubated with digoxygenin peroxidase for 30 min. at RT. The slides were washed again in 1×PBS and then incubated with 0.8

mg/ml diaminobenzidine and 0.01% hydrogen peroxide in 0.1 M Tris, pH 7.2. The sections were counterstained with methyl green and coverslipped using Permount mounting media (Fisher). For negative controls, slides were incubated without TDT.

Analysis of Mitochondrial Activity

Mitochondrial status was examined in two ways:

1. Changes in the potential difference of the mitochondrial membrane were monitored by confocal microscopy using the fluorescent cation rhodamine-123 (Rh-123) according to the method of Lemasters *et al.*[40] Rh-123 enters the mitochondria and is retained as a function of the electrochemical and proton gradient[41]; as the mitochondrial membrane potential decreases, the marker is progressively lost. Cells were plated at a density of $2–3\times10^3/cm^2$ onto non-fluorescent glass cover slips (Fisher) or cover glass chambers (Nunc) in media as previously described. After treatment with inducers, Rh-123 at a final concentration of 0.5–1.0 μM was added to the culture medium for 30 min at 37°C in the dark. The cells were washed several times in 1×PBS to remove excess dye and a laser scanning confocal system was used to analyze Rh-123 localization. Confocal imaging was performed using a Meridian confocal microscope (Lansing, MI) at 20× magnification. An argon-ion laser beam (700 mW) was used for single fluorescein emission after excitation at 450–490 nm with emission at 515–550 nm. At each session, 3–5 similar fields of cells, containing at least 20 cells apiece, were counted, photomicrographic images were recorded, and temporal alterations in fluorescence intensity were calculated automatically with a digital imaging processor by a computer assisted image analyzing system (Meridian, Lansing MI).
2. Mitochondrial activity was also estimated by enzyme activity. 3-(4,5-dimethylthiazol-2-yl)-2,5-diphenyltetrazolium (MTT; 5 mg/ml; Sigma) was added to cells and left at 37°C for several hours. MTT is a water-soluble salt that is cleaved by active mitochondrial dehydrogenases present in living cells,[42] resulting in the formation of an insoluble purple precipitate. This precipitate was solubilized and quantified by spectrophotometer (OD_{570}) yielding a measure of absorbance as a function of concentration of converted dye.

RESULTS

CoQ_{10} Can Protect Differentiated but not Undifferentiated Cells against Some Cytotoxic Insults

We first determined that CoQ_{10} at various concentrations does not have any adverse effect on cell death (data not shown). Next we used several parameters to examine the amount and type of cell death induced by various inducers. To examine if CoQ_{10} can protect cells from cell death inducers, we exposed cells to ceramide (10 μM), cycloheximide (50 μg/ml), ethanol (1%–2.5%) or in the case of PC12 cells,

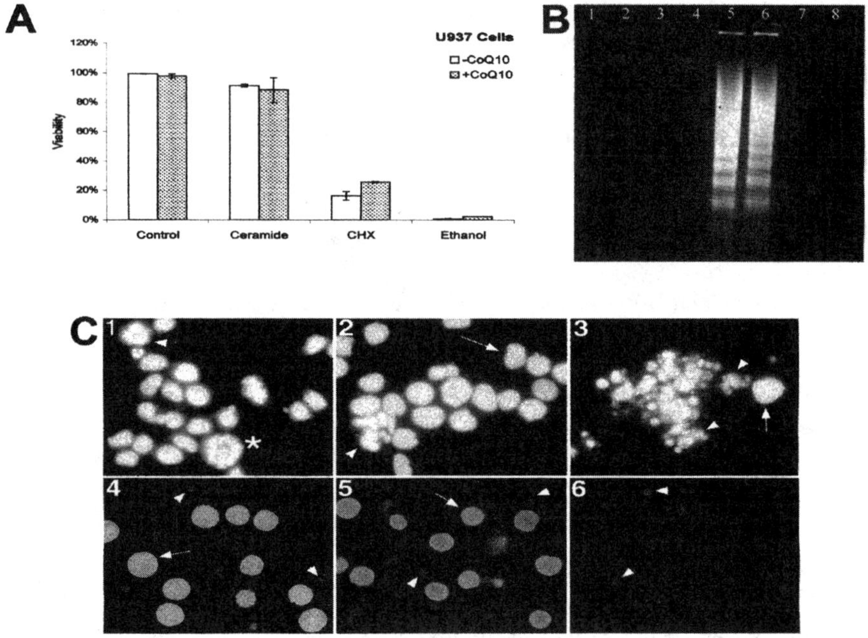

FIGURE 1. A: CoQ_{10} does not protect U937 cells against induction of cell death by a variety of toxins. Cells were treated with 10 μM ceramide, 50μg cycloheximide or 2.5% ethanol ± 100 μM CoQ_{10} as described in METHODS. 18 hrs post treatment, cells were collected and assessed for viability. Cycloheximide and ethanol were toxic to U937 cells. CoQ_{10} did not protect U937 cells against any of the toxins ($p >> 0.05$). **B**: CoQ_{10} does not prevent apoptosis in U937 cells. Cells were treated with inducers as described in **1A**. 18 hrs post treatment cells were collected and low molecular weight DNA was isolated and electrophoresed on a 2% agarose gel as described in METHODS. The fragmentation of DNA into bands differing by approximately 180bp was visualized by ethidium bromide staining. Untreated U937 cells ± CoQ_{10} (Lane 1, no CoQ_{10}; Lane 2, CoQ_{10}) show no DNA fragmentation. Ceramide treated cells (Lanes 3–4) show low levels of DNA fragmentation. Cycloheximide treated cells exhibit DNA ladders in both lanes (Lanes 5–6) regardless of the presence (Lane 6) of CoQ_{10}. In ethanol-treated cells (Lanes 7–8), mortality was very great and DNA fragmentation was a faster event, so that at 24 hours no laddering was evident. **C:** U937 cells normally die by nuclear fragmentation. Fluorescence microscope images of U937 cells (800× magnification) illustrate Hoechst 33258 stained cells (1–3) and cells stained with LIVE/DEAD® kit (4–6) from Molecular Probes (Eugene, OR). 1. Untreated cells show minimal nuclear fragmentation (*arrowhead*). A dividing cell was also present in the untreated U937 control (*asterisk*). 2. Cells treated with 10 μM ceramide show infrequent nuclear fragmentation by Hoechst staining, which does not vary significantly from control (arrow indicates a live cell). 3. Cells treated with 50 μg/ml cycloheximide exhibit severe nuclear fragmentation as shown by Hoechst staining (*arrowheads*) with infrequent live cells (*arrows*). 4. LIVE/DEAD assay of untreated cells shows brightly colored (live) cells (*arrows*) and more darkly shaded gray (dead) cells (*arrowhead*). 5. Ceramide-treated cells show little variation from untreated cells also when analyzed by LIVE/DEAD Assay. Brightly colored cells (*arrows*) represent more living cells than dead (*arrowheads*). 6. LIVE/DEAD assay of cells treated with cycloheximide also shows more dead cells (*gray*) than brightly colored living cells.

NGF depletion. The amount of cell death was measured by different parameters as described above.

We found that U937 cells are differentially sensitive to the several toxicants. While ceramide is not particularly toxic, cycloheximide and ethanol are significant cell death inducers. When we treated U937 cells with these toxicants in the presence of 100 μM CoQ_{10}, killing was not seriously diminished as determined by trypan blue exclusion (FIG. 1A); this result correlated with DNA fragmentation assay as shown in FIGURE 1B. Untreated U937 cells, in the absence or presence of CoQ_{10}, do not exhibit nuclear fragmentation, and neither do cells treated with ceramide. The characteristic DNA ladder is seen in cells treated with cycloheximide both in the absence and presence of CoQ_{10}. Ethanol treated cells often reveal extensive nuclear fragmentation as well (unpublished observations). However, because of the extreme toxicity of this compound, nuclear fragmentation is a fast event and by 18 hours, there is little remaining nuclear material visible by gel electrophoresis. Nuclear fragmentation was corroborated by Hoechst and LIVE/DEAD staining which revealed fragmented nuclei and extensive cell death as a result of treatment with cycloheximide, but not following ceramide treatment (FIG. 1C). Furthermore neither nuclear fragmentation or cell death was prevented by CoQ_{10}.

B16F10 melanoma cells were sensitive to CHX which resulted in a slight but significant ($p << 0.05$) decrease in cell viability (FIG. 2). B16F10 cells were not at all sensitive to ceramide. Similar to U937 cells, CoQ_{10} failed to protect B16F10 cells.

Differentiated PC12 cells are equally sensitive to ceramide and NGF withdrawal ($p = 0.05$ for both), resulting in a 10–15% loss of viability after treatment for 24 hours (FIG. 3a). Both cycloheximide and ethanol were more toxic and resulted in a 20%–40% loss of viability after 24 hours respectively ($p < 0.01$ for both). Cell killing by these agents was apoptotic as evidenced by nuclear fragmentation clearly visible

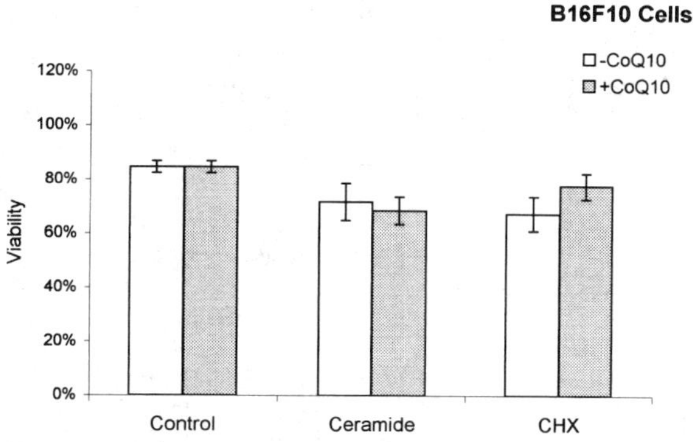

FIGURE 2. CoQ_{10} does not protect B16F10 cells from induced cell death. B16F10 cells were treated with 10 μM ceramide or 50 μg/ml cycloheximide ±100 μM CoQ_{10}. 24 hours later cells were scraped, and cell viability was determined as described in METHODS. B16F10 cells are not sensitive to ceramide ($p > 0.05$). Although B16F10 cells are sensitive to CHX ($p \leq 0.01$), there is no difference in viability in the presence of CoQ_{10} as compared with cells cultured in its absence ($p > 0.05$).

by Hoechst staining after NGF withdrawal (FIG. 3B-2) and ceramide treatment (FIG. 3B-4). Nuclear fragmentation after cycloheximide treatment is less obvious (FIG. 3B-5), but treatment with ethanol resulted in massive and rapid (within 12 hours) nuclear fragmentation (data not shown). The apoptotic nature of ceramide-induced cell death was further confirmed by *in situ* TUNEL labeling (FIG. 3B-6). While untreated

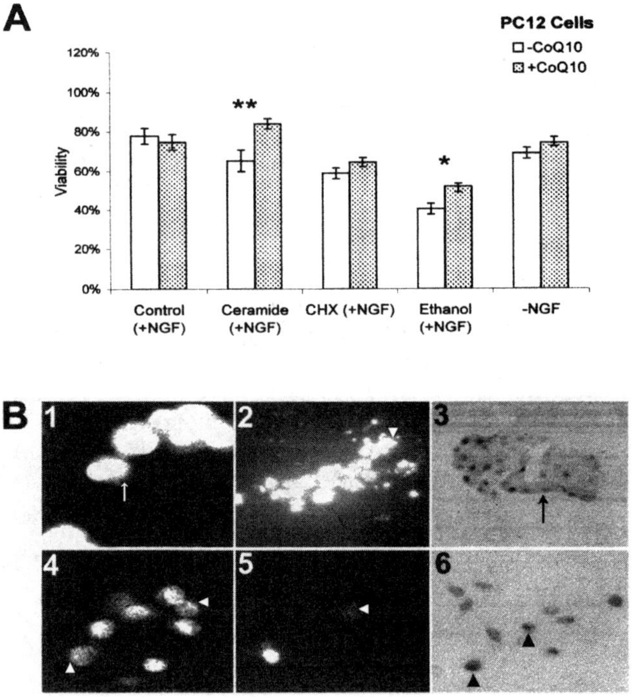

FIGURE 3. **A:** CoQ_{10} protects PC12 cells from certain types of induced cell death. Cultures of neuronal PC12 cells were exposed to 10 μM ceramide, 50 μg/ml cycloheximide, 2.5% ethanol or NGF withdrawal in the presence or absence of 100 μM CoQ_{10}. After 24 hours, cells were collected and assessed for viability as described in METHODS. CoQ_{10} offers significant protection ($p << 0.05$) to cells treated with ceramide or ethanol. ($*p \leq 0.05$, $**p \leq 0.01$) **B:** Cell death in neuronal PC12 cells is apoptotic (*arrowheads*). Fluorescence microscope images represent Hoechst 33258 stained cells (1–2, 4–5). Phase contrast images represent *in situ* TUNEL staining (3,6). 1. Magnified image (1200×) of untreated neurons does not show nuclear fragmentation. 2. Cells exposed to NGF withdrawal exhibit nuclear fragmentation after 24 hours (magnification 1200×). 3. TUNEL assay shows little staining of untreated cells (magnification 400×). 4. Cells treated with ceramide show some nuclear fragmentation by Hoechst staining (magnification 1200×). 5. Nuclear fragmentation is also present in cells treated with cycloheximide (magnification 1200×). 6. Magnified image (800×) of ceramide treated cells shows strong positive staining for apoptosis (dark cells) by TUNEL assay. Similar positive TUNEL staining is evident when cells are exposed to cycloheximide, ethanol, or NGF withdrawal (data not shown).

PC12 cells were clearly TUNEL negative (FIG. 3B-3), consistent, heavy positive TUNEL staining was evident in all experimental treatments (data not shown).

In contrast to the highly proliferative U937 cells, differentiated PC12 cells were differentially protected against apoptosis by CoQ_{10}. CoQ_{10} protected PC12 cells against ceramide or ethanol, significantly alleviating the loss of viability ($p << 0.05$). Such protection is not apparent when cells are exposed to either cycloheximide or NGF withdrawal.

CoQ_{10} Protection of Cells Correlates with Alteration of Mitochondrial Activity

To determine whether CoQ_{10} protection is mediated by alterations in mitochondrial membrane polarity, we treated PC12 cells as above, adding rhodamine 123 (Rh-123) during the last half-hour of treatment. PC12 cells were again treated with toxicants in the absence or presence of CoQ10 (in the presence of NGF), this time measuring average fluorescence/field using confocal microscopy after both the first hour of treatment and 24 hours later (described in METHODS). In cells treated with cell death inducers but without CoQ_{10}, we see a differential drop in mitochondrial membrane potential. When membrane potential is examined one hour after treatment (FIG. 4A) we find significant loss in potential over the control after ceramide (p < 0.01) and ethanol ($p < 0.05$) treatment but no difference in potential after NGF withdrawal ($p > 0.1$). This trend continues after 24 hours (FIG. 4B). At this time, ceramide exposed cells continue to show a loss in membrane potential similar to that seen after 1 hour ($p < 0.01$) while cells treated with NGF withdrawal show a slight decrease in mitochondrial membrane potential compared to controls ($p < 0.05$) and there is a further drop over the control values in ethanol treated cells ($p < 0.01$).

Addition of CoQ_{10} alone to cells had no effect after the first hour of treatment, nor was there any effect when CoQ_{10} was added to cells treated with NGF withdrawal for one hour (FIG. 4A). However, when CoQ_{10} was added to cells exposed to exogenous ceramide, we found that the loss of mitochondrial membrane potential (fluorescence/field) was ameliorated ($p < 0.01$) as compared to cells treated with ceramide alone, although mitochondrial membrane potential was not recovered to levels observed in control untreated cells. Cells treated with ethanol in the presence of CoQ_{10} also demonstrated similar short-term protection against loss of mitochondrial membrane potential. We found less mitochondrial membrane potential loss ($p < 0.05$) after the first hour in cells treated with ethanol and CoQ_{10} as compared to cells treated with ethanol alone, although there was no increase in mitochondrial membrane potential over values found for treatment with CoQ_{10} alone.

Addition of CoQ_{10} alone to cells caused an increase ($p < 0.05$) in mitochondrial membrane potential in control cells after 24 hours (FIG. 4B). However, addition of CoQ_{10} had no significant effect on loss of mitochondrial membrane potential in ceramide treated cells after 24 hours ($p > 0.05$). Interestingly, although loss of mitochondrial membrane potential was prevented in cells treated with short term (1 hour) ethanol in the presence of CoQ_{10}, when cells were treated with ethanol in the presence of CoQ_{10} for a longer term (24 hours), loss of mitochondrial membrane potential was not prevented. In contrast, when PC12 cells were exposed to NGF withdrawal in the presence of CoQ_{10} for one hour, there was no significant increase in mitochondrial membrane potential as compared either to cells treated with NGF withdrawal or to control cells. However, after 24 hours we observed a significant in-

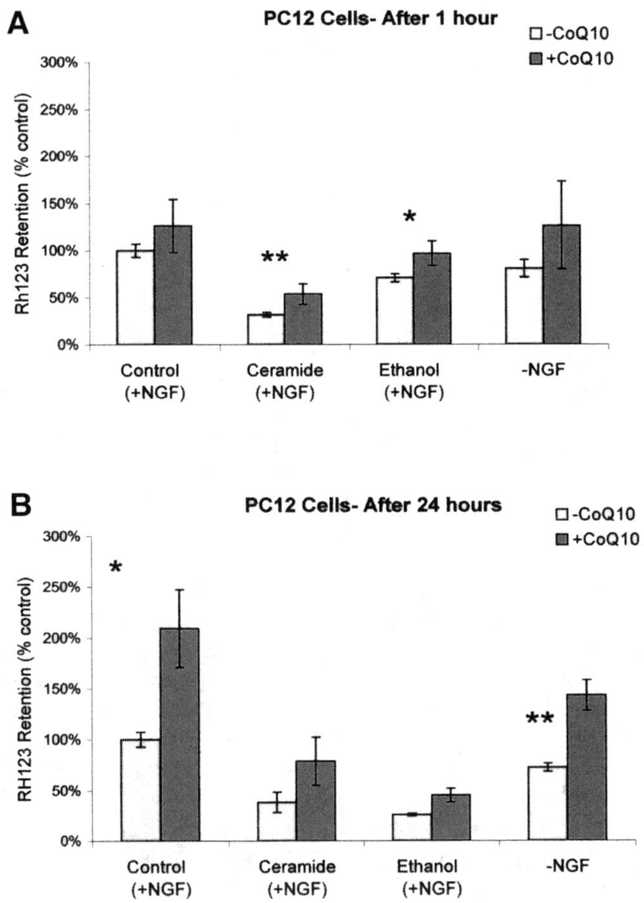

FIGURE 4. A: Induction of cell death results in an immediate loss of mitochondrial membrane potential, which can be offset by CoQ_{10}. Using confocal microscopy as described in Methods, we calculated average fluorescence/field and examined activity following induction of cell death. CoQ_{10} prevented loss of mitochondrial potential following treatment with ceramide ($p << 0.05$) and ethanol ($p = 0.05$). Loss of potential was not significant in cells exposed to NGF withdrawal, and there were no significant changes in mitochondrial potential in the presence of CoQ_{10}. **B:** After 24 hours, control cells treated with CoQ_{10} alone showed increased mitochondrial potential ($p \leq 0.05$), but cells treated with both CoQ_{10} and either ceramide or ethanol showed no alterations at this time relative to cells treated in the absence of CoQ_{10}. Nevertheless, CoQ_{10} did prevent loss of mitochondrial potential in PC12 cells following 24 hours of NGF withdrawal ($p \leq 0.005$). (Asterisks indicate significance of $+CoQ_{10}$ relative to $-CoQ_{10}$ $*p \leq 0.05$, $**p \leq 0.01$).

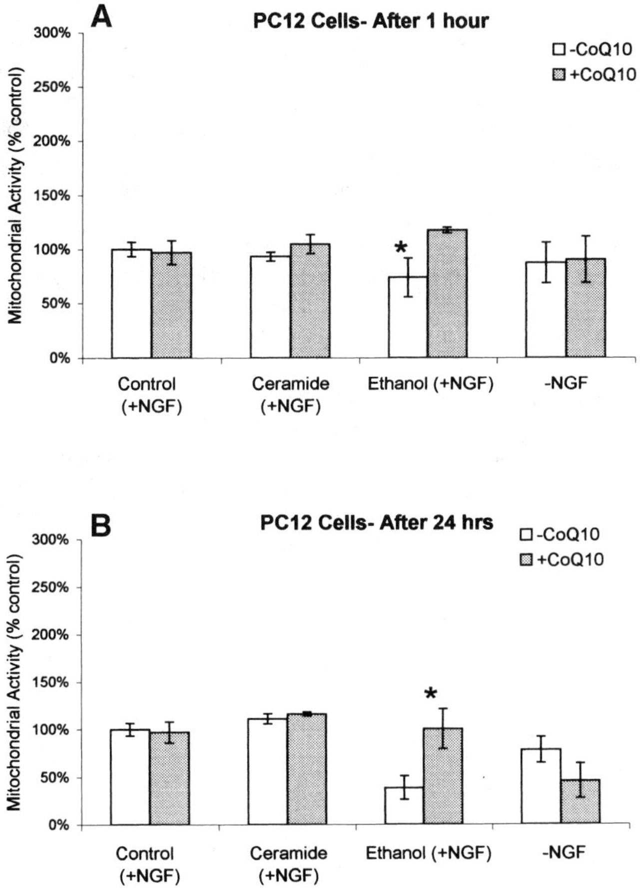

FIGURE 5. **A:** Mitochondrial enzyme activity was selectively disrupted by cell death induction in PC12 cells as measured by MTT assay (described in METHODS). Treatment with CoQ_{10} did not alter mitochondrial enzyme activity in untreated cells, nor did ceramide (10 μM) or NGF withdrawal result in alterations to mitochondrial enzyme activity. However within the first hour of treatment with ethanol (2.5%), enzyme activity was decreased below control. This loss was prevented by treatment with CoQ_{10} ($p < 0.05$). In addition, mitochondrial enzyme activity was unaffected following treatment with either ceramide or NGF withdrawal even in the presence of CoQ_{10}. **B:** After 24 hours, neither untreated, nor ceramide treated cells exhibit losses in mitochondrial enzyme activity although losses are seen in cells treated with either ethanol or NGF withdrawal. Although this loss was unaffected by the presence of CoQ_{10} following NGF withdrawal, following ethanol treatment this loss was somewhat ameliorated ($p < 0.05$). ($*p \leq 0.05$).

crease in mitochondrial membrane potential in cells deprived of NGF but provided with CoQ_{10} as compared to cells deprived of NGF alone ($p < 0.005$).

Finally, mitochondrial activity and viability was also estimated by dehydrogenase activity using MTT (described in METHODS). Living cells contain the enzyme mitochondrial dehydrogenase, which is able to cleave the water-soluble salt MTT resulting in an insoluble purple precipitate. For this experiment we measured mitochondrial activity in PC12 cells following treatment similar to that described above. After 1 hour of treatment (FIG. 5a), we found no differences in mitochondrial enzyme activity as compared to untreated cells following addition of ceramide or withdrawal of NGF, but a significant ($p < 0.05$) loss of mitochondrial enzyme activity following ethanol treatment. Interestingly, after 24 hours (FIG. 5b), there was still no loss in mitochondrial enzyme activity in ceramide treated cells, however, both ethanol exposed ($p << 0.05$) and NGF depleted cells ($p < 0.05$) showed decreased mitochondrial activity.

When mitochondrial enzyme activity was measured in cells treated with CoQ_{10} alone, we found no differences in enzyme activity after either short (1 hour) or long (24 hour) term treatment. Similarly, mitochondrial enzyme activity was not affected in cells treated with either ceramide or NGF withdrawal in the presence of CoQ_{10} after either short or long term treatment. However, CoQ_{10} provided significant protection against loss of mitochondrial enzyme as compared to untreated controls, when PC12 cells were treated with ethanol, both within the first hour of incubation and after 24 hours ($p < 0.05$).

DISCUSSION

CoQ_{10} protects some cells against cytotoxic insults. Our results suggest a protective role of CoQ_{10} against specific toxins in specific cell types; just as cell death stimuli do not affect different cell lines equivalently, neither does CoQ_{10} serve as a universal protective agent. Some highly proliferative cells, such as U937, are sensitive to a variety of cytotoxic stimuli (CHX and ethanol, but not ceramide), and are not protected by CoQ_{10}. Other proliferative cells, such as B16F10, are slightly sensitive to both ceramide and CHX ($p \leq 0.05$) and yet are not protected by CoQ_{10}. Neuronal cells, which are non-proliferative, are sensitive to most cell death inducers; they are dependent on the hormone NGF and they die in response to a variety of other cytotoxic stimuli ($p < 0.05$). These cells exhibit the most consistent protection afforded by CoQ_{10} after ceramide or ethanol treatment but show little or no protection against CHX or NGF withdrawal.

These results suggest that the toxicities of ceramide and ethanol are similar to each other but different from the toxicity produced by administration of CHX or withdrawal of NGF. They also suggest that specific toxins may act differently in different cell types, or that the detection of apoptosis varies with cell type. U937 cells treated with ceramide exhibit no significant loss in viability when examined by trypan blue exclusion. However when apoptosis is analyzed by gel electrophoresis, DNA fragmentation into small oligonucleosomal fragments is evident starting at 18 hours. Ethanol-treated cells show significant loss of viability by trypan blue exclusion at 18 hours, but DNA fragmentation is detectable much earlier, by 6 hours (data not shown). This suggests that the induction of apoptosis is an extended process and that its kinetics may differ significantly depending on mode of induction. In addition, since cell death induced by CHX in the presence of CoQ_{10} has varying effects on the three cell types chosen, we suggest that more than one pathway may be used by a specific inducer or that regulatory and feedback mechanisms are sufficiently different to produce apparently different results, depending on the time of analysis. The varying ability of CoQ_{10} to suppress apoptosis may reflect alternative uses for CoQ_{10} within a cell. Although CoQ_{10} is widely thought of as an antioxidant whose presence is mainly necessary for mitochondrial redox activities, it may have additional functions.

The effect of CoQ_{10} can be mediated by alterations in mitochondrial activity. Our results point to a probable signaling pathway that may be common to specific cell death stimuli: the involvement of mitochondria. Mitochondria participate in the control of apoptotic nuclear disintegration.[43] Mitochondrial participation can be seen prior to nuclear apoptosis (chromatin condensation and DNA fragmentation), by the reduction of the inner mitochondrial transmembrane potential ($\Delta\Psi_m$) mediated by the opening of mitochondrial permeability transition (MPT) pores.[44] Hyperproduction of superoxide anions, outflow of matrix calcium and glutathione, disruption of mitochondrial biogenesis, and release of soluble intermembrane proteins are important consequences of MPT.[45] Swelling of the mitochondria also releases an apoptosis-inducing protein (possibly a protease) that is normally sequestered in the intermembrane space[46] and cytochrome c is also released upon mitochondrial swelling.[47]

We found a general enhancement of mitochondrial viability and activity in the presence of CoQ_{10} in addition to a protection by CoQ_{10} of mitochondrial viability and activity after exposure to ceramide or ethanol. This consequence may explain the protective effect of CoQ_{10} against cell death stimuli, since apoptosis often starts in the mitochondria, resulting in release of cytochrome c and consequent activation of cysteine-aspartate proteases (caspases).[48] Thus CoQ_{10} could interact with the apoptotic machinery in the mitochondria, preventing the loss of mitochondrial potential by sequestering ROS generated in response to cell death signals and thus blocking cell death. This effect is correlated with the significant loss of mitochondrial membrane potential exhibited by PC12 cells within the first hour of treatment with either ceramide or ethanol. This loss in membrane potential was not evident in the presence of CoQ_{10}. Furthermore, we found that after 24 hours of NGF withdrawal, PC12 cells also exhibited a loss in mitochondrial potential, which was prevented by the presence of CoQ_{10}. Thus a change in mitochondrial membrane potential may be a means by which CoQ_{10} elicits its protective effect, but protection of mitochondrial potential does not guarantee survival. This may be because protection of mitochondrial potential is not necessarily an immediate event and can take up to 24 hours to be observed. However by the time CoQ10's effect on mitochondrial membrane potential is obtained, it is superfluous since the cell death signal has already been initiated.

Our results also indicate that CoQ_{10} does not appear to work at the level of mitochondrial activity in most cases of cell death, regardless of significant changes in mitochondrial membrane potential. Although we reported that CoQ_{10} significantly increased mitochondrial membrane potential in cells exposed to short term ceramide treatment and long term NGF withdrawal, its presence had no significant effect on mitochondrial activity (as measured by MTT assay). Thus, it appears that mitochondrial dehydrogenases are retained and remain capable of accepting electrons after depolarization. This implies a substantial limit to the use of MTT as an early assay for cell death. Other investigators have shown that ceramide causes a disruption of the inner mitochondrial transmembrane potential that precedes the nuclear signs of apoptosis in various cell types.[49–50] Ceramide may result in increased ROS generation,[51] and oxidative stress may cause changes in mitochondrial membrane potential.[52] Thus, CoQ_{10} may in this instance act as an antioxidant.

CoQ_{10}'s protective role in ethanol induced toxicity may also be indirect. It is only following ethanol treatment that PC12 cells exhibit rapid and sustained loss of cell viability, mitochondrial membrane potential and mitochondrial enzyme activity, which are ameliorated by CoQ_{10}. Previous investigators have indicated that CoQ_{10} protects against glutamate neurotoxicity *in vivo*[54] and that N-methyl D-aspartate (NMDA) is a key mediator of glutamate induced cytotoxicity.[53] Ethanol has also been shown to inhibit NMDA receptors.[34] Taking all these observations together, we may postulate that any benefit gained by treatment with CoQ_{10} may also result from some sort of indirect protection of mitochondrial enzyme activity.

The greatest effect of CoQ_{10} on NGF withdrawal is an increase in mitochondrial membrane potential, evident 24 hours after treatment. Differentiating neurons depend on NGF[36]; withdrawal of it activates the c-Jun NH_2 terminal kinase (JNK) group of MAP kinases[55] and the endogenous ceramide signaling pathway.[56] Since the ceramide signaling pathway is activated 18–24 hours after NGF withdrawal (our unpublished observations), the late depolarization of the mitochondria may directly result from ceramide generation, as we see after 1 hour of administration of exogenous ceramide. These results suggest that the prevention of mitochondrial membrane potential loss is the most common effect of CoQ_{10} at the cellular level and perhaps is the more important means of preventing cell death by these inducers. However, mitochondrial membrane potential and mitochondrial enzyme activity are not necessarily coupled during induced cell death.

The variations in the induction, frequency, and reversibility of apoptosis in the three cell types may be explained by a number of different factors. It appears that in limited circumstances, CoQ_{10} can prevent apoptosis. Depending on both cell type and stimulus, the route to apoptosis may vary. Finally, where CoQ_{10} is effective, the effect may derive from directly stabilizing mitochondrial membrane potential or by indirectly protecting against loss in mitochondrial enzyme activity. However, apoptosis may be achieved independently of a protection of the mitochondria. It will be of interest in the future to determine those circumstances in which CoQ_{10} can be useful as a therapeutic supplement, and the means by which it acts.

ACKNOWLEDGMENTS

This research was supported in part by NIH Grants R13AG15846 (for the meeting) and SK04-AG00631, as well as by a CUNY Collaboration Initiative Grant.

REFERENCES

1. KRUMAN, I., Q. GUO & M.P. MATTSON. 1998. Calcium and reactive oxygen species mediate staurosporine-induced mitochondrial dysfunction and apoptosis in PC12 cells. J. Neurosci. Res. **51:** 293–308.
2. HA, H.C., J.D. YAGER, P.A. WOSTER & R.A. CASERO JR. 1998. Structural specificity of polyamines and polyamine analogues in the protection of DNA from strand breaks induced by reactive oxygen species. Biochem. Biophys. Res. Commun. **244:** 298–303.
3. HRUSZKEWYCZ, A.M. & D.S. BERTOLD. 1990. The 8-hydroxyguanine content of isolated mitochondria increases with lipid peroxidation. Mutat. Res. **244**: 123–128.
4. ZHANG, J.R. & A. SEVANIAN. 1991. Effect of vitamin E on arachidonic acid peroxidation and its binding to Chinese hamster V79 cell DNA. Biochim. Biophys. Acta **1085:** 159–166.
5. PARK, J.W. & K.A. FLOYD. 1992. Lipid peroxidation products mediate the formation of 8-hydroxydeoxyguanosine in DNA. Free Radic. Biol. Med. **12:** 245–250.
6. JACOTOT, E., P. CONSTANTINI, E. LABOUREAU, N. ZAMZAMI, S.A. SUSIN & G. KROEMER. 1999. Mitochondrial membrane permeabilization during the apoptotic process. Ann. N.Y. Acad. Sci. **887:** 18–30. This volume.
7. ZWIZINSKI, C.W. & H.H. SCHMID. 1992. Peroxidative damage to cardiac mitochondria identification and purification of modified adenine nucleotide translocase. Arch. Biochem. Biophys. **294:** 178–183.
8. KATO, T., S. YONEDA, T. KAKO, M. KOKETSU, I. HAYANO & T. FUJINAMI. 1990. Reduction in blood viscosity by treatment with coenzyme Q_{10} in patients with ischemic heart disease. Int. J. Clin. Pharmacol. Ther. Toxicol. **28:** 123–126.
9. TOPI, P.L., A. DAVINI, G. SQUARCINI. 1989. Efficiency of ubidecarenone in the treatment of patients with cardiac insufficiency. Minerva Cardiangiol. **37:** 255–258.
10. LANGSJOEN, P.H., K. FOLKERS, K. LYSON, K. MURATSU, T. LYSON & P. LANGSJOEN. 1988. Effective therapy with coenzyme Q_{10} for cardiomyopathy [abstract]. Klin. Wochenschr. **66:** 583.
11. FOLKERS, K., P. LANGSJOEN, Y. NARA, K MURATSU, J. KOMOROWSKI, P.C. RICHARDSON & T.H. SMITH. 1988. Biochemical deficiencies of Coenzyme Q10 in HIV-infection and exploratory treatment. Biochem. Biophys. Res. Commun. **153:** 888–896.
12. DENSNUELLE, J., F. PELLISSIER, G. SERRATRICE, J. POUGET & D.M. TURNBULL. 1989. Kearns-Sayer syndrome: mitochondrial encephalomyopathy caused by deficiency of the respiratory chain. Rev. Neurol. (Paris). **145:** 842–850.
13. BLIZNAKOV, E. 1973. Effect of stimulation of the host defense system by Coenzyme Q_{10} on dibenzpyrene-induced tumors and infection with friend leukemia virus in mice. Proc. Natl. Acad. Sci. USA **70:** 390–394.
14. FOLKERS, K., R. BROWN, W.V. JUDY, M. MORITA. 1993. Survival of cancer patients on therapy with Coenzyme Q_{10}. Biochem. Biophys. Res. Commun. **192:** 241–245.
15. JOLLIET, P., N. SIMON, J. BARRE, J.Y. PONS, M. BOUKEF, B.J. PANIEL & J.P. TILEMENT. 1998. Plasma coenzyme Q_{10} concentrations in breast cancer: prognosis and therapeutic consequences. Int. J. Clin. Pharmacol. Ther. **36:** 506–509.
16. RAWN, D.J. 1989. Oxidative Phosphorylation. Biochemistry. Neil Patterson Publishers, Burlington, NC.
17. MITCHELL, P. 1975. Proton-motive redox mechanism of the cytochrome b-c1 complex in the respiratory chain: proton-motive ubiquinone cycle. FEBS Lett. **56:** 1–6.
18. SCHULZE-OSTHOFF K., A.C. BAKKER, B. VANHAESEBROECK, J. BEYAERT, W.A. JACOB & W. FIERS 1992. Cytotoxic activity of tumor necrosis factor is mediated by early damage of mitochondrial functions. Evidence for the involvement of mitochondrial radical generation. J. Biol. Chem. **267:** 5317–5323.
19. FORSMARK-ANDREE, P. & L. ERNSTER. 1994. Evidence for a protective effect of endogenous ubiquinol against oxidative damage to mitochondrial protein and DNA during lipid peroxidation. Mol. Aspects Med. **15**(Suppl.)**:** 73–81.

20. STOCKER, R., V.W. BOWRY & B. FREI. 1991. Ubiquinol-10 protects human low- density lipoproteins more efficiently against lipid peroxidation than does α-tocopherol. Proc. Natl. Acad. Sci. USA **88:** 1646–1650.
21. MOHR, D., V.W. BOWRY & R. STOCKER. 1993. Dietary supplementation with Coenzyme Q_{10} results in increased levels of ubiquinol-10 within circulating lipoproteins and increased resistance of human low-density lipoprotein to the initiation of lipid peroxidation. Biochim. Biophys. Acta **1126:** 247–254.
22. FREI, B., M.C. KIM & B.N. AMES. 1990. Ubiquinol-10 is an effective lipid-soluble antioxidant at physiological concentrations. Proc. Natl. Acad. Sci. USA **87:** 4879–4883.
23. FERRARI, G., C.Y. YAN & L.A. GREENE. 1995. N-acetylcysteine (D- and L-stereoisomers) prevents apoptotic death of neuronal cells. J. Neurosci. **15:** 2857–2866.
24. IIMURA, O., F. VRTOVSNIK, F. TERZI & G. FRIEDLANDER. 1997. HMG-CoA reductase inhibitors induce apoptosis in mouse proximal tubular cells in primary culture. Kidney Int. **52:** 962–972.
25. BARROSO, M.P., C. GOMEZ-DIAZ, J.M.VILLALBA, M.I. BURON, G. LOPEZ-LLUCH & P. NAVAS. 1997. Plasma membrane ubiquinone controls ceramide production and prevents cell death induced by serum withdrawal. J. Bioenerg. Biomembr. **29:** 259–267.
26. SUN, I.L., E.E. SUN & F.L. CRANE. 1992. Stimulation of serum-free cell proliferation by Coenzyme Q. Biochem. Biophys. Res. Commun. **189:** 8–13.
27. CRANE, F.L., I.L. SUN, R.A. CROWE, F.J. ALCAIN & H. LOW. 1994. Coenzyme Q_{10}, plasma membrane oxidase and growth control. Mol. Aspects Med **15** (Suppl.): s1–s11.
28. CAMILLERI-BROET, S., H. VANDERWERFF, E. CALDWELL & D. HOCKENBERRY. 1998. Distinct alterations in mitochondrial mass and function characterize different models of apoptosis. Exp. Cell Res. **239:** 277–292.
29. HANNUN, Y.A. 1996. Functions of ceramide in coordinating cellular responses to stress. Science **274:** 1855–1859.
30. HAIMOVITZ-FRIEDMAN, A., R.N. KOLESNICK & Z. FUKS. 1997. Ceramide signaling in apoptosis. Br. Med. Bull. **53:** 539–553.
31. HANNUN, A. 1994. The sphingomyelin cycle and the second messenger function of ceramide. J. Biol. Chem. **269:** 3125–3128.
32. YOUNG, D.A. 1969. Glucocorticoid action on rat thymus cells. Interrelationships between carbohydrate, protein and adenine nucleotide metabolism and cortisol effects on these functions in vitro. J. Biol. Chem. **244:** 2210–2217.
33. MATSUE, H., D. EDELBAUM, A.C. HARTMANN, A. MORITA, P.R. BERGSTRESSER, H.YAGITA, K. OKUMURA & A. TAKASHIMA. 1999. Dendritic cells undergo rapid apoptosis in vitro during antigen specific interaction with $CD4^+$ T cells. J. Immunol. **162:** 5287–5298.
34. MILLER, M.W. 1996. Mechanisms of ethanol induced neuronal death during development: from the molecule to behavior. Alcohol Clin. Exp. Res. **20** (Suppl. 8): 128A–132A.
35. KUROSE, I., H. HIGUCHI, S. KATO, S. MIURA & H. ISHII. 1996. Ethanol induced oxidative stress in the liver. Alcohol Clin. Exp. Res. **20**(Suppl. 1): 77A–85A.
36. BATISTATOU, A. & L. GREENE. 1991. Aurintricarboxylic acid rescues PC12 cells and sympathetic neurons from cell death caused by nerve growth factor deprivation: correlation with suppression of endonuclease activity. J. Cell Biol. **115:** 461–471.
37. MIYASHITA, T. & J.C. REED. 1993. Bcl-2 oncoprotein blocks chemotherapy-induced apoptosis in a human leukemia cell line. Blood **81:** 151–157.
38. KAGAN, T. & Z. ZAKERI. 1998. Detection of apoptotic cells in the nervous system. Methods in Molecular Medicine. J. Harry & H.A. Tilson, Eds. **22:** 105–123.

39. ZAKERI, Z., D. QUAGLINO & H.S. AHUJA. 1994. Apoptotic cell death in the mouse limb and its suppression in the hammertoe mutant. Dev. Biol. **165:** 294–297.
40. LEMASTERS, J.J., J. DIGIUSEPPI, A.L. NIEMINEN & B. HERMAN. 1987. Blebbing, free $Ca2^+$ and mitochondrial membrane potential preceding cell death in hepatocytes. Nature **325**: 78–81.
41. JOHNSON, L.V., M.L. WALSH, B.J. BOCKUS & J.B. CHEN. 1981. Monitoring of relative mitochondrial membrane potential I living cells by fluorescence microscopy. J. Cell Biol. **88:** 526–535.
42. MOSSMAN, T. 1983. Rapid colorimetric assay for cellular growth and survival: application to proliferation and cytotoxicity assays. J. Immunol. Methods **65:** 55–63.
43. SKULACHEV, V.P. 1996. Why are mitochondria involved in apoptosis? Permeability transition pores and apoptosis as selective mechanisms to eliminate superoxide- producing mitochondria and cell. FEBS Lett. **397:** 7–10.
44. MACOUILLARD-POULLETIER DE GANN, M.A. BELAUD-ROTUREAU, P. VOISIN, N. LEDUCQ, F. BELLOC & P. CANIONI. 1998. Flow cytometric analysis of mitochondrial activity in situ: application to acetylceramide induced mitochondrial swelling and apoptosis. Cytometry **33:** 333–339.
45. PASTORINO, J.G., G. SIMBULA, K. YAMAMOTO, P.A. GLASCOTT JR., R.J. ROTHMAN & J.L. FARBER. 1996. The cytotoxicity of tumor necrosis factor depends on induction of the mitochondrial permeability transition. J Biol. Chem. **271:** 29792–29798.
46. SUSIN, S.A., N. ZAMZAMI, & M. CASTEDO. 1997. The central executioner of apoptosis: multiple connections between protease activation and mitochondria in Fas/APO-1/CD95- and ceramide-induced apoptosis. J. Exp. Med. **186:** 25–37.
47. YANG, J.C. & G.A. CORTOPASSI. 1998. dATP causes specific release of cytochrome c from mitochondria. Biochem Biophys Res Commun. **250:** 454–457.
48. ZAMZAMI, N., S.A. SUSIN, P. MARCHETI, T. HIRSCH, I. GOMEZ-MONTERREY, M. CASTEDO & G. KROEMER. 1996. Mitochondrial control of nuclear apoptosis. J. Exp. Med. **183:** 1533–1544.
49. ARORA, A.S., B.J. JONES, T.C. PATEL, S.F. BRONK & G.J. GORES. 1997. Ceramide induces hepatocyte cell death through disruption of mitochondrial function in the rat. Hepatology. **25:** 958–963.
50. FRANCE-LANORD, V., B. BRUGG, P.P. MICHEL, Y. AGID & M. RUBERG. 1997. Mitochondrial free radical signal in ceramide dependent apoptosis: a putative mechanism for neuronal death in Parkinson's disease. J. Neurochem. **69:** 1612–1621.
51. QUILLET-MARY, A., J.P. JAFFREZOU, V. MANSAT, C. BORDIER, J. NAVAL & G. LAURENT. 1997. Implication of mitochondrial hydrogen peroxide generation in ceramide induced apoptosis. J. Biol Chem. **272:** 21388–21395.
52. SATOH, T., Y. ENOKIDO, H. AOSHIMA, Y. UCHIYAMA & H. HATANAKU. 1997. Changes in mitochondrial membrane potential during oxidative stress induced apoptosis in PC12 cells. J. Neuro. Res. **50:** 413–420.
53. BALAZS, R., O.S. JORGENSEN & N. HACK. 1988. N-methyl D-aspartate promotes the survival of cerbellar granule cells in culture. Neuroscience **27:** 437–451.
54. FAVIT, A., F. NICOLETTI, U. SCAPAGNINI & P.L. CEACHICO. 1992. Ubiquinone protects cultured neurons against spontaneous and excitotoxin induced degeneration. J. Cereb. Blood & Met. **12:** 638–645.
55. XIA, K., N.K. MUKHOPADHYAY, R.C. INHORN, D.L. BARBER, P.E. ROSE, R.S. LEE, R.P. NARSIMHAN, A.D. D'ANDREA, J.D. GRIFFIN & T.M. ROBERTS. 1996. The cytokine activated tyrosine kinase JAK2 activates Raf1 in a p21ras dependent manner. Proc. Natl. Acad. Sci. USA **93:** 11681–11686.
56. DOBROWSKY, R.T., M.H.WERNER, A.M. CASTELLINO, M.V. CHAO & Y.A. HANNUN. 1994. Activation of the sphingomyelin cycle through the low affinity neurotrophin receptor. Science **265:** 1596–1599.

Neither Caspase-3 nor DNA Fragmentation Factor Is Required for High Molecular Weight DNA Degradation in Apoptosis

P. ROY WALKER,[a] JULIE LEBLANC, CHRISTINE CARSON, MARIA RIBECCO, AND MARIANNA SIKORSKA

Apoptosis Research Group, Institute for Biological Sciences, National Research Council, Ottawa, Ontario, Canada K1A 0R6

ABSTRACT: In this paper, we show that there is a two-step process of DNA fragmentation in apoptosis; DNA is first cleaved to large fragments of 50–300 kb that are subsequently cleaved to smaller oligonucleosomes in some, but not all cells. Significantly, only the first stage is considered essential for cell death since some cells, for example human MCF7 breast carcinoma cells and human NT2 neuronal cells, do not show this behavior but still display normal nuclear morphological apoptotic changes. In cells that usually produce small fragments blocking the second (internucleosomal) stage of DNA fragmentation prevents neither nuclear condensation nor apoptosis. We are beginning to understand why the extent of DNA fragmentation during apoptosis varies enormously and why it appears to be a function of the cell type not the inducer. Presumably, this reflects the content of not only endonuclease activit(ies) but also on the ability of the cells to activate caspases, particularly caspase-3, and other proteases that may be involved in endonuclease activation. Since NT2 cells activate caspase-3, but do not correctly process DFF45,[b] other factors must also impinge on the inevitability of that process.

INTRODUCTION

DNA fragmentation in apoptosis is a two step process in which the DNA is first cleaved to large fragments of 50–300 kb that are subsequently cleaved to smaller oligonucleosomes (the DNA ladder) in some, but not all cells.[1,2] Significantly, only the first stage is considered essential for cell death since some cells, for example human MCF7 breast carcinoma cells and human NT2 neuronal cells, never produce ladders, but show normal nuclear morphological apoptotic changes. In cells that usually do

[a]Address for correspondence: Apoptosis Research Group, Institute for Biological Sciences, National Research Council, Bldg. M54, Montreal Road, Ottawa, ON, Canada K1A 0R6; 613-993-2892 (voice); 613-990-7963 (fax).
e-mail: roy.walker@nrc.ca

[b]ABBREVIATIONS: CAD, caspase-activated DNase (mouse); CAGE, conventional agarose gel electrophoresis; DFF45, DNA fragmentation factor; CPAN, caspase activated nuclease (human homolog of CAD); DEVDamc, aspartate-glutamate-valine-aspartate-acetoxymethylchlorine; DME, Dulbecco's modified essential medium; DNase, deoxyribonuclease; ICAD, inhibitor of CAD (ICAD and DFF45 are synonymous); HMW DNA, high molecular weight deoxyribonucleic acid; MEM, minimal essential medium; PFGE, Pulsed field gel electrophoresis; RFU, relative fluorescence units; DEVDfmk, Aspartate-glutamate-valine-aspartate-fluoromethylketone.

produce ladders blocking the second (internucleosomal) stage of DNA fragmentation does not prevent either nuclear condensation or apoptosis.[3–5]

Recently, a cytoplasmic endonuclease that is activated specifically during apoptosis by caspase-3 has been described in human and mouse tissue.[6–9] The enzyme, CAD (caspase activated DNase[6,7]) and its human homolog, CPAN (caspase-activated nuclease [9]), is kept in an inactive state in the cytoplasm by the tight binding of an inhibitor (ICAD, inhibitor of CAD [6]) or DFF45 (DNA fragmentation factor [10]). Active enzyme is only produced when ICAD is cleaved by activated caspase-3. Cleavage of ICAD by caspases is a two-stage process with a N-terminal site being cleaved initially followed by secondary cleavage nearer the C-terminus. Although other caspases can cleave the N-terminal site only caspase-3 can catalyze the second cleavage step.[11] The active CAD/CPAN subunit is only released following the second cleavage and is then believed to enter the nucleus to catalyze internucleosomal DNA fragmentation. This mechanism is appealing since it links the caspase proteolytic cascade(s) occurring during apoptosis with DNA degradation, one of the major events of cell death in all cells.

However, most cell deaths are now considered to be independent of caspase-3.[12,13] For example, there are numerous examples of cultured cells which still undergo apoptosis when caspase-3 activity is blocked [4,12,14–16] the most notable example being MCF-7 cells that are defective in caspase-3.[17] Cells derived from caspase-3 knockout mice are not resistant to apoptosis and caspase-3 knockout mice also have comparatively few *in vivo* defects in apoptosis.[12] In addition, DFF45 knockout mice develop normally, even though some cells (thymocytes and macrophages) appear to be defective in internucleosomal DNA fragmentation.[18] This suggests that whilst a caspase-3-DFF45/ICAD-CAD/CPAN pathway of endonuclease activation may be required in some cells for internucleosomal DNA fragmentation, it raises doubts about whether it can be the only degradative mechanism operating during apoptosis and whether it is necessary at all for high molecular weight (HMW) DNA fragmentation.

So far, the CAD/CPAN endonuclease has only been shown to be involved in generating small oligonucleosomal fragments, both in the various knockout animals as well as in the *in vitro* target-nuclei assay.[7,9] To try to determine whether activation of CAD/CPAN is essential for the critical first stage of HMW DNA fragmentation we have studied apoptosis in MCF7 cells that carry a mutated caspase-3 gene and are unable to correctly process DFF45/ICAD resulting in a failure to activate CAD/CPAN. In addition, we have studied NT2 cells which, although they do activate caspase-3 during apoptosis, undergo only HMW DNA fragmentation. We show that both of these cell types undergo the initial, HMW stage of DNA fragmentation and exhibit characteristic nuclear morphological changes independent of caspase-3 activity and without activating a cytoplasmic CAD/CPAN-like enzyme. From these studies, we conclude that the elements of DNA fragmentation that are essential for apoptosis are catalyzed by an endogenous nuclear enzyme(s) and that neither caspase-3 nor CAD/CPAN cytoplasmic activities are essential for either cell death or the biochemical and morphological changes that occur in the nucleus during apoptosis.

MATERIALS AND METHODS

Cell Culture Conditions and Induction of Apoptosis

MCF7, human breast adenocarcinoma cells (ATCC HTB 22), were cultured in MEM supplemented with 10% fetal bovine serum and 100 nM bovine insulin, as described previously.[3] Human NT2 cells (Stratagene, San Diego, CA) were propagated in DME supplemented with 10% fetal bovine serum and 40 μg/ml gentamycin.[3] Apoptosis was induced in both cell types by serum withdrawal. The induction of apoptosis was monitored by assessing nuclear morphology and plasma membrane permeability of the cells. Thus, 70% confluent cultures grown on glass coverslips were subjected to serum withdrawal to induce apoptosis, in the presence or absence of 10 μM zDEVDfmk, an inhibitor or caspase-3. Samples were taken and the live cells were detected by staining with 10 μM Hoechst 33342 and apoptotic cells with ethidium homodimer. The coverslips were rinsed briefly in PBS, fixed for 5 minutes in 3% paraformaldehyde (J.B. EM Services, Inc., Dorval, Quebec), rinsed in PBS and mounted onto glass slides. Cells were examined on an Olympus BX-50 microscope equipped with epifluorescence optics and photographed onto Kodak Ektachrome P1600 film. For the time courses, nuclear morphology was scored for four hundred cells per sample, as either normal (uncondensed, blue nuclei labeled by Hoechst alone), necrotic (uncondensed red nuclei, labeled by Hoechst and by ethidium homodimer due to breakdown of plasma membrane integrity) or apoptotic (condensed blue and red nuclei).

DNA Electrophoresis

The conditions for pulsed field gel electrophoresis (PFGE), the handling of cells, their immobilization in agarose and deproteinization of the embedded cells were as described previously.[3,19] Typically 2×10^6 cells from each time point were immobilized in a mini plug and loaded on the gel. The PFGE electrophoresis conditions were as described[19] and permitted the ordered separation of fragments from 100 bp to greater than 1Mb. In some experiments, plugs prepared as described above were loaded onto 0.8% agarose gels and subjected to conventional agarose gel electrophoresis (CAGE) as described previously.[3]

Preparation of Cytoplasmic Extracts and Digestions of Target Nuclei

Cells were harvested and disrupted in 100 μl of an extraction buffer consisting of 10 mM HEPES-NaOH, pH 7.0, 40 mM β-glycerophosphate, 50 mM NaCl , 2mM $MgCl_2$, 5 mM EGTA by 3 cycles of freezing and thawing with homogenization during each thawing cycle. The suspension was centrifuged at $100{,}000\times g$ for 60 min in a Beckman Ultra centrifuge and the supernatant recovered. Target nuclei were prepared from normal Jurkat cells since nuclei from these cells do not contain a nuclease capable of internucleosomal cleavage. The cells were harvested, washed in PBS, resuspended in a buffer consisting of 0.25 M sucrose, 15 mM Tris-HCl, pH 7.4, 60 mM KCl, 15 mM NaCl, 2mM EDTA, 0.5 mM EGTA, 0.5 mM spermidine, 0.15 mM spermine, 15 mM mercaptoethanol and 0.15% NP40 and incubated on ice for 5 minutes. The extracts were centrifuged at 1,000 rpm for 5 min, washed and resuspended in extraction buffer such that 10 µl contained 2×10^6 nuclei. 50 µl of cytoplasmic extract was added and the digestions were carried out for 2 h at 37°C. The nuclei were then embedded in plugs, deproteinzed and the DNA subjected to CAGE as described above.

DFF45 Processing

Western blot analysis was used to detect processing of DFF45. Cells were treated as indicated in the Figure legends, harvested and extracted with RIPA buffer as described previously.[3] 100 µg of protein from each extract was resolved on a 12% (v/v) SDS-PAGE gel, transferred onto Nylon and probed with anti-DFF45/ICAD polyclonal antibody (1 µg/ml, N-terminal peptide antibody, Upstate Biotechnology, Lake Placid, NY). Detection was via an alkaline phosphatase-conjugated goat anti-rabbit secondary antibody.

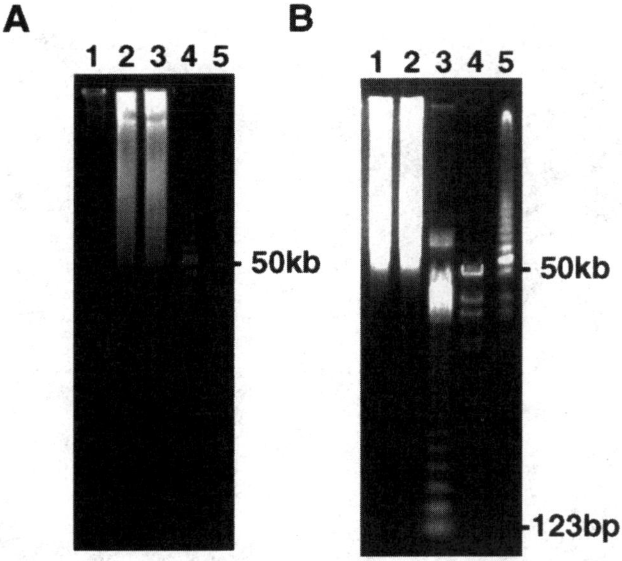

FIGURE 1. PFGE analysis of DNA degradation in apoptotic MCF7 (**A**) and NT2 (**B**) cells. Cells were harvested at 72 h after serum withdrawal, embedded in agarose and processed for PFGE analysis. **A:** *lane 1,* cells from control cultures; *lane 2,* cells at 72 h after serum withdrawal; *lane 3,* cells treated with 10 µM zDEVDfmk 30′ prior to serum withdrawal (72h); lane 4, high range DNA size markers; lane 5, lambda DNA ladder (monomer size, 50Kb as indicated). **B:** *lane 1,* cells at 72 h after serum withdrawal; *lane 2,* cells treated with 10 µM zDEVDfmk 30′ prior to serum withdrawal (72 h); lane 3, 123 bp ladder (monomer size indicated); *lane 4,* HindIII digest of lambda DNA; *lane 5,* high range DNA size markers (lambda DNA ladder monomer size of 50 Kb indicated).

RESULTS

Biochemical and Morphological Characterization of Apoptosis in MCF7 and NT2 Cells

To establish definitively that both MCF7 and NT2 cells, which produce only high molecular weight fragments during apoptosis, still showed the typical morphological nuclear changes of chromatin condensation and formation of apoptotic bodies we examined the cells using DNA-binding fluorochromes. Both MCF7 and NT2 cells were induced to undergo apoptosis by subjecting them to serum withdrawal. Analysis of DNA fragmentation by PFGE showed that, as expected, HMW DNA fragments appeared in MCF7 cells following 72 h of serum withdrawal (FIG. 1A, lane 2). Moreover, there were no DNA fragments smaller than 50 kb in DNA extracted from apoptotic MCF7 cells at any stage of apoptosis. Similar results were obtained when NT2 cells were subject to serum withdrawal (FIG. 1B, lane 1) with the appearance of DNA fragments of 50–300 kb without any internucleosomal DNA fragmentation.

At 48 h after serum withdrawal approximately 15–20% of the MCF7 cells showed the typical features of apoptosis (FIG. 2A,B). These cells were shrunken and the

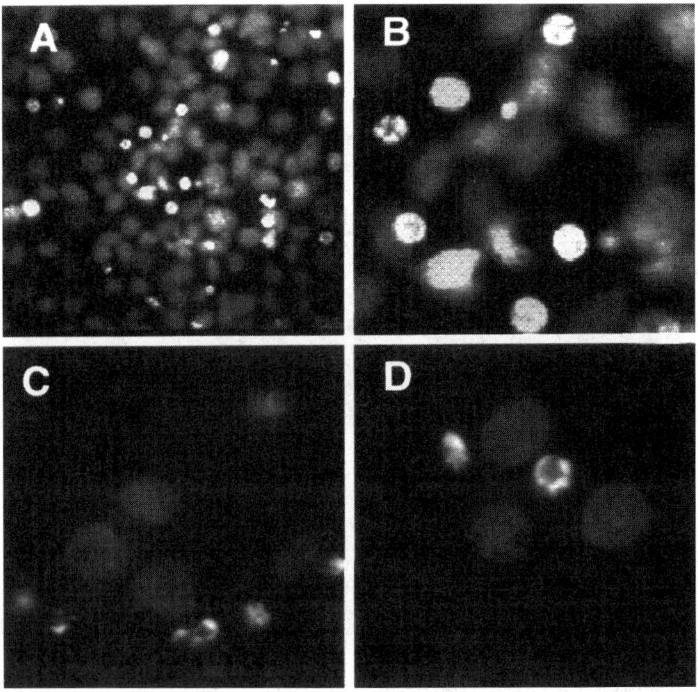

FIGURE 2. Fluorescence microscopy of apoptotic MCF7 (panels **A**, **B**) and NT2 (panels **C**, **D**) cells. Cells were harvested at 48 h after serum withdrawal, fixed and stained with Hoechst 33342.

bright Hoechst staining showed clear condensation of nuclear chromatin with the formation of apoptotic bodies (panel B). Similarly, the NT2 cells, which were approximately 50% apoptotic by 48 h, also showed many cells with typical apoptotic nuclear morphology including condensed chromatin and apoptotic bodies (FIG. 2, panels C and D). Thus, neither MCF7 nor NT2 cells degrade their DNA into internucleosomal fragments, yet show typical nuclear morphological changes, confirming that HMW DNA fragmentation is sufficient to cause nuclear collapse and chromatin condensation, in agreement with previous studies.[1,2,4]

Caspase-3 Activity in MCF7 and NT2 Cells

Given that the CAD/CPAN endonuclease, which has been implicated in internucleosomal DNA fragmentation but has not yet been studied in relation to HMW DNA fragmentation, requires caspase-3 for activation we next investigated whether caspase-3 is activated in cells producing only HMW fragments (TABLE 1). There was no detectable caspase-3 activity in MCF7 cells following serum withdrawal since the cells lack a functional caspase-3 protein due to a deletion in the gene.[17, 20] An increase in caspase-3 like activity (i.e. DEVDase activity) was readily detected in NT2 cells at 24 h after serum withdrawal, increasing from 1050 RFU/10^6 cells in the control to 5306 RFU/10^6 cells in 24 h serum-withdrawn cells (TABLE 1).

We next investigated whether the caspase-3 inhibitor zDEVDfmk had any effect on cell death or DNA degradation in these cells. Addition of the inhibitor at 10 μM to the NT2 cells, 30 min prior to serum withdrawal, abolished all DEVDase activity (TABLE 1). Despite this, the only effect of the inhibitor was to slightly delay, but not prevent cell death (FIG. 3A). Significantly, PFGE analysis of DNA extracted from the inhibitor-treated cells showed the same pattern of HMW DNA fragmentation as in the untreated cells (FIG. 1B, lane 3). Thus, inhibition of DEVDase activity does not influence the nature of the degradative process (i.e. pattern of DNA fragmentation). As expected, zDEVD-fmk did not prevent the MCF7 cells from undergoing apoptosis (FIG. 3B) although the rate of entry of inhibitor-treated MCF7 cells into the DNA degradative phase of the process was also slightly delayed (FIG. 3B). However, the pattern of fragmentation remained unchanged (FIG. 1A, lane 3). Since MCF7 cells have no functional caspase-3 either another zDEVD-fmk inhibitable caspase may be affecting the rate of entry of the cells into the DNA degradative stage of apoptosis or the inhibitor may have some non-specific effects on cellular metabolism, thereby slowing the process.

TABLE 1. Caspase-3 activity in NT2 and MCF7 cells

	NT2[a]	MCF7
control	1050	ND[b]
24h-serum	5306	ND
24h-serum + 10 μM DEVD	134	ND

[a]Activity is expressed as RFU/10^6 cells.
[b]ND = not detected.

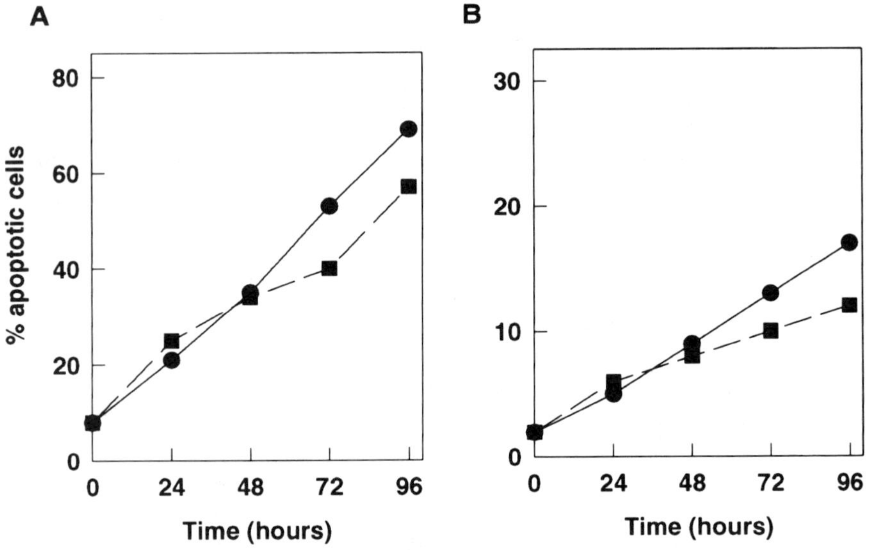

FIGURE 3. Time courses of the appearance of apoptotic cells following serum withdrawal of (**A**) NT2 and (**B**) MCF7 cells. Prior to serum withdrawal, cells were either untreated (*circles*) or treated for 30 min with 10 μM zDEVDfmk (*squares*). The appearance of apoptotic nuclei were scored as described in MATERIALS AND METHODS.

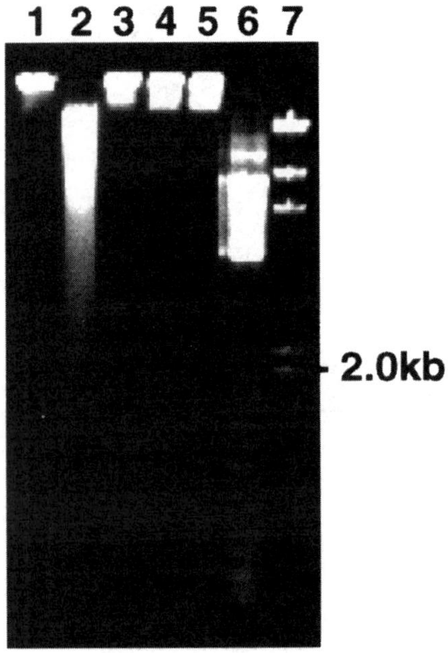

FIGURE 4. Cytoplasmic endonuclease activities. Cellular extracts from apoptotic MCF7 and NT2 cells (48 h after serum withdrawal) and anti-FAS–treated Jurkat cells were incubated for 2 h with normal Jurkat cell nuclei and processed for CAGE. *Lane 1,* freshly isolated, unincubated nuclei; *lane 2,* Jurkat cell extract; *lane 3,* MCF7 cell extract; *lane 4,* NT2 cell extract; *lane 5,* nuclei incubated without any added extract; *lane 6,* 123bp DNA ladder; *lane 7,* HindIII digest of lambda DNA. The position of the 2.0 Kb size marker is indicated at the right of the gel.

Cytoplasmic Endonuclease Activities in MCF7 and NT2 Cells

Cytoplasmic extracts were prepared from serum-withdrawn, apoptotic MCF7 and NT2 cells, as well as anti-FAS treated human Jurkat cells as a positive control. As shown in FIGURE 4, the cellular extract prepared from Jurkat cells treated for 4 h with anti-FAS antibody (lane 2) was able to degrade DNA in the target nuclei to low molecular weight fragments (producing the same pattern usually seen in apoptotic Jurkat cells). However, cellular extracts from either apoptotic MCF7 or NT2 cells did not contain any activity capable of degrading DNA in target nuclei (lanes 3 and 4, respectively), indicating the absence of a caspase-3 activated CAD/CPAN-like cytoplasmic activity in these cells.

ICAD/DFF45 Processing in NT2 Cells

Active CAD/CPAN is released upon degradation of ICAD/DFF45 by caspase-3. To detect ICAD/DFF45 processing cell extracts were taken at suitable times after treatment with apoptotic inducers, resolved on acrylamide gels and subject to western blotting with a polyclonal antiDFF45 antibody. MCF7 cells have been shown to be unable to correctly process DFF45, producing a fragment of 30 kD due to partial processing at the N-terminal site in the absence of caspase-3 as described by Tang and Kidd.[11] Although the NT2 cells activate caspase-3 they failed to process DFF45. The antibody recognized the long and short forms of DFF45/ICAD and a band of about 40 kD appeared with time in the serum-withdrawn cells. However, this band does not correspond to cleavage of the protein at a known caspase cleavage site and may reflect processing by other proteases or the emergence of a non-specific band. There were no detectable smaller fragments that would reflect correct processing at both caspase cleavage sites.

DISCUSSION

These data shows clearly that HMW DNA fragmentation is all that is required for cells to complete apoptosis normally, including the destruction of the nucleus and the packaging of the fragmented chromatin into apoptotic bodies. Furthermore, we demonstrate that neither caspase-3 nor the CAD/ICAD(DFF45) endonuclease system is required for HMW DNA fragmentation. Thus, nuclear shrinkage, chromatin condensation and the formation of apoptotic bodies appear to proceed normally without a functional caspase-3 or the cytoplasmic endonuclease CAD/CPAN.

Recent work has shed some light on the role that CAD/CPAN might play in apoptosis. Significantly, CAD/CPAN is not expressed ubiquitously in human tissues and cells[21] even though all cells are believed to be capable of apoptosis. Cells such as Hela, which do not express CAD, showed little or no internucleosomal DNA fragmentation when undergoing apoptosis. Transfection of these cells with the CAD gene restored internucleosomal DNA fragmentation. These observations suggest that CAD is the, or one of the, endonuclease(s) that mediates internucleosomal DNA fragmentation. Since transfection of caspasc-3–deficient MCF7 cells with caspase-3 restores the ability of these cells to undergo internucleosomal DNA fragmentation when stimulated to undergo apoptosis,[17] it is evident that caspase-3 drives the pathway of internucleosomal DNA fragmentation via activation of CAD/CPAN. Consistent with this, cells from knockout mice that are lacking DFF45 cannot undergo internucleosomal DNA fragmentation.[18] Taken together, this work supports the existence of such a pathway. However, this pathway is independent of the one producing HMW DNA fragmentation. The endonuclease responsible for this initial stage of fragmentation appears to be nuclear, but has not yet definitively been identified, although a chromatin-bound enzyme, DNAseY has recently been identified in our laboratory.[22]

The enzyme responsible for HMW DNA degradation is activated in MCF7 cells in the absence of caspase-3. Moreover, the caspase inhibitor zDEVDfmk, at concentrations sufficient to completely block caspase-3 activity, did not prevent the activation of the endonuclease responsible for HMW DNA fragmentation in NT2 cells. Therefore, caspase-3 and other caspases that metabolize the DEVD-amc substrate, such as caspases 1, 2, 4, 6–8, 10[23] are not required for the activation of this endonuclease. Moreover, since caspase-6 and 7 are believed to be activated by caspase-3[24] it is unlikely that they too could be involved. This suggests that the onset of DNA fragmentation, which is initiated by the formation of single-strand breaks,[19] may be caspase-independent. It appears, therefore, that caspases are superimposed over an underlying series of degradative events that constitute the fundamental mechanism of apoptosis. The fundamental mechanism drives the essential plasma membrane, mitochondrial and nuclear events that ensure the death of the cell. Indeed, in situations where caspases are involved in the process they do not determine commitment

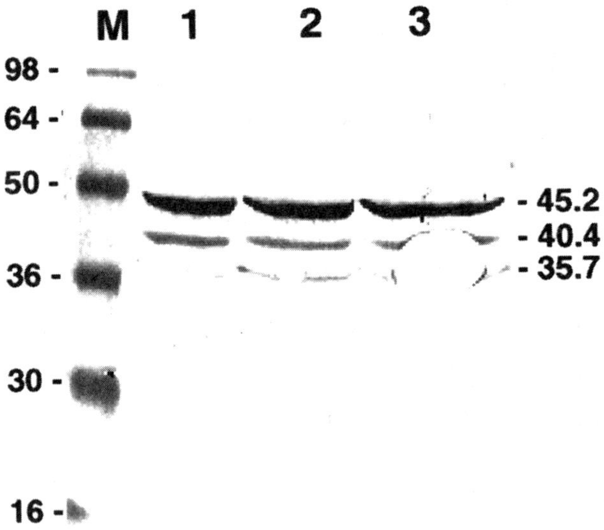

FIGURE 5. Processing of DFF45 in apoptotic NT2 cells. RIPA extracts were prepared from cells at 0, 24, and 48 h after serum withdrawal (*lanes 1–3,* respectively) and subjected to electrophoresis and western blotting. The size marker lane (lane M) was used to compute the sizes of the observed proteins.

to death.[25] The caspases serve two purposes in the process. They amplify apoptotic signals and trigger events that assist in the "clean-up" process by stimulating downstream degradative events, such as internucleosomal DNA fragmentation. Such mechanisms may have evolved to cope with the increased complexity of mammalian cells. The size of the genome of a mammalian cell is many fold that of *C. elegans* providing much larger amounts of DNA to be disposed of. DNA fragmentation to the HMW stage involves the introduction of about 50,000 dsbs into the genome compared with about 3×10^6 which accumulate during internucleosomal DNA fragmentation. In some cells, CAD/CPAN may play a role in determining the extent of DNA degradation during the later stages of the process. Thus CAD may play a role in the "mitochondrial accelerator" mechanism whereby the rate of apoptosis is increased in some cells.[26]

Finally, we are beginning to understand the basis for the enormous variation in the extent of DNA fragmentation during apoptosis varies enormously and why it appears to be a function of the cell type not the inducer. Presumably, this reflects the content of not only endonuclease activit(ies) but also on the ability of the cells to activate caspases, particularly caspase-3, and other proteases that may be involved in endonuclease activation.[27] Since NT2 cells activate caspase-3, but do not correctly process DFF45 (FIG. 5) other factors must also impinge on the inevitability of that process.

ACKNOWLEDGMENT

The authors wish to thank Dr. Siyaram Pandey for helpful discussions and advice.

REFERENCES

1. SIKORSKA, M. & P.R. WALKER. 1997. Endonuclease activities and apoptosis. *In* Physiological Cell Death. R.A. Lockshin, J.L. Tilly & Z. Zakeri, Eds. New York. Wiley-Liss, Inc. pp. 211–242.
2. WALKER, P.R. & M. SIKORSKA. 1997. New Aspects of the Mechanism of DNA fragmentation in Apoptosis. Biochem. Cell Biol. **75:** 7–19.
3. PANDEY, S., P.R. WALKER & M. SIKORSKA. 1997. Identification of a novel 97 kDa endonuclease capable of internucleosomal DNA cleavage. Biochemistry **36:** 711–720.
4. SANE, A.T. & R. BERTRAND. 1998. Distinct Steps in DNA fragmentation pathway during camptothecin-induced apoptosis involved caspase-, benzoylcarbonyl- and n-tosyl-l-phenylalanylchloromethyl ketone-sensitive activities. Cancer Res. **58:** 3066–3072.
5. COHEN, G.M., X.M. SUN, H. FEARNHEAD, M. MACFARLANE, D.G. BROWN, R.T. SNOWDEN & D. DINSDALE. 1994. Formation of large molecular weight fragments of DNA is a key committed step of apoptosis in thymocytes. J. Immunol. **153:** 507–516.
6. SAKAHIRA, H., M. ENARI & S. NAGATA. 1998. Cleavage of CAD inhibitor in CAD activation and DNA degradation during apoptosis [see comments]. Nature **391:** 96–99.
7. ENARI, M., H. SAKAHIRA, H. YOKOYAMA, K. OKAWA, A. IWAMATSU & S. NAGATA. 1998. A caspase-activated DNase that degrades DNA during apoptosis, and its inhibitor ICAD [see comments]. Nature **391:** 43–50.

8. Mitamura, S., H. Ikawa, N. Mizuno, Y. Kaziro & H. Itoh. 1998. Cytosolic nuclease activated by caspase-3 and inhibited by DFF-45. Biochem. Biophys. Res. Commun. **243:** 480–484.
9. Halenbeck, R., H. MacDonald, A. Roulston, T.T. Chen, L. Conroy & L.T. Williams. 1998. CPAN, a human nuclease regulated by the caspase-sensitive inhibitor DFF45. Current. Biology **8:** 537–540.
10. Liu, X., H. Zou, C. Slaughter & X. Wang. 1997. DFF, a heterodimeric protein that functions downstream of caspase-3 to trigger DNA fragmentation during apoptosis. Cell **89:** 175–184.
11. Tang, D. & V.J. Kidd. 1998. Cleavage of DFF-45/ICAD by multiple caspases is essential for its function during apoptosis. J. Biol. Chem **273:** 28549–28552.
12. Woo, M., R. Hakem, M.S. Soengas, G.S. Duncan, A. Shahinian, D. Kagi, A. Hakem, M. McCurrach, W. Khoo, S.A. Kaufman, G. Senaldi, T. Howard, S.W. Lowe & T.W. Mak. 1998. Essential contribution of caspase 3/CPP32 to apoptosis and its associated nuclear changes. Genes Dev. **12:** 806–819.
13. Kuida, K., T.S. Zheng, S. Na, C. Kuan, D. Yang, H. Karasuyama, P. Rakic & R.A. Flavell. 1996. Decreased apoptosis in the brain and premature lethality in CPP32-deficient mice. Nature **384:** 368–372.
14. Lavoie, J.N., M. Nguyen, R.C. Marcellus, P.E. Branton & G.C. Shore. 1998. E4orf4, a novel adenovirus death factor that induces p53-independent apoptosis by a pathway that is not inhibited by zVAD-fmk. J. Cell Biol. **140:** 637–645.
15. D'Mello, S.R., F. Aglieco, M.R. Roberts, K. Borodezt & J.W. Haycock. 1998. A DEVD-inhibited caspase other than CPP32 is involved in the commitment of cerebellar granule neurons to apoptosis induced by K^+ deprivation. J. Neurochem. **70:** 1809–1818.
16. Walker, P. R. & M. Sikorska, Unpublished observations.
17. Janicke, R.U., M.L. Sprengart, M.R. Wati & A.G. Porter. 1998. Caspase-3 is required for DNA fragmentation and morphological changes associated with apoptosis. J. Biol. Chem. **273:** 9357–9360.
18. Zhang, J.H., X.S. Liu, D.C. Scherer, L. Vankaer, X.D. Wang & M. Xu. 1998. Resistance to DNA fragmentation and chromatin condensation in mice lacking the DNA fragmentation factor 45. Proc. Natl. Acad. Sci. USA **95:** 12480–12485.
19. Walker, P. R., J. Leblanc & M. Sikorska. 1997. Evidence that DNA fragmentation in apoptosis is initiated and propagated by single-strand breaks. Cell Death & Differentiation **4:** 506–515.
20. Janicke, R.U., P. Ng, M.L. Sprengart & A.G. Porter. 1998. Caspase-3 is required for alpha-fodrin cleavage but dispensable for cleavage of other death substrates in apoptosis. J. Biol. Chem **273:** 15540–15545.
21. Mukae, N., M. Enari, H. Sakahira, Y. Fukuda, J. Inazawa, H. Toh & S. Nagata. 1998. Molecular cloning and characterization of human caspase-activated DNase. Proc. Natl. Acad. Sci. USA **95:** 9123–9128.
22. Liu, Q.Y., S. Pandey, R.K. Singh, W. Lin, M. Ribecco, H. Borowy Borowski, B. Smith, J. Leblanc, P.R. Walker & M. Sikorska. 1998. DNaseY: a rat DNaseI-like gene coding for a constitutively expressed chromatin-bound endonuclease. Biochemistry **37:** 10134–10143.
23. Samejima, K., S. Tone, T.J. Kottke, M. Enari, H. Sakahira, C.A. Cooke, F. Durrieu, L.M. Martins, S. Nagata & S.A. Kaufman. 1998. Transition from caspase-dependent to caspase-independent mechanisms at the onset of apoptotic execution. J. Cell Biol. **143:** 225–239.
24. Hirata, H., A. Takahashi, S. Kobayashi, S. Yonehara, H. Sawai, T. Okazaki, K. Yamamoto & M. Sasada. 1998. Caspases are activated in a branched protease cascade and control distinct downstream processes in Fas-induced apoptosis. J. Exp. Med. **187:** 587–600.

25. OHTA, T., T. KINOSHITA, M. NAITO, T. NOZAKI, M. MASUTANI, T. TSURUO & A. MIYAJIMA. 1997. Requirement of the caspase-3/CPP32 protease cascade for apoptotic death following cytokine deprivation in hematopoietic cells. J. Biol. Chem. **272:** 23111–23116.
26. BOSSY WETZEL, E., D.D. NEWMEYER & D.R. GREEN. 1998. Mitochondrial cytochrome c release in apoptosis occurs upstream of DEVD-specific caspase activation and independently of mitochondrial transmembrane depolarization. EMBO J. **17:** 37–49.
27. LIU, Q. Y., M. RIBECCO, S. PANDEY, P. R. WALKER & M. SIKORSKA. 1999. Apoptosis-relevant biochemical and structural features of DNaseI-like family of nucleases. Ann. N.Y. Acad. Sci. **887:** 60–76. This volume.

Apoptosis-related Functional Features of the DNaseI-like Family of Nucleases

QING Y. LIU, MARIA RIBECCO, SIYARAM PANDEY, P. ROY WALKER, AND MARIANNA SIKORSKA[a]

Apoptosis Research Group, Institute for Biological Science, National Research Council of Canada, Ottawa, Ontario, Canada K1A 0R6

ABSTRACT: Rat DNaseY[b] and its human homolog DHP2 are members of a new family of DNaseI-like endonucleases. They contain all the conserved amino acid residues to engage a DNaseI-like catalytic activity and the same molecular mechanisms of DNA hydrolysis. The sequence similarity can be extended to other families of nucleases, such as FEN-1, DNA polymerases, RNaseH and exonuclease III, involved in the ion-dependent hydrolysis of nucleic acids. Their unique features include the NLS signals that place them in the nuclei and a high content of positively charged amino acid residues that results in their high affinity for the substrate. Their properties are consistent with a role in the early stage DNA degradation during apoptosis. The caspase-DFF45/CIDE-CPAN pathway is most likely involved in the second stage of internucleosomal DNA degradation. However, cells express constitutively multiple transcripts encoding DNA degrading enzymes and related molecules, hence they have the molecular diversity to engage the self-destructive pathway appropriate to a given trigger.

[a]Address for correspondence: Apoptosis Research Group, Institute for Biological Sciences, National Research Council of Canada, 1200 Montreal Road, Bldg. M-54, Ottawa, Ontario, K1A0R6; 613-993-5916 (voice); 613-990-7963 (fax).

e-mail: marianna.sikorska@nrc.ca

[b]ABBREVIATIONS: T4RNaseH, phage T4 ribonuclease H; CAD, caspase-activated deoxyribonuclease; CIDE- A and B, cell death-inducing DFF45-like effectors A and B, respectively; CPAN, caspase activated nuclease (human homolog of CAD); DFF45, deoxyribonucleic acid fragmentation factor (inhibitor of CAD and CPAN); DHP1 and DHP2, human deoxyribonuclease I homologous protein 1 and 2 (synonymous with DNAS1L2 and DNAS1L3), respectively; DNA, deoxyribonucleic acid; DNaseI, deoxyribonuclease I; DNaseII, deoxyribonuclease II; DNaseX and DNL1L, X chromosome-linked deoxyribonuclease; DNaseY and DNase γ, deoxyribonuclease I-like nuclease (rat homolog of DNAS1L3/ DHP2); ERNaseHI, *E. coli* ribonuclease HI; EEXOIII, *E. coli* exonuclease III; ER, endoplasmic reticulum; FEN-1, flap-1 endonuclease; HEXOIII, human exonuclease III; H-, M- or XFENI, human, mouse or Xenopus flap-1endonuclease, respectively; GAPDH, glyceraldehyde 3-phosphate dehydrogenase; LS- DNase, liver and spleen deoxyribonuclease; MAR, matrix attachment region; NLS, nuclear localization signal; NUC18, 18 kD nuclease; ORF, open reading frame; PCR, polymerase chain reaction; PFGE, pulsed field gel electrophoresis; RNaseH, ribonuclease H; SPRAD2, *S. pombe* DNA repair endonuclease; TUNEL, terminal deoxynucleotidyl transferase-mediated deoxyuridine 5'-triphosphate nick end-labeling; TaqPol, *Thermus aquaticus* DNA polymerase; TdT, terminal deoxynucleotidyl transferase; T4 kinase, phage T4 polynucleotide kinase; T5Exo, phage T5D15 gene for 5' exonuclease; VM26, teniposide topoisomerase II inhibitor.

MOLECULAR MECHANISMS OF DNA FRAGMENTATION DEFINES FUNCTIONAL FEATURES OF APOPTOTIC ENDONUCLEASE(S)

Chromatin degradation is considered to be a universal biochemical hallmark of apoptosis and it is one of the most thoroughly studied apoptotic events. DNA fragmentation in apoptosis is primarily a two-stage process and probably involves more than one endonuclease.[1,2] At the early stage of nuclear disassembly the chromatin is cleaved into large domains of 50–300 kilobase pairs (kb). This step is most likely catalysed by an endonuclease that resides at the matrix attachment regions (MARs) where DNA binds to the nuclear scaffold [2,3] hence, such enzyme would be expected to be tightly associated with chromatin. The second stage involves more extensive DNA degradation and usually produces oligonucleosomal DNA fragments. However, since blocking oligonucleosomal cleavage does not protect cells from death,[4–6] only the high molecular weight stage of DNA fragmentation is considered to be essential for apoptosis.

A great deal has been learned about the mechanism of DNA degradation and yet, the actual endonuclease responsible for this process have remained elusive. Candidate apoptotic endonucleases have been identified based upon biochemical properties, mainly the cation and pH requirements needed to recreate, *in vitro*, the pattern of DNA fragments generated in cells undergoing apoptosis. In most cell types, DNA degradation occurs at or around neutral pH, therefore, endonucleases involved in apoptosis would be expected to have a basal activity at neutral pH in the presence of magnesium ions and to be further activated by calcium ions. Zinc ions have been shown to be inhibitory, at least for the activities responsible for the secondary stages of DNA fragmentation. The fragment ends almost universally have been found to be 3'-OH and 5'-P.

Over the years more than 20 enzymatic activities have been biochemically characterized and implicated in apoptosis. In recent years several of these enzymes have actually been cloned. These include DNaseI,[7,8] NUC18,[9] DNaseII,[10,11] CAD,[12] CPAN,[13,14] as well as a family of DNaseI-like genes, i.e., DNL1L,[15,16] DNaseX,[17] DHP1/DNAS1L2 and DHP2/DNAS1L3,[18,19] DNaseY,[20] DNase γ,[21] and LS-DNase.[22]

Many of the biochemical properties ascribed to the apoptotic endonuclease(s) are found in DNaseI and this enzyme has, indeed, been implicated in apoptosis.[7] However, the criticism leveled against DNaseI is that it is primarily a secretory protein being released into the alimentary tract and bloodstream. In contrast, the enzyme responsible for the initial stages of DNA fragmentation, would be expected to reside in the nucleus and be tightly-associated with chromatin. More recently it has become apparent that there is a family of DNaseI-like genes. In this paper we discuss apoptosis-relevant structural features and biochemical properties of the DNaseI-like family of endonucleases. One of these enzymes has all of the properties described above that are required for the initial stages of DNA cleavage during apoptosis, including the ability to localize to the nucleus.

DNASEI-LIKE GENE FAMILY

The first human DNaseI-like gene (DNL1L), found to be X chromosome-linked and termed DNaseX,[17] was cloned by Parrish *et al.*[15] It encodes a 302 amino acid protein with 39% identity to human DNaseI and exhibits highest expression in heart and muscle tissues.[15–17] The gene was shown to be at least 10.2 kb in length and consisted of 10 exons, compared to 9 in human DNaseI. The gene is differentially spliced in the 5'-untranslated region resulting in a transcript with either two or three noncoding exons, in contrast to only one in the DNaseI gene. However, in both genes 6 out of 8 coding exons have identical intron-exon boundaries, indicating that both genes may have evolved from the same ancestor.[17]

Rodriguez *et al.*[18] identified two additional human DNaseI-like cDNAs, DHP1 (DNaseI Homologous Protein 1) and DHP2 (DNaseI Homologous Protein 2), predicted to encode protein variants of 299 and 305 amino acids, with 56 and 46% sequence identity to DNaseI, respectively. The genes are mapped to human chromosomes 16 and 3, respectively, and were shown to be expressed in brain (DHP1) and liver (DHP2). Alignment of the predicted protein sequences for DHP1 and DHP2 with DNAseI reveals extensive similarities between these proteins, including a predicted signal peptide, conserved catalytic, calcium-binding and DNA-binding residues as well as the cysteine residues essential for the formation of disulfide bridges. The same proteins were cloned by Zeng *et al.*[19]

We have isolated and sequenced a large 30 kb contig from a rat genomic library containing a 17 kb gene encoding a 310 amino acid protein which we named DNaseY.[20] Analysis of the amino acid sequence revealed that this protein also has a typical eukaryotic signal peptide of 25 amino acids which includes a charged N terminus, a hydrophobic core and a probable cleavage site "Ala-Leu-Ser".[23] Cleavage of this signal peptide would produce a 33 kD mature protein. We found this 33 kD active endonuclease to be constitutively present in nuclei of a wide variety of cells and tissues. The DNaseY gene contained the same number of exons as DNaseI, separated by much larger introns, resulting in a gene of >17 kb compared to <4 kb gene for DNaseI.[20] The DNaseY has 42% identity to DNaseI, including the signal peptide, conserved critical active site residues, the essential disulfide bridges and the calcium binding site. Sequence analysis of the cDNA showed it to be a rat homolog of human DNAS1L3.[18,19]

Two additional laboratories reported the cloning of cDNAs encoding DNAseI-like proteins.[21,22] Shiokawa *et al.*[21,24] have isolated an endonuclease from nuclei of apoptotic thymocytes, cloned its cDNA and named it DNase γ. This protein is identical to DNaseY described by us.[20] The authors detected the expression of this protein in spleen, lymph nodes, thymus and liver using the Northern blotting technique and in kidney and testis by RT-PCR amplification. Baron *et al.*[22] cloned human and murine DNAseI-like cDNAs, termed LS-DNase, and show its highest expression in macrophage populations of liver and spleen tissues, although it is also expressed in many other tissues. This protein is the same as those identified by Rodriguez *et al.*[18] and Zeng *et al.*[19]

The studies cited above describe the identification of a new family of endonucleases which, although unique, are highly homologous to DNaseI. A comparison of their genomic organization, such as location and sequence of intron/exon boundaries, points to their common ancestry. For example, the DNaseY gene had the same number of exons, but due to interruption by long introns the gene is much larger than DNaseI.[20] This suggests that this gene family may have arisen from an early DNaseI gene duplication and later evolved and adopted a distinct function.

STRUCTURAL AND FUNCTIONAL FEATURES OF DNASEI-LIKE ENDONUCLEASES

Catalytic Properties

Biochemical studies revealed that the product of the rat DNaseI-like gene, DNaseY enzyme, is a p33 endonuclease with biochemical properties very similar to DNaseI.[20,21,24] Thus, like DNaseI, the p33 enzyme is capable of cleaving both single and double stranded DNA in the activity gel and plasmid digestion assays (FIG. 1). It requires both Ca^{2+} and Mg^{2+} for full activity, although it is active in the presence of Mg^{2+}, Co^{2+} or Mn^{2+} alone and, like DNaseI, it is fully inhibited by Zn^{2+} (FIG. 1A). It is active over a broad pH range with optimum pH of 7–8 (FIG. 1B). Furthermore, at neutral pH, it produces DNA fragments with the same 3'-OH ends that are typically observed in apoptotic cells and which form the basis of the widely used TUNEL assay.[25]

The catalytic properties of DNaseI have been related to its well-defined secondary structure and mutagenesis studies which identified all critical amino acid resi-

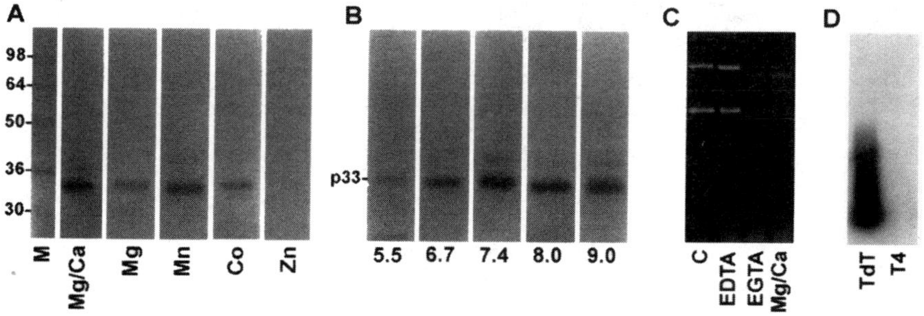

FIGURE 1. Biochemical properties of DNaseY. The enzyme was purified from the nuclei of 5123tc cells and its activity assayed as described by Liu *et al.*[20] **A:** *Cation requirements.* The activity of DNaseY (indicated by a dark band at the bottom of the gel) was assayed in the presence of 5 mM $MgCl_2$ and 2 mM $CaCl_2$ (lane Ca/Mg), 5 mM $MgCl_2$ and 2 mM EGTA (lane Mg), 1 mM $MnCl_2$ (lane Mn), 1 mM $CoCl_2$ (lane Co), or 1 mM $ZnCl_2$, 5 mM $MgCl_2$ and 2 mM $CaCl_2$ (lane Zn). **B:** *The pH profile.* The following buffers containing, 5 mM $MgCl_2$ and 2 mM CaC12, were used in this assay: 100 mM sodium acetate buffer for pH 5.5 and 50 mM Tris-HCl buffers of pH 6.7 to 9.0. The enzymatic activity is indicated by the presence of dark band. **C:** *Plasmid DNA digestion.* 1 µg of plasmid DNA and approximately 0.5 µg of DNaseY were incubated for 1 h in 25 µl of 10 mM Tris-HCl buffer (pH 7.4), containing either 2 mM EDTA (EDTA), 5 mM $MgCl_2$ and 2mM EGTA (EGTA) or 5mM $MgCl_2$ and 1mM CaC12 (Mg/Ca). Plasmid DNA incubated without the enzyme is shown in lane C. DNA was resolved on an agarose gel and stained with ethidium bromide. Undigested plasmid is represented by two white bands, lanes C and EDTA, and digested plasmid by the white smear, lanes EGTA and Mg/Ca. **D:** *Detection of free 3'-hydroxyl ends.* Plasmid DNA was incubated with DNAseY in the presence of 5mM $MgCl_2$ and 1mM $CaCl_2$ and the digested DNA was end-labeled using radioactive nucleotides and either terminal deoxynucleotidyl transferase (TdT) or T4 polynucleotide kinase (T4). After electrophoresis the gel was dried and autoradiographed overnight. Incorporation of radioactive nucleotides is indicated by the black smear in lane TdT.

dues involved in the process of DNA hydrolysis.[26–30] This knowledge is essential for meaningful sequence comparisons between the two gene families. The alignment of the DNaseY sequence with all known DNaseI and DNaseI-like proteins is shown in FIGURE 2. All active site residues are highly conserved in these proteins. They are Y78, E80, D235, H156 and H275 in DNaseY and they correspond to Y76, E78, D212, H134 and H252 of bovine DNaseI. Amongst these residues, E80, H156 and Y78 (which is a key residue important for the recognition of the geometry of the minor groove and for DNA binding) form the first active site triad, whereas D235 and H275 constitute the second active site. The disulfide bridge essential for DNaseI activity is absolutely conserved in all these proteins. The two residues, E39 and D190 (E39 and D168 in bovine DNaseI), which serve as a magnesium ligand site, are also conserved. It has been shown by Jones *et al.*[30] that magnesium ions are indeed critical for DNaseI activity. Mutation of D168S in bovine DNaseI almost completely abolished its activity. Rat DNaseY does not cut circular plasmids in the present of EDTA either, indicating its absolute requirement for the magnesium ions (FIG. 1C).[20] In addition, DNaseI contains two Ca^{2+} binding sites and, although the catalytic metal is actually magnesium,[28,30] they are important for the structural integrity and the folding of the protein. Amino acid residues D101, D109, F111 and E114, which are conserved in the DNaseI-like and DNaseI proteins, form the first calcium-binding site, whereas D220, T225, T227 and T230 form the second site, although, T227 is conserved only in the DNaseI proteins. Taken together the above sequence comparison suggests that the proteins of these two gene families might have a similar tertiary structure which accounts for the fact that their biochemical properties and the mechanism of DNA hydrolysis are identical.

The DNaseI-DNA interaction is inhibited by the binding of monomeric actin to the enzyme.[31,32] Several residues that comprise the DNaseI actin-binding site are not conserved in DNaseI-like proteins (FIG. 2). However, it should be stressed that the sequence similarity does not necessarily guarantee functional equivalence. For example, a conservative substitution of V67 to I67 found in the rat DNaseI sequence results in an almost 1000-fold lower affinity for actin than human DNaseI.[33,34] The following substitutions are found in the actin-binding site of DNaseY and DHP2: D53, V67 and A114 of DNaseI are replaced by E53, I69 and F116 in DNaseY and DHP2. It has been shown by Baron *et al.*[22] that the activity of the human homolog of DNaseY is not regulated by actin.

Magnesium-binding Sites

Most nucleases and polynucleotidyl transferases have an absolute requirement for divalent cations, usually magnesium, and contain acidic amino acids at their active sites that coordinate the metal.[35] Well-characterized examples include the flap endonuclease-1 family (FEN1), DNA polymerase, RNaseH as well as several restriction endonucleases.[30,35,36] Their amino acid sequences are aligned in FIGURE 3. The remarkable feature of this sequence comparison is the presence of seven acidic residues corresponding to those of T4 RNaseH, whose secondary structure has been well characterized. T4 RNaseH contains 9 such residues and coordinates two magnesium ions.[37] Of the proteins aligned in FIGURE 3, the *E. coli* RNaseHI is a shorter protein of 155 amino acids and coordinates only one magnesium ion in its active site.[36] Therefore the sequence similarity of *E. coli* RNaseHI with other nucleases in

```
                                                  ▼E39
RDNASEY     LRLCSFNVRSFGESKKENHNAMDIIVKIIKRCDLILLMEIKDSNNNICPMLMEKLNGNSR 60
DHP2        MRICSFNVRSFGESKQEDKNAMDVIVKVIKRCDIILVMEIKDSNNRICPILMEKLNRNSR
RDNASEI     LRIAAFNIRTFGDTKMSNATLSSYIVKILSRYDIAVVQEVRDTHLVAVGKLLDELNRDIP
MDNASEI     LRIAAFNIRTFGETKMSNATLSVYFVKILSRYDIAVIQEVRDSHLVAVGKLLDELNRDKP
HDNASEI     LKIAAFNIQTFGETKMSNATLVSYIVQILSRYDIALVQEVRDSHLTAVGKLLDNLNQDAP
BDNASEI     LKIAAFNIRTFGETKMSNATLASYIVRIVRRYDIVLIQEVRDSHLVAVGKLLDYLNQDDP
DHP1        LRIGAFNIQSFGDSKVSDPACGSIIAKILAGYDLALVQEVRDPDLSAVSALMEQINSVSE
HDNASEX     FRICAFNAQRLTLAKVAREQVMDTLVRILARCDIMVLQEVVDSSGSAIPLLLREINRFDD
             .. .** . .  .*           . ... *. .. *. *        *.  .*

                          Y78▼ ▼E80                 !       ! !  !
RDNASEY     RSTTYNYVISSRLGRNTYKEQYAFLYKEKLVSVKAKYLYHDYQDG-DTDVFSREPFVVWF 120
DHP2        RGITYNYVISSRLGRNTYKEQYAFLYKEKLVSVKRSYHYHDYQDG-DADVFSREPFVVWF
RDNASEI     D--NYRYIISEPLGRKSYKEQYLFVYRPSQVSVLDSYHYDDGCEPCGNDTFSREPAIVKF
MDNASEI     D--TYRYVVSEPLGRKSYKEQYLFVYRPDQVSILDSYQYDDGCEPCGNDTFSREPAIVKF
HDNASEI     D--TYHYVVSEPLGRNSYKERYLFVYRPDQVSAVDSYYYDDGCEPCGNDTFNREPAIVRF
BDNASEI     N--TYHYVVSEPLGRNSYKERYLFLFRPNKVSVLDTYQYDDGCESCGNDSFSREPAVVKF
DHP1        H--EYSFVSSQPLGRDQYKEMYLFVYRKDAVSVVDTYLYPD-----PEDVFSREPFVVKF
HDNASEX     SG-PYSTLSSPQLGRSTYMETYVYFYRSHKTQVLSSYVYND-----EDDVFAREPFVAQF
                *  . *  ***  * * * ....         * * *       * * *** .  *

                                               ▼H156
RDNASEY     QAPFT--------------------AAKDFVIVPLHTTPETSVKEIDELADVYTDVRRRW 180
DHP2        QSPHT--------------------AVKDFVIIPLHTTPETSVKEIDELVEVYTDVKHRW
RDNASEI     FSPYT--------------------EVREFAIVPLHSAPTEAVSEIDALYDVYLDVRQKW
MDNASEI     FSPYT--------------------EVQEFAIVPLHAAPTEAVSEIDALYDVYLDVWQKW
HDNASEI     FSRFT--------------------EVREFAIVPLHAAPGDAVAEIDALYDVYLDVQEKW
BDNASEI     SSHST--------------------KVKEFAIVALHSAPSDAVAEINSLYDVYLDVQQKW
DHP1        SAPGTGERAPPLPSRRALTPPPLPAAAQNLVLIPLHAAPHQAVAEIDALYDVYLDVIDKW
HDNASEX     SLPSN--------------------VLPSLVLVPLHTTPKAVEKELNALYDVFLEVSQHW
                                         . ...**..*     *.. * .*. .*  .*

                     ▼D190                         !    ! !  !    ▼D235
RDNASEY     KAENFIFMGDFNAGCSYVPKKAWKNIRLRTDPNFVWLIGDQEDTTVKKSTSCAYDRIVLR 240
DHP2        KAENFIFMGDFNAGCSYVPKKAWKNIRLRTDPRFVWLIGDQEDTTVKKSTNCAYDRIVLR
RDNASEI     GLEDIMFMGDFNAGCSYVTSSQWSSIRLRTSPIFQWLIPDSADTTAT-STHCAYDRIVVA
MDNASEI     GLEDIMFMGDFNAGCSYVTSSQWSSIRLRTSPIFQWLIPDSADTTVT-STHCAYDRIVVV
HDNASEI     GLEDVMLMGDFNAGCSYVRPSQWSSIRLWTSPTFQWLIPDSADTTAT-PTHCAYDRIVVA
BDNASEI     HLNDVMLMGDFNADCSYVTSSQWSSIRLRTSSTFQWLIPDSADTTAT-STNCAYDRIVVA
DHP1        GTDDMLFLGDFNADCSYVRAQDWAAIRLRSSEVFKWLIPDSADTTVG-NSDCAYDRIVAC
HDNASEX     QSKDVILLGDFNADCASLTKKRLDKLELRTEPSFHWVIADGEDTTVRASTHCTYDRVVLH
               . ...*****.*. .       . * .   * *.* *  ***    . *.***.*
                          |____________________________________|

                                              ▼H275
RDNASEY     GQEIVNSVVPRSSGVFDFQKAYELSEEEALDVSDHFPVEFKLQSSRAFTNSRKSVSLKKK 300
DHP2        GQEIVSSVVPKSNSVFDFQKAYKLTEEEALDVSDHFPVEFKLQSSRAFTNSKKSVTLRKK
RDNASEI     GALLQAAVVPSSAVPFDFQAEYRLTNQMAEAISDHYPVEVTLRKT---------------
MDNASEI     RALLQAAVVPNSAVPFDFQAEYGLSNQLAEAISDHYPVEVTLRKI---------------
HDNASEI     GMLLRGAVVPDSALPFNFQAAYGLSDQLAQAISDHYPVEVMLK-----------------
BDNASEI     GSLLQSSVVPGSAAPFDFQAAYGLSNEMALAISDHYPVEVTLT-----------------
DHP1        GARLRRSLKPQSATVHDFQEEFGLDQTQALAISDHFPVEVTLKFHR--------------
HDNASEX     GERCRSLL--HTAAAFDFPTSFQLTEEEALNISDHYPVEVELKLSQAHSVQPLSLTVLLL
                   .   .    .*   . * .  *  .***.***  *

RDNASEY     KKGSRS------ 312
DHP2        TKSKRS------
RDNASEI     ------------
MDNASEI     ------------
HDNASEI     ------------
BDNASEI     ------------
DHP1        ------------
HDNASEX     LSLLSPQLCPAA
```

FIGURE 2. Sequence comparison of DNaseI and DNaseI-like proteins. The alignment was done using the ClustalX program. " * " - conserved residues; " . " - conserved substitutions; " ! " - the residue below is involved in calcium binding. Arrow heads indicate magnesium binding and active site residues, the numbers beside the arrow heads indicate the position of these residues in this alignment. The nuclear localization signal sequences are underlined. Cysteine residues that are involved in disulfide bridge formation are also indicated. The N-glycosylation sites and the residues involved in actin binding in all the DNase I sequences are boxed. RDNASEY - rat DNaseY; DHP2 - human DNaseI homologous protein 2; RNASEI - rat DNaseI; MDNASEI - mouse DNaseI; HDNASEI - human DNaseI; BDNSAEI - bovine DNaseI; DHP1 - human DNaseI homologous protein1; HDNASEX - human DNaseX.

this alignment was limited to short segments that include the *E. coli* catalytic residues D10, E48 and D70, corresponding to T4 RNaseH residues D19, D132 and D155, respectively. It has been demonstrated that the acidic residues are clustered around the magnesium ions and their carboxylate oxygen atoms coordinate magnesium by forming hydrogen bonds to water molecules.[37] These residues and their interactions are essential for the molecular mechanisms of substrate hydrolysis as demonstrated by mutagenesis studies.[35]

```
HFENI       29)RKVAIDASMSIYQFLIAVR-QGGDVLQNEEGETTSHLMGMFYRTIRMMENGIKPVY-VFD
MFENI       29)RKVAIDASMSIYQFLIAVR-QGGDVLQNEEGETTS-LMGMFYRTIR-MENGIKPVY-VFD
XFENI       29)RKVAIDASMCIYQFLIAVR-QDGNTLQNEEGETTSHLMGMFYRTIRMVEHGIKPVY-VFD
SPRAD2      29)RKVAIDASMSLYQFLIQVRSQDGQQLMNEQGETTSHLMGMFYRTLRIVDNGIKPCF-VFD
TaqPol      13)RVLLVDGHHLAYRTFFALK--GLTTSRGEPVQAVYGFAKSLLKALK--EDGYKAVFVVFD
T5Exo       21)NLMIVDGTNLGFRFKHNNS--K-KPFASSYVSTIQSLAKSYSARTT---------IVLGD
T4RNaseH    14)GICLIDFSQIALSTALVNF-PDKEKINLSMVRHLILNSIKFNVKKAKTL-GYTKIVLCID
ERNaseHI     5)VEIFTDGSCLGN--------PGPG-------------------------GYGAILR---
                    *D10

HFENI          GKPPQLKSGELAKRSERRAEAEKQLQQAQAAGAEQEVEKFTKRLVKVTKQHNDECKHLLS
MFENI          GKPPQLKSGELAKRSERRAEAEKQLQQAQEAGMEEEVEKFTKRLVKVTKQHNDECKHLLS
XFENI          GKPPQMKSGELAKRSERRAEAEKLLEAAEEAGEVENIEKFTKRLVKVTKQHNEECKKLLT
SPRAD2         GKPPTLKSGELAKRVARHQKAREDQEETKEVGTAEMVDRFAKRTVKVTRQHNDEAKRLLE
TaqPol         AKAPS-FRHEAYEA-YKAGRAPTPEDFP--------------------RQLALI-KELVD
T5EXO          KGKSV-FRLEHLPE-YKGNRDEKY------AQRTEEEKALDEQFFEYLKDA-----FELC
T4RNaseH       NAKSGYWRRDFAYY-YKKNR-------------GKAREESTWDWEGYFESSHKVIDELKA
ERNaseHI       ---------------YRGREKTF--------------------SAGYTRTT---------

HFENI          LMGIPYLDAPS-EAEASCAALVKA----GKVYAAATEDMDCLTFGSPVLMRHLTASEAKK
MFENI          LMGIPYLDAPS-EAEASCAALAKA----GKVYAAATEDMDCLTFGSPVLMRHLTASEAKK
XFENI          LMGIPYVDAPC-EAEATCAALVKA----GKVYAAATEDMDALTFGTPVLLRHLTASEAKK
SPRAD2         LMGIPFVNAPC-EAEAQCAALARS----GKVYAAASEDMDTLCFQAPVLLRHLTFSEQRK
TaqPol         LLGFTRLEVPGYEADDVLATLAKKAEKEGYEVRILTADRDLYQL---VSDRVAVLHPEGH
T5Exo          KTTFPTFTIRGVEADDMAAYIVKLIGHLYDHVWLISTDGDWDTL---LTDKVSRFS---F
T4RNaseH       YMPYIVMDIDKYEADDHIAVLVKKFSLEGHKILIISSDGDFTQLHKYPNVKQWSPMHKKW
ERNaseHI       ----------NNRMELMAA-IVALEALKEHCEVILSTDSQYVRQGITQWIHNWKKRGWKT
                             *E48                D70*

HFENI          LPIQEFHLSRILQELGLNQ-EQFVDLCILLG---------SDYCESIRGIGPK(244
MFENI          LPIQEFHLSRVLQELGLNQ-EQFVDLCILLG---------SDYCESIRGIGAK(242
XFENI          LPIQEFHLNRVIQDIGITH-EQFVDLCILLG---------SDYCETIRGIGPK(244
SPRAD2         EPISEYNIEKALNGLDMSV-EQFVDLCILLG---------CDYCEPIRGVGPA(245
TaqPol         LITPEW----LWEKYGLRP-EQWVDFRALVGDP-------SDNLPGVKGIGEK(203
T5Exo          TTRREYHLRDMYEHHNVDDVEQFISLKAIMGDL-------GDNIRGVEGIGAK(215
T4RNaseH       VKIK------------SGSAEIDCMTKILKGDK-------KDNVASVKVRSDF(212
ERNaseHI       ADKKPVKNVDLWQRLDAALGQHQIKWEWVKGHAGHPENERCDELARAAAMNPT(145
                                                        *D134
```

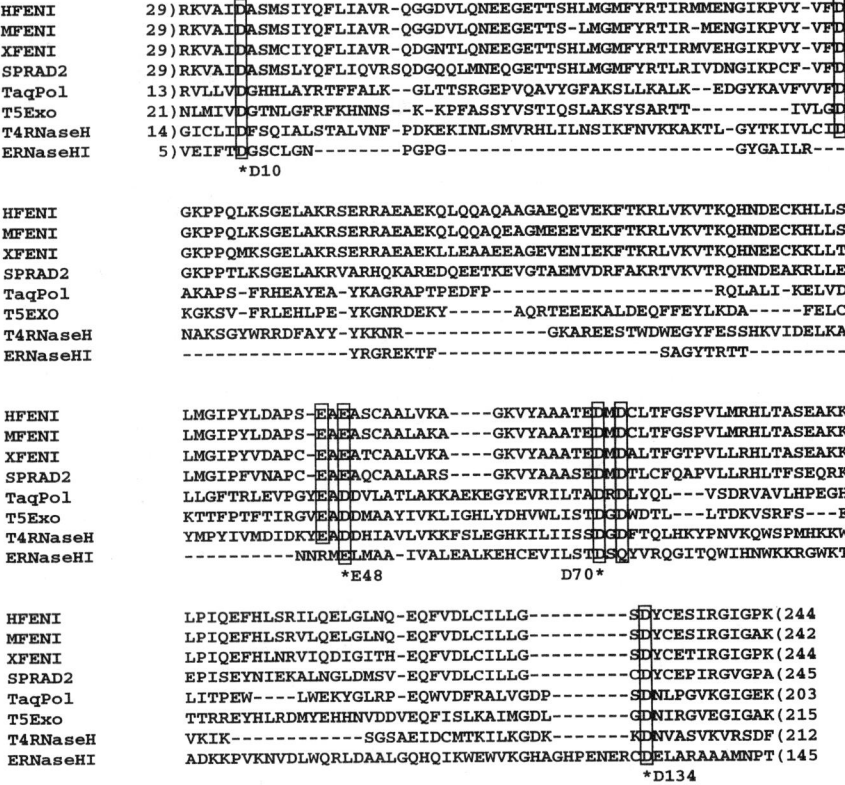

FIGURE 3. Amino acid sequence alignment of different nucleases showing conserved residues identified as Mg^{2+} - ligands. The alignment was done using the ClustalV program with some adjustments made by eye based on the known X-ray crystal structures and results of mutagenesis studies. The beginning and end of each sequence used in the alignment are indicated by the number at each end. The common residues that are possible Mg^{2+}- ligands are boxed. The residues also conserved in *E. coli* RNaseH1 are numbered and indicated by " * ". H, M or XFENI - human, mouse or Xenopus flap-1 endonuclease; SPRAD2 - S. pombe DNA repair endonuclease; TaqPol - *Thermus aquaticus* DNA polymerase; T5Exo - phage T5D15 gene for 5' exonuclease; T4RNaseH - phage T4 RNaseH; ERNaseHI - *E. coli* RNaseHI.

T4 RNaseH, *E. coli* RNaseHI and the multi-functional DNA-repair enzyme exonuclease III have similar enzymatic properties and a similar protein folding structure around the metal-binding site. Furthermore, the structural motif of β strands around the magnesium- binding sites of the *E. coli* RNaseHI is similar to that of DNaseI.[36–39] The sequence alignment of DNaseI, DNaseY and *E. coli* exonuclease III is presented in FIGURE 4. It shows that the exonuclease III protein shares some structural and functional similarity with the DNaseI-like enzymes. Although the Y78-E80-H156 active site triad is missing in the exonuclease III and H156 is replaced by Y109, the exonuclease III contains a D229-H259 and D151 and E34 pairs, which are exact structurally equivalent of D235-H275, D190 and E39 in the DNaseI-like family (FIG. 2). Taken together, it is fair to conclude that all the enzymes involved in the hydrolysis of nucleic acids share a functionally appropriate protein folding structure, encompassing at least one magnesium-coordinating active site, which supports a common mechanisms of substrate hydrolysis.

Intracellular Localization

The endonuclease activities responsible for DNA fragmentation during apoptosis are generally considered to reside constitutively in the nucleus, since nuclei isolated

```
BDNASEI    LKIAAFNIRTFGETKMSNATLASYIVRIVRRYDIVLIQEVR-DSHLVAVGK
RDNASEY    LRLCSFNVRSFGESKKENHNAMDIIVKIIKRCDLILLMEIK-DSNNNICPM
EEXOIII    MKFVSFNINGL---RARPHQLEAIVEK--HQPDVIGLQETKVHDDMFPLEE
           ... .**.  .   .        .  .  .. *..  *  .  .  .  .
                 N7                          E34
BDNASEI    L--LDYLNQDDPN--TYHYVVSEPLGRNSYKERYLFLFRPNKVSVLDTYQY
RDNASEY    L--MEKLNGNSRRSTTYNYVISSRLGRNTYKEQYAFLYKEKLVSVKAKYLY
EEXOIII    VAKLGYNVFYHGQ--KGHYGVA-------------LLTKETPIAVRRGFPG
           .                *  ..                * .*  ..*   .

BDNASEI    DDGCESCGNDSFSREPAVVKFSSHSTKVKEFAIVALHSAPSDAVAEI-NSL
RDNASEY    HDYQDGD-TDVFSREPFVVWFQAPFTAAKDFVIVPLHTTPETSVKEI-DEL
EEXOIII    -DDEEAQRRIIMAEIPSLLGNVTVI----NGYFPQGESRDHPIKFPAKAQF
            *  ...     . * ..    *       Y109..  .         .

BDNASEI    YDV---YLDVQQKWHLNDVMLMGDFNAD---------------------CS
RDNASEY    ADV---YTDVRRRWKAENFIFMGDFNAG---------------------CS
EEXOIII    YQNLQNYLETELKRD-NPVLIMGDMNISPTDLDIGIGEENRKRWLRTGKCS
            .    * .   ..  .   ..*** *                      **
                                   D151
BDNASEI    YVTS-SQWSSIRLRT---SSTFQWLIPDSADTTA----TSTNC------AY
RDNASEY    YVPK-KAWKNIRLRT---DPNFVWLIGDQEDTTV---KKSTSC------AY
EEXOIII    FLPEEREWMD-RLMSWGLVDTFRHANPQTADRFSWFDYRSKGFDDNRGLRI
           ..     *   ** .      *     . *        *     .

BDNASEI    DRIVVAGSLLQSSVVPGSAAPFDFQAAYGLSNEMALAISDHYPVEVTLT-
RDNASEY    DRIVLRGQEIVNSVVPRSSGVFDFQKAYELSEEEALDVSDHFPVEFKLQS
EEXOIII    DLLLASQPLAECCVETGID--------YEIRSMEKP--SDHAPVWATFRR
           * ..           *              * .         *** **    .
           D229                                   H259
```

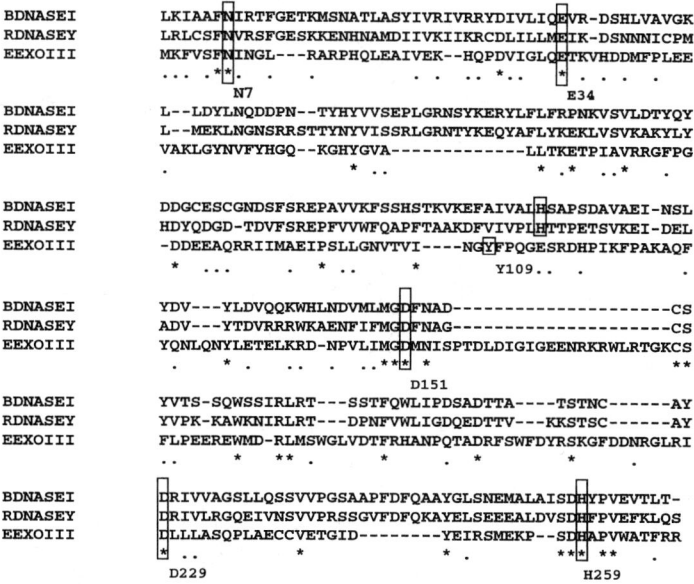

FIGURE 4. Amino acid sequence comparison of bovine DNaseI, rat DNaseY and E. coli exonuclease III. The alignment was done using the ClustalV program with some adjustments made by eye according to the known X-ray crystal structures. " * " - marks conserved residues and ". " - conserved substitutions. The conserved residues involved in the catalytic activity of exonuclease III are boxed. The numbers under the alignment indicate their positions in the *E. coli* exonuclease III sequence. BDNASEI - bovine DNaseI. RDNASEY- rat DNase Y. EEXOIII - *E. coli* exonuclease III.

from normal tissues can reproduce all of the aspects of DNA fragmentation seen during apoptosis.[40] Analysis of the amino acid sequence of DNAseY (FIG. 2) shows that, although the protein contains a typical eukaryotic signal peptide, it also contains two nuclear localization signals (NLS), one of which is a bipartite nuclear targeting sequence, starting 14 amino acids after the signal peptide cleavage site, and the other one is located at the C-terminal end of the protein, strongly resembling the SV40-type of NLS.[41] Of the proteins aligned in FIGURE 2 only rat DNaseY and its human homolog DHP2 contain this type of NLS.

The presence of both a signal peptide, which usually determines ER processing and export from the cell, and a nuclear localization signal in the same protein is unusual. The N-terminal hydrophobic signal peptide of DNaseY was shown to be cleaved in the microsomes and the mature 33 kD active DNaseY is smaller than the 36 kD pro-protein.[20] This suggested that the protein must pass through the ER membranes before being imported into the nucleus. It also implies that after the signal peptide is cleaved the mature protein must be released back to the cytoplasm and then transported to the nucleus through the nuclear pore. A precedent of a protein containing both a functional N-terminal hydrophobic signal peptide and a NLS is the Hepatitis B virus precore protein.[42] In this case, after removal of the signal peptide in the ER, 70% of the mature protein is released back to the cytoplasm and is then translocated to the nucleus. This nuclear targeting is due to the presence of the NLS in the viral protein. The complicated translocation process of the Hepatitis B virus precore protein and DNaseY is likely to be inefficient although the two NLSs of DNaseY may act cooperatively to improve this efficiency. Indeed, it has been reported that multiple NLSs contribute to the efficient and quick transport of a protein into the nucleus.[43]

The two N-glycosylation sites for the addition of mannose-6-phosphate found in all DNaseI sequences are neither conserved in DNaseY nor in DHP2 proteins (FIG. 2). Although the glycosylation is not required for DNaseI activity since the active protein can be synthesized in bacteria,[44] it might be essential for the proper protein targeting. The attachment of mannose-6-phosphate is believed to target proteins into the lysosomal compartment.[45] Thus, it is unlikely that either DNaseY or DHP2 proteins would be targeted to the lysosomes.

Both DNaseY and DHP2 have exceptionally high isoelectric points (pI) of 9.46 and 9.78, respectively. These high pI values are caused by the high content of positively charged residues which are distributed throughout the mature protein. Besides the positively charged amino acids in the NLS, they both have 23 more positively charged residues than the other DNaseI-like proteins listed in FIGURE 2. The pI values of all other known DNaseI-like proteins are near or lower than 5.0. It has been suggested that positively charged residues are required for high affinity-binding to the substrate, they interact with the phosphate groups on both strands of the helix around the minor groove of the DNA.[29,36] Consistently, the p33 DNaseY protein has been found in the nuclei of the variety of cells. Following the disruption of nuclear architecture by high salt conditions (i.e., buffer containing 4 M NaCl), the nuclease remains tightly bound to DNA.[20]

The Activation Mechanism

The activation mechanism of p33/DNaseY during apoptosis is currently being studied in our lab. However, it is clear that since the enzyme is present constitutively in nuclei, it must be prevented from gaining access to DNA and therefore must undergo some modification(s) to modulate its endonuclease activity in normal cells. Alternatively, changes in the chromatin structure itself, due to proteolysis of chromatin-associated proteins, may render DNA accessible to the endonuclease at early stages of apoptosis. Degradation of nuclear matrix and DNA-associated proteins have been observed at early stages of apoptosis and may support the latter hypothesis.[46]

ENDONUCLEASES IMPLICATED IN OLIGONUCLEOSOMAL DNA CLEAVAGE

In the later stage of apoptosis, cells usually undergo more extensive DNA degradation and produce oligonucleosomal DNA fragments. As many as $2–3 \times 10^6$ double-strand breaks are introduced into the DNA during the secondary stage of fragmentation, requiring a substantial increase in degradative activity. This could be achieved by either further activation of the same endonuclease or by the cooperation of additional endonuclease(s). Recent cloning of mouse and human homologs of caspase-activated nucleases, termed CAD and CPAN, respectively, confirms the existence of such enzymatic activities.[12–14] They are regulated by a caspase-sensitive inhibitor ICAD or DFF45,[12,47] which binds to the newly synthesized CAD or CPAN in the cytoplasm and inhibits its activity. After the cleavage of DFF45 by caspase3 the active CPAN is released and, presumably, gains access to the nuclear DNA and degrades it further.[12,13] The CPAN and DFF45 proteins share 57% sequence similarity in a 58 amino acid N-terminal region, likely involved in the homophilic interaction of these two proteins. Although the endonucleolytic activity of CPAN is supported by either Mg^{2+} or Mn^{2+} and is inhibited by Zn^{2+},[13] there is no significant sequence homology with the DNaseI-like enzymes (data not shown). However, comparison of the C-terminal region of CPAN with that of the human exonuclease III (FIG. 5), revealed that some sequence alignment is possible, especially with the seven residues of exonuclease III that are involved in cation binding and DNA cleavage.[39] This suggests that the DNAse-I like enzymes and CPAN may share a similar mechanism of DNA hydrolysis.

Two cDNAs, encoding a new family of caspase independent death effectors, CIDE-A and CIDE-B, have recently been cloned by Inohara *et al.*[48] The two proteins are highly related and share homology with the N-terminal region of DFF45. Their overexpression induces internucleosomal DNA fragmentation and cell death that can be inhibited by DFF45, but not by caspase inhibitors. However, CIDEs themselves are not nucleases. Instead, they may function as signaling components to regulate the activity of other endonuclease(s), potentially CPAN. Neither DNaseY nor its human homolog show any sequence similarity with the CIDEs and DFF45 that would be indicative of potential interactions, such as the homophilic interaction between the CPAN and DFF45 (data not shown). Therefore, it is unlikely that the function of DNaseY is controlled by these molecules.

```
HEXOIII   62)LKICSWNVDGLRAWIKKKGLDWVKEEAPDILCLQETKCSENKLPAELQEL
CPAN      82)-QGYVSDIRRFLSAFHEPQVGLI-QAAQQLLCDEQAPQRQRLLADLLHNV
              .    ..  . . .    .. . . * ..** ...   .  *.  *...

HEXOIII      PG-------LSHQYWSAPSDKEGYSGVGLLSRQCPLKV-SY---------
CPAN         SQNIAAETRAEDPPWFEGLESRFQSKSGYLRYSCESRIRSYLREVSSYPS
             .          .  *    .    *  * *   *  .. **

HEXOIII      GIGDEEHDQEGRVIVA---EFDSFVLVTAYVPNAGRGLVRLEYRQRWDEA
CPAN         TVGAEAQEEFLRVLGSMCQKLRSMQYNGSYFDRGAKGGSRLCTPEGW---
              .* * ...  **. .    . *     .*   ...*  **   . *

HEXOIII      FRKFLKGLASRKPLVLCGDLNVAHEEIDLRNPKGNKKNAGFTPQERQGFG
CPAN         FS-----------CQGPFDMDSCL-SRHSINP--------YSNRESRILF
             *                 * .      .  **        .. .* . .

HEXOIII      ELLQAVPLADSFRHLYPNTPYAYTFWTYMMNARSKNVGWRLDYFLL----
CPAN         STWNLDHIIEKKRTIIP------TLVEAIKEQDGREV----DWEYFYGLL
                .   . .  * . *      *.   . .  ...*    *   .

HEXOIII      ----SHSLLPALCDSKIRSKALGSDHCPITLYLAL(319
CPAN         FTSENLKLVHIVC------------HKKTTHKLNC(317
                 .  *.  .*            *   *  *
```

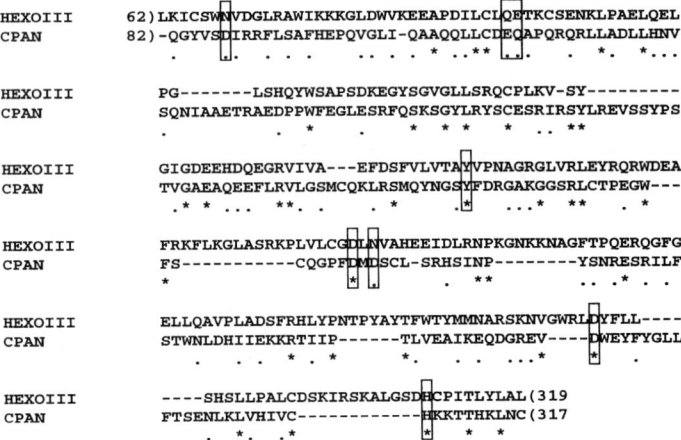

FIGURE 5. Comparison of the C-terminal region of CPAN with human exonuclease III. The alignment was done using the ClustalV program with some adjustments made by eye. " * " - conserved residues, " . " - conserved substitutions. The conserved residues involved in the catalytic activity of exonuclease III are boxed. The beginning and end of each sequence used in the alignment are indicated by the number at each end.

Cyclophilin A, B, and C which share sequence homology with NUC18 have also been implicated in the genome degradation during apoptosis.[9,49] Since they were shown to require both Ca^{2+} and Mg^{2+} for nuclease activity, we did pairwise comparison of each member of the cyclophilin family with all known metal ion-requiring nucleases. No sequence homology in either the metal binding site or in the catalytic site residues was found in the alignment (data not shown). Given the high level of conservation of these residues in all other nuclease it is unlikely that cyclophilins share with them the same mechanism of catalysis. However, since the active residues of the cyclophilin family have not yet been determined, it is possible they may belong to a completely new class of enzymes with a different mechanism of DNA hydrolysis.

Barry and Eastman[50] have implicated DNaseII in the internucleosomal DNA degradation following intracellular acidification of apoptotic. This enzyme is distinct from the DNaseI class of endonucleases. It functions under acidic condition, in the absence of metal ions and it cleaves DNA to produce 3'-P and 5'-OH oligonucleotides.[51] This enzyme does not contain the NLS signal and is believed to function in the lysosomes. The recent cloning of DNaseII gene enabled us to compare its amino acid sequence[10,11,52] with other nucleases. No homology was found (data not shown). Although a histidine residue may also be involved in its active site,[53] this enzyme probably engages a totally different catalytic mechanism.

CONSTITUTIVE EXPRESSION OF APOPTOSIS-RELATED TRANSCRIPTS

Multiple transcripts encoding a number of DNA-degrading enzymes and related molecules are constitutively expressed in cells. Shown in FIGURE 6 is the expression pattern of several of these transcripts, i.e., DNaseI, DNaseY, CPAN, DFF45, Casp3 and CIDE A, in four different immortalized human cell lines. Transcripts for both DNaseI and DNaseY were present all four cell lines. Their expression was very low, with the exception of the human neuroblastoma NT2 cells which produced a higher level of the DNaseY transcript. The expression of CPAN was also detected in all cells, including the lymphocytic Jurkat cells, in which, according to the published data,[12–14] this DNA-degradation pathway is particularly significant. However, CAD/CPAN is not universally expressed.[14] All cells expressed a significant level of the DFF45 transcript. The expression of casp3 was the same in these cells, although the PCR product detected in MCF-7 cells was shorter than predicted from the designed primers, i.e. 739 bp rather then 864 bp. This 125 bp deletion in the caspase-3 gene alters its ORF to produce a non-functional protein. This phenomenon was also reported by Jänicke *et al.*[54] A low level of the CIDE A transcript was present only in the NT2 cells and, none of the cell lines expressed CIDE B.

In order to try to establish a correlation between expression of endonucleases in these cells and the apoptotic outcome, we assessed their relative sensitivities to VM26 (topoisomerase II inhibitor) and analyzed the DNA fragmentation pattern that the treatment produced (FIG. 7). During a 48-h treatment with the drug, 65–75% of the Jurkat, DU145 and NT2 cells, but only 25% of MCF-7 cells lost their membrane integrity according to trypan blue exclusion assay (data not shown). Thus, the MCF-7 cells were the most treatment resistant. The DNA fragmentation pattern was

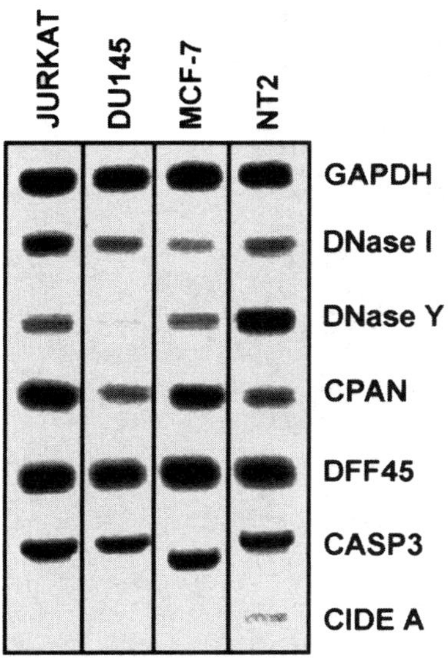

FIGURE 6. RT-PCR amplification of apoptosis-regulating factors. Total RNA was isolated from MCF-7 (ATCC HTB 22), DU145 (ATCC HTB 81), Jurkat (ATCC T1B152) and NT2 (Stratagene, San Diego, CA), propagated according to the conditions suggested by the supplier. The RNA isolation and cDNA synthesis was performed as described by Liu *et al.*[20] Gene specific primer sets are listed in TABLE 1. PCR amplifications were performed with one denaturation cycle at 94°C for 5 min, followed by 35 cycles (25 cycles for GAPDH) of 94°C for 1 min, 60°C for 50 sec and 72°C for 1 min. The PCR products were analyzed by electrophoresis on 1% agarose, visualized by ethidium bromide staining and photographed with a CCD camera (Ultra Lum Inc., Carson, CA).

TABLE 1. PCR primers

Gene	Sequences of primer sets	PCR product
GAPDH	f: GTG AAG GTC GGT GTC AAC G r: CAT ACT CAG CAC CAG CAT C	271bp
Casp3	f: AAG GTA TCC ATG GAG AAC ACT GAA AAC r: AAC CAC CAA CCA ACC ATT TCT	864 bp
CPAN	f: TTC CCG GCT GTG CCT GTA CGA GGA T r: CTC TCA CAG CTG TAT CTC AGA TAG CC	332 bp
DFF45	f: CTG GCC ATT GAT AAG TCC CTG A r: CTG GCC ACA TTC TTC CAC TTC A	239 bp
CIDE A	f: GAG CCC TCA TCA GGC CCC TGA r: GAA GTC CTT GGG GTT CAG CCT G	398 bp
DNaseY	f: AAC TCA CGA AGA AGC ACG AC r: TGT CAT AGG CAC AGC TGG TG	473 bp
DNaseI	f: GTT GCT GTT GGG AAG CTA CTG G r: CAC CTC CAC TGG GTA ATG GTC	639 bp

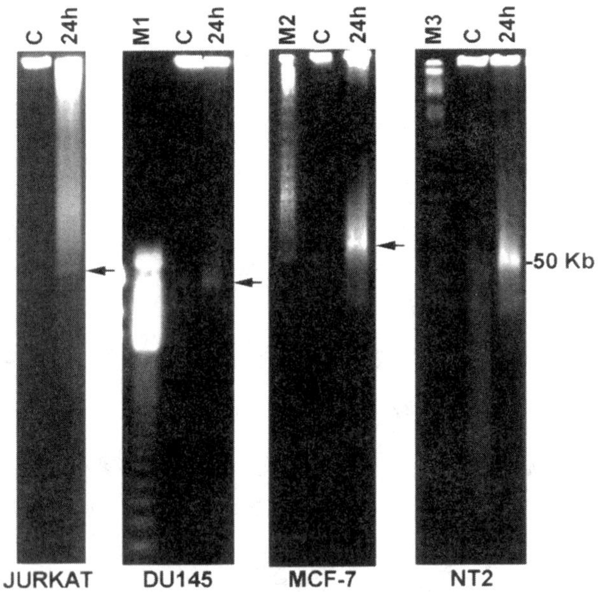

FIGURE 7. DNA fragmentation pattern in VM26-treated cells. 70% confluent MCF7, NT2, DU145 and Jurkat cells were treated with 10 μM VM26 for 24 hours. The DNA damage was assessed by PFGE.[55] C - control untreated cells; 24h - VM26 treated cells; M1, M2 and M3 are DNA molecular size markers, 123 bp DNA ladder, lambda DNA ladder, and yeast chromosomes, respectively.

very similar in all cells, i.e, they all produced large fragments (10–50 kb), but not DNA ladders. However, the amounts of fragmented DNA in DU1435 cells was very low in comparison to the other three cell lines. This was despite the fact that a very significant percentage of these cells were trypan blue positive (approximately 45–50%). Therefore, it seems that no reliable prediction can be made from the transcript profiles, with the exception of the MCF-7 cells where drug resistance may be linked to the non-functional caspase-3. There seems to be no obvious reasons, for example, why DU145 cells do not fragment their DNA since they do express transcripts encoding nucleases?

In summary, it seems that cells may use one or more enzymes, from a pool of constitutively expressed endonucleases, to effect DNA degradation in apoptosis. If a DNaseI-like enzyme is involved it is likely to be DNaseY, since it has all of the required properties and is localized in the nucleus. The CAD/CPAN pathway is used by some, but not all cells, to complete oligonucleosomal DNA fragmentation. However, many cells do not undergo internucleosomal DNA cleavage and do not activate a cytoplasmic nuclease.[56] There may also be a role for DNaseII-like enzymes if cells undergo acidification to a level sufficient for their activation.

ACKNOWLEDGMENTS

The authors wish to thank Brandon Smith for PCR primer design, Joanne Chartier and Wei Lin for their helpful assistance of in the preparation of this manuscript.

REFERENCES

1. Walker, P.R. & M. Sikorska. 1997. New aspects of the mechanism of DNA fragmentation in apoptosis. Biochem. Cell Biol. **75:** 287–299.
2. Sikorska, M. & P.R. Walker. 1998. A comprehensive evaluation of apoptosis and programmed cells death. *In* When Cells Die. R.A. Lockshin, J. L. Tilly & Z. Zakeri, Eds.: 211–242. Wiley-Liss Inc. New York.
3. Lagarkova M.A., O.V. Iarovaia & S.V. Razin. 1995. Large-scale fragmentation of mammalian DNA in the course of apoptosis proceeds via excision of chromosal DNA loops and their oligomers. J. Biol. Chem. **270:** 20239–20241.
4. Cohen, G.M., X.M. Sun, R.T. Snowden, D. Dinsdale & D.N. Skilleter. 1992. Key morphological features of apoptosis may occur in the absence of internucleosomal DNA fragmentation. Biochem. J. **286:** 331–334.
5. Brown, D.G., X.M. Sun & G.M. Cohen. 1993. Dexamethasone-induced apoptosis involves cleavage of DNA to large fragments prior to internucleosomal fragmentation. J. Biol. Chem. **286:** 3037–3039.
6. Walker, P.R., V.M. Weaver, B. Lach, J. Leblanc & M. Sikorska. 1994. Endonuclease activities associated with high molecular weight and internucleosomal DNA fragmentation in apoptosis. Exp. Cell Res. **213:** 100–106.
7. Peitsch, M.C. , B. Polzar, H. Stephan, T. Crompton, H.R. MacDonald, H.G. Mannherz & J. Tschopp. 1993. Characterization of the endogenous deoxyribonuclease involved in nuclear DNA degradation during apoptosis (programmed cell death). EMBO J. **12:** 371–377.
8. Liu, Q.Y., M. Ribecco, X. Hou, P.R. Walker & M. Sikorska. 1997. DNaseI primary transcript is alternatively spliced in both normal and apoptic cells: no evidence of up-regulation in apoptosis. DNA Cell Biol. **16:** 911–918.

9. MONTAGUE J., M. GAIDO, C. FRYE & J. CIDLOWSKI. 1994. A calcuim-dependent nuclease from apoptotic rat thymocytes is homologous with cyclophilin. Recombinant cyclophilins A, B, and C have nuclease activity. J. Biol. Chem. **269:** 18877–18880.
10. YASUDA, T., H. TAKESHITA, R.IIDA, T. NAKAJIMA, O. HOSOMI, Y. NAKASHIMA & K. KISHI. 1998. Molecular cloning of the cDNA encoding human deoxyribonuclease II. J. Biol. Chem. **273:** 2610–2616.
11. KRIESER, R.J. & A. EASTMAN. 1998. The cloning and expression of human deoxyribonuclease II. J. Biol. Chem. **273:** 30909–30914.
12. ENARI, M., H. SAKAHIRA, H. YOKOYAMA, K. OKAWA, A. IWAMATSU & S. NAGATA. 1998. A caspase-activated DNase that degrades DNA during apoptosis, and its inhibitor ICAD. Nature **391:** 43–50.
13. HALENBECK, R., H. MACDONALD, A. ROULSTON, T.T. CHEN, L. CONROY & L.T. WILLIAMS. 1998. CPAN, a human nuclease regulated by the caspase-sensitive inhibitor DFF45. Current Biol. **8:** 537–540.
14. MUKAE, N., M. ENARI, H. SAKAHIRA, Y. FUKUDA, J. INAZAWA, H. TOH & S. NAGATA. 1998. Molecular cloning and characterization of human caspase-activated Dnase. Proc. Natl. Acad. Sci. USA **95:** 9123–9128.
15. PARRISH, J.E., A. CICCODICOLA, M. WEHHERT, G.F. COX, E. CHEN & D.L. NELSON. 1995. A muscle-specific DNase I-like gene in human Xq28. Hum. Mol. Genet. **4:** 1557–1564.
16. PERGOLIZZI,R., V. APPIERTO, A. BOSSETTI, G.L. DEBELLIS & I. BIUNNO. 1996. Cloning of a gene encoding a DNase I-like endonuclease in the human Xq28 region. Gene. **168:** 267–270.
17. COY, J.F., I. VELHAGEN, R. HIMMELE, H. DELIUS, A. POUSTKA & H. ZENTGRAF. 1996. Isolation, differential splicing and protein expression of a DNase on the human X chromosome. Cell death differ. **3:** 199–206.
18. RODRIGUEZ, A. M., D. RODIN, H. NOMURA, C.C. MORTON, S. WEREMOWICZ & M.C. SCHNEIDER. 1997. Identification, localization, and expression of two novel human genes similar to deoxyribonuclease I. Genomics **42:** 507–513.
19. ZENG, Z., D. H. PARMELEE, H. HYAW, T.A. COLEMAN, K. SU, J. ZHANG, R. GENTZ, S. RUBEN, C. ROSEN & Y. LI. 1997. Cloning and characterization of a novel human DNase. Biochem. Biophys. Res. Commun. **231:** 499–504.
20. LIU, Q.Y., S. PANDEY, R.K. SINGH, W. LIN, M. RIBECCO, H. BOROWY-BOROWSKI, B. SMITH, J. LEBLANC, P.R. WALKER & M. SIKORSKA. 1998. DNaseY: a rat DNaseI-like gene coding for a constitutively expressed chromatin-bound endonuclease. Biochemistry **37:** 10134–10143.
21. SHIOKAWA, D. & S. TANUMA. 1998. Molecular cloning and expression of a cDNA encoding an apoptotic endonuclease DNase γ. Biochem J. **332:** 713–720.
22. BARON, W.F., C.Q. PAN, S.A. SPENCER, A.M. RYAN, R.A. LAZARUS & K.P. BAKER. 1998. Cloning and characterization of an actin-resistant Dnase I-like endonuclease secreted by macrophages. Gene **215:** 291–301.
23. VON HEIJNE, G. 1984. How signal sequences maintain cleavage specificity. J. Mol. Biol. **173:** 243–251.
24. SHIOKAWA, D., H. OHYAMA, T. YAMADA & S. TANUMA. 1997. Purification and properties of DNase γ from apoptotic rat thymocytes. Biochem. J. **326:** 675– 681.
25. GAVRIELI, Y., Y. SHERMAN & S.A. BEN-SASSON. 1992. Identification of programmed cell death in situ via specific labeling of nuclear DNA fragmentation. J. Cell Biol. **119:** 493–501.
26. SUCK, D.,C. OEFNER & W. KABSCH. 1984. Three-dimensional structure of bovine pancreatic DNase I at 2.5 Å resolution. EMBO J. **3:** 2423–2430.
27. SUCK, D & C. OEFNER. 1986. Structure of DNase I at 2.0 Å resolution suggests a mechanism for binding to and cutting DNA. Nature **321:** 620–625.

28. OEFNER, C. & D. SUCK. 1986. Crystallographic refinement and structure of DNase I at 2 Å resolution. J. Mol. Biol. **192:** 605–632.
29. WESTON, S., A. LAHM & D. SUCK. 1993. X-ray structure of the Dnase I-d(GGTATACC)$_2$ complex at 2·3 Å resolution. J. Mol. Biol. **226:** 1237–1256.
30. JONES, S.J., A.F. WORRALL & B.A. CONNOLLY. 1996. Site-directed mutagenesis of the catalytic residues of bovine pancreatic deoxyribonuclease I. J. Mol. Biol. **264:** 1154–1163.
31. KABSCH, W., H.G. MANNHERZ, D. SUCK, E.F. PAI & K.C. HOLMES. 1990. Atomic structure of the actin:DNase I complex. Nature **347:** 37–44.
32. ULMER, J.S., A. HERZKA, K.J. TOY, D.L. BAKER, A.H. DODGE, D. SINICROPI, S. SHAK & R.A. LAZARUS. 1996. Engineering actin resistant human DNase I for treatment of cystic fibrosis . Proc. Natl. Acad. Sci. USA **93:** 8225–8229.
33. LACKS, S.A. 1981. Deoxyribonuclease I in mammalian tissues: specificity of inhibition by actin. J. Biol. Chem. **256:** 2644–2648.
34. POLZAR, B. & H.G. MANNHERZ. 1990. Nucleotide sequence of a full length cDNA clone encoding the deoxyribonuclease I from the rat parotid gland. Nucleic Acids Res. **18:** 7151.
35. SHEN, B., J.P. NOLAN, L.A. SKLAR & M.S. PARK. 1997. Functional analysis of point mutations in human flap endonuclease-1 active site. Nucleic Acids Res. **25:** 3332–3338.
36. KATAYANAGI, K., M. MIYAGAWA, M. MATSUSHIMA, M. ISHIKAWA, S. KANAYA, M. IKEHARA, T. MATSUZAKI & K. MORIKAWA. 1990. Three-dimensional structure of ribonuclease H from E. coli. Nature **347:** 306–309.
37. MUESER, T.C., N.G. NOSSAL & C.C. HYDE. 1996. Structure of bacteriophage T4 RNase H, a 5' to 3' RNA-DNA and DNA-DNA exonuclease with sequence similarity to the RAD2 family of eukaryotic proteins. Cell **85:** 1101–1112.
38. KASHIWAGI, T., D. JEANTEUR, M. HARUKI, K. KATAYANAGI, S. KANAYA & K. MORIKAWA. 1996. Proposal for new catalytic roles for two invariant residues in Escherichia coli ribonuclease HI. Protein Engineering **9:** 857–867.
39. MOL, C.D., C. KUO, M.M. THAYER, R.P. CUNNINGHAM & J.A. TAINER. 1995. Structure and function of the multifunctional DNA-repair enzyme exonuclease III. Nature **374:** 381–386.
40. WALKER, P.R., S. PANDEY & M. SIKORSKA. 1995. Degradation of chromatin in apoptosis. Cell death differ. **2:** 97–104.
41. DINGWALL, C. & R.A. LASKEY. 1991. Nuclear targeting sequences--a consensus? Trends Biochem. Sci. **16:** 478–481.
42. OU, J.H., C.T. YEH & T.S. YEN. 1989. Transport of hepatitis B virus precore protein into the nucleus after cleavage of its signal peptide. J. Virol. **63:** 5238–5243.
43. SILVER, P. & H. GOODSON. 1989. Nuclear protein transport. Crit. Rev. Biochem. Mol. Biol. **24:** 419–435.
44. WORRALL, A.F. & B.A. CONNOLLY. 1990. The chemical synthesis of a gene coding for bovine pancreatic DNase I and its cloning and expression in Escherichia coli. J. Biol. Chem. **265:** 21889–21895.
45. PFEFFER, S.R. 1991. Targeting of proteins to the lysosome. In Curr. Topics Microbiol. Immunol. **170:** 43–65.
46. WEAVER, V.M., C.E. CARSON, P.R. WALKER, N. CHALY, B. LACH, Y. RAYMOND, D.L. BROWN & M. SIKORSKA. 1996. Degradation of nuclear matrix and DNA cleavage in apoptotic thymocytes. J. Cell Sci. **109:** 45–56.
47. LIU, X., H. ZOU, C. SLAUGHTER. & X. WANG. 1997. DFF, a heterodimeric protein that functions downstream of caspase-3 to trigger DNA fragmentation during apoptosis. Cell **89:** 175–184.

48. INOHARA, N., T. KOSEKI, S. CHEN, X. WU & G. NÚÑEZ. 1998. CIDE, a novel family of cell death activators with homology to the 45 kDa subunit of the DNA fragmentation factor. EMBO J. **17:** 2526–2533.
49. MONTAGUE, J.W., F.M. HUGHES JR. & J.A. CIDLOWSKI. 1997. Native recombinant cyclophilins A, B, and C degrade DNA independently of peptidylprolyl cis-tras-isomerase activity. J. Biol. Chem. **272:** 6677–6684.
50. BARRY, M.A. & A. EASTMAN. 1993. Identification of deoxyribonuclease II as an endonuclease involved in apoptosis. Arch. Biochem. Biophys. **300:** 440-450.
51. BERNARDI, G. 1971. Spleen acid deoxyribonuclease. In The Enzymes. P.D. Boyer, Ed. **4:** 271–287. Academic Press. New York.
52. BAKER, K.P., W.F. BARON, W.J. HENZEL & S.A. SPENCER. 1998. Molecular cloning and characterization of human and murine DNase II. Gene **215:** 281–289.
53. LIAO, T. 1985. The subunit structure and active site sequence of porcine spleen deoxyribonuclease. J. Biol. Chem. **260:** 10708–10713.
54. JÄNICKE, R.U., M.L. SPRENGART, M.R. WATI & A.G. PORTER. 1998. Caspase-3 required for DNA fragmentation and morphological changes associated with apoptosis. J. Biol. Chem. **273:** 9357–9360.
55. WALKER, P.R., L. KOKILEVA, J. LEBLANC & M. SIKORSKA. 1993. Detection of the initial stages of DNA fragmentation in apoptosis. Biotechniques **15**: 1032–1041.
56. WALKER, P.R., J. LEBLANC, C. CARSON, M. RIBECCO & M. SIKORSKA. 1999. Neither caspase-3 nor DNA fragmentation factor is required for high molecular weight DNA degradation in apoptosis. Ann. N.Y. Acad. Sci. **887:** 48–59. This volume.

Does the Oxidative/Glycolytic Ratio Determine Proliferation or Death in Immune Recognition?

M. KAREN NEWELL,[a,c] MARY-ELLEN HARPER,[a] KAREN FORTNER,[a] JULIE DESBARATS,[a] ALICIA RUSSO,[a] AND SALLY A. HUBER[b]

[a]*Division of Immunobiology, Department of Medicine, University of Vermont College of Medicine, Burlington, Vermont 05405, USA*

[b]*Division of Immunobiology, Department of Pathology, University of Vermont College of Medicine, Burlington, Vermont 05405, USA*

ABSTRACT: Here we discuss the possibility that the way cells utilize fuel(s) for energy confers the properties that can be recognized by the immune system and, reciprocally, that recognition by the immune system can alter the balance of the cell's energy metabolism. We propose that immune recognition, of somatic cells via MHC can alter the their energy metabolism and induce a metabolic shift. We demonstrate the reciprocal relationship that inducing a shift in metabolism toward glycolysis by supplying glucose and insulin results in the upregulation of immunologically recognizable molecules such as cell surface Fas. Thus, immune recognition can induce metabolic deviation. Metabolic deviation can result in altered immune recognition and ultimately in cell proliferation, cell differentiation, or cell death.

INTRODUCTION

The immune system functions to recognize and discriminate between self and nonself. Immune responses include "innate", non-antigen specific recognition, and "acquired," antigen-specific recognition. Innate immunity includes physical barriers and secretions, the influx of neutrophils, and the release of bradykinins and platelet activating factors at the site of damage. Acquired immunity includes production of cytokines, T cell–mediated cytotoxicity, and B cell–mediated antibody production. The innate or acquired immune response can result in at least four fates for the recognized or affected cells (see FIG. 1):

1. the cell will be tolerated and left unchanged;
2. the cell will be induced to divide;
3. the cell will be induced to differentiate; or
4. the cell will be induced to die.

The purpose of this chapter is to discuss the possibility that the way cells utilize fuel(s) for energy confers the properties that can be recognized by the immune system and, reciprocally, that recognition by the immune system alters the balance of the cell's energy metabolism.

[c]Present address for correspondence: Department of Biology, University of Colorado, Colorado Springs, CO 80918, USA. 719-262-3256 (voice); 719-262-3047 (fax).
e-mail: mnewell@mail.uccs.edu

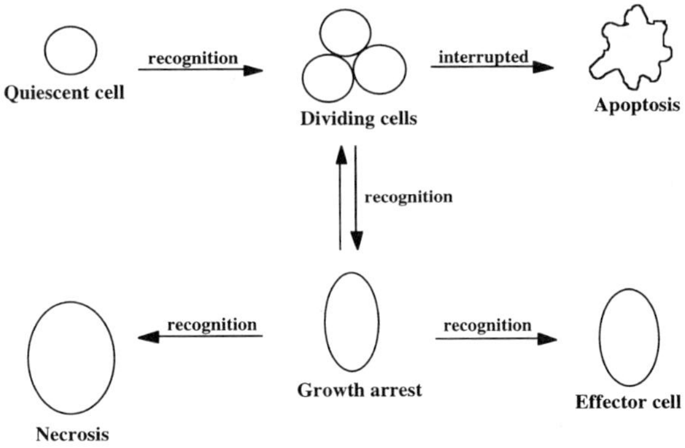

FIGURE 1. The consequences for immune recognition. Immune recognition can result in at least four fates for the recognized cell: 1) the cell will be tolerated and left unchanges; 2) the cell will be induced to divide; 3) the cell will be induced to differentiate; or 4) the cell will be induced to die.

MOLECULAR MECHANISMS OF IMMUNE RECOGNITION: MAJOR HISTOCOMPATIBILITY COMPLEX

Over 60 years ago, the Major Histocompatibility Complex (MHC)–encoded molecules were identified as the cell surface molecules that control rejection of tumor cells (reviewed in ref. 1). Subsequently, these molecules were shown to be responsible for graft rejection between genetically distinct members of the same species. The mechanism for both phenomena has been attributed to antigen-specific T cell recognition and effector function, which occur only when MHC molecules are engaged. Our work and the work of others have demonstrated that MHC molecules also serve as signal transducing receptors.[2–4] MHC engagement can result in cell activation, division, or death depending on cell type, position within the cell cycle, and activation state.[5] Therefore, MHC molecules control the rejection of allografted tumor cells[1] or transplants by T cell recognition and/or by MHC-mediated death. Reciprocally, alterations in MHC-mediated signaling in tumor cells may be a mechanism for inducing immunological tolerance of tumor cells.

ENERGY METABOLISM

Rapid cell division is a feature of some tumor cells,[6] developing tissues, and of adult tissues capable of regenerating, including hematopoietic cells, skin cells, and regenerating liver cells.[7,8] These cell types are distinguished by a marked shift away from mitochondrial and toward cytoplasmic ATP production.[6] During rapid cell division, the ATP required by the cell is produced by a relatively increased rate of glycolysis in the cytosol and there is less reliance on mitochondria as a source of ATP. It seems paradoxical that rapidly dividing cells depend heavily on glycolysis to generate ATP since the cell's energy demands are high. In non-dividing cells the mitochondria produce about 90% of cellular ATP by ATP synthetase.[9] Mitochondrial ATP synthesis is fueled by the oxidation of a variety of energy substrates in the cell (including fatty acids, glucose, and amino acids, such as glutamine). ATP is produced by coupling electron transport to the phosphorylation of ADP by ATP synthase.[10] Coincident to the transfer of electrons (or reducing equivalents) through the components of the electron transport chain, protons are pumped into the intermembrane space of the mitochondria,[11] producing an electrochemical gradient, or protonmotive force, across the inner membrane. Protonmotive force directly fuels mitochondrial ATP synthesis, since the activity of ATP synthase depends on the movement of protons from the intermembrane space to the matrix. Research in the last decade has convincingly shown that the coupling of substrate oxidation to ADP 2 phosphorylation in mitochondria is far from perfect. The main mechanism underlying imperfect coupling is a proton leak which is estimated to account for 25–35% of the resting energy expenditure of cells.[9] Leak results in the loss of potential energy when protons return to the matrix through a mechanism that bypasses ATP synthase. Thus, energy substrates are oxidized and oxygen is consumed by the mitochondria, but ATP is not produced and the energy can be released as heat. It is important to remember that regardless of the efficiency of mitochondrial ATP synthesis, mitochondrial energy expenditure is proportional to oxygen consumption, as oxygen is the final electron acceptor in the electron transport chain. The application of metabolic control analysis to studies of mitochondrial proton leak have shown that the amount of energy "wasted" through leaks is carefully controlled by cells.[12] Results have demonstrated that when leak is low when ATP need is high, and leak is high when ATP need is low.[12–14] Leak therefore has a tonic effect on overall oxygen consumption and has been proposed to be an important contributor to basal metabolic rate.[14]

Cancer cells have an altered balance between glycolytic and mitochondrial metabolism. This was first described by Warburg when he discovered that some cancer cells had a respiratory deficiency that played an important role in tumorigenesis.[15] The studies that followed were focused on identifying the genes responsible for the cancer phenotype, and studies of the mitochondria were few until Clayton in 1967[16] and Hadler in the 1970s[17] proposed a relationship between the changes in mitochondrial function and carcinogenesis.[17] Cancer cells exhibit a variety of differentiation states which parallel states that occur in normal cells. These include the highly differentiated cell, with high rates of glucose oxidation and decreased rates of cell division, to the highly dedifferentiated cell, which is highly glycolytic and exhibits a high rate of cell division. We suggest that these states are important in maintaining the immunological "privilege" of tumor cells.

IMMUNE RECOGNITION, FAS AND FAS LIGAND, AND CELL DEATH

During tissue regeneration and wound healing, a marked increase in cell proliferation is accompanied by a metabolic shift toward glycolysis, as demonstrated by increased glucose transporter expression, glucose uptake, glucose-6-phosphatase and tissue hexokinase activity, and lactate production.[18–23] Wound healing can be accelerated by the topical application of ordinary granulated sugar or sugar paste.[24–28] More recently, wound healing in burn patients has been accelerated by the infusion of high dose glucose plus insulin.[29] Thus, the proliferating cells of a healing wound are primarily glycolytic, and supplying glucose and insulin to these cells can speed the healing process. The mechanism by which this phenomenon occurs remains poorly understood.

We have shown that increasing glucose and insulin concentrations in cultures of tumor cells induced increased cell proliferation, and upregulated cell surface Fas expression (A. Russo, M.-E. Harper & M.K. Newell, manuscript in preparation). However, Fas, a member of the nerve growth factor/tumor necrosis factor family, is widely known as a death effector molecule. Fas engagement can also lead to accelerated proliferation in some cells including human peripheral blood T cells[31] and mouse memory T cells (J. Desbarats & M.K. Newell, submitted for review). The ligand for Fas (Fas Ligand, FasL) is expressed by cells of the immune system including activated T and B cells, NK cells, and PMNs, many of which infiltrate healing wounds. Notably, when PMNs are suppressed or absent, the consequence is impaired wound healing.

We suggest that immune system recognition and engagement of Fas on wounded tissues may accelerate proliferation and promote healing. We have demonstrated that cell surface Fas expression correlates with increased glucose utilization, and a shift toward glycolytic metabolism tumor cells and pancreatic islet β cells (M.-E. Harper & M.K. Newell, manuscript in preparation). This suggests that increasing insulin and glucose promotes cell division. When glucose levels drop, Fas-dependent cell death occurs and is necessary for the wound to heal.

On the basis of these data, we have proposed that Fas expression is a function of increasing rates of cytosolic glycolysis and that Fas is retained intracellularly in cells in which oxidative metabolism is uncoupled from glycolysis.We suggest that in glycolytic cells Fas engagement can lead to cell proliferation when glucose is present and usable (i.e. glucose transporters are being expressed, insulin is available, and the cells are not insulin resistant). Conversely, in glycolytic cells Fas engagement in the absence of sufficient glucose or in insulin resistant cells (such as during type II diabetes) may trigger apoptotic cell death. Interestingly, *c-myc*, a proto-oncogene which, like Fas can trigger either death or proliferation, has recently been described to induce a glucose-dependent apoptotic pathway, suggesting that the metabolic regulation of cell death may be a general phenomenon that operates via multiple mechanisms.

MODEL

In summary, we propose that immune recognition, of somatic cells via MHC can alter the cell's energy metabolism and induce a metabolic shift. We have demonstrated a reciprocal relationship that inducing a shift in metabolism toward glycolysis by supplying glucose and insulin results in the upregulation of immunologically recognizable molecules such as cell surface Fas. Thus, immune recognition can induce metabolic deviation. Metabolic deviation can result in altered in immune recognition and ultimately, in cell proliferation, cell differentiation, or cell death. Therefore, immune recognition can result in four outcomes. First, recognition can result in immunological tolerance. This occurs in immune privileged tissues, such as the brain, eye, and Sertoli cells of the testis. Tumors are also tolerated by the immune system. Second, recognition can result in cell division that occurs during "programmed cell division" during development or antigen-specific expansion of lymphocytes. The recognition that promotes cell division could provide protection to developing fetal tissues and tumor cells. Third, recognition can result in "post-mitotic" terminal differentiation of cells into "effectors." Recognition results in differentiation of cells to form specialized tissues with specific cellular functions. Finally, recognition can result in cell death. Cell death provides mechanisms for growth control, for the regulation of the type of immune responses that is needed, and for protection of the whole organism from damage and/or tumor cell growth.

REFERENCES

1. SNELL, G. 1981. Studies in histocompatibility. Science **213:** 172–177.
2. NEWELL, M.K., L.B. JUSTEMENT, K. LEHMANN, K. CALDWELL, D.M.F. COOPER, J.H. FREED & J.C. CAMBIER. 1987. Do class 11 major histocompatibility molecules function as signal transducers during B lymphocyte activation? *In* H-2 Antigens: Genes, Molecules, Functions. C. S. David, Ed. Plenum Press. New York. pp. 531–540.
3. CAMBIER, J.C., M.K. NEWELL, L.B. JUSTEMENT, K. LEACH, J.C. MCGUIRE & Z.Z. CHEN. 1987. Ia binding ligands and CAMP stimulate nuclear translocation of PKC in B lymphocytes. Nature (London) **327:** 629–632.
4. BISHOP, G.A. & G. HAUGHTON. 1986. Induced differentiation of a transformed clone of Ly-l$^+$ B cells by clonal T cells and antigen. Proc. Natl. Acad. Sci. USA **83:** 7410–7414.
5. ANDRÉ, P., J.C. CAMBIER, T.K. WADE, T. RAETZ & W.F. WADE. 1994. Distinct structural compartmentalization of the signal transducing functions of major histocompatibility complex class II (Ia) molecules. J. Exp. Med. **179**: 763–768.
6. BAGGETTO, L.G. 1992. Deviant energy metabolism of glycolytic cancer cells. Biochimie **74:** 959–974.
7. BRAND, K.A. & U. HERMFISSE. 1997. Aerobic glycolysis by proliferatingcells: a protective strategy against reactive oxygen species. FASEB J. **11:** 388–395.
8. GREINER, E.F., M. GUPPY & K. BRAND. 1994. Glucose is essential for proliferation and the glycolytic enzyme induction that provokes a transition to glycolytic energy production. J. Biol. Chem. **269:** 31484–31490.
9. BRAND, M.D., E.K. CHIEN, E.K. AINSCOW, D.F.S. ROLFE & R.K. PORTER. 1994. The causes and functions of mitochondrial proton leak. Biochim. Biophys. Acta **1187:** 132–139.
10. HARPER, M.E. 1997. Obesity research continues to spring leaks. Clin. Invest. Med. **20:** 239–244.
11. HIMMS-HAGEN, J. 1992. Brown Adipose Tissue. J.B. Lippincott. Philadelphia.
12. BRAND, M.D., M.-E. HARPER & H.C. TAYLOR. 1993. Control of the effective P/O ration of oxidative phophorylation in liver mitochondria and hepatocytes. Biochem. J. **291:** 739–748.
13. HAFNER, R.P., G.C. BROWN & M.D. BRAND. 1990. Analysis of the control of respiration rate, phosphorylation rate, proton leak rate and protonmotive force in isolated mitochondria using the "top down" approach of metabolic control theory. **188:** 313–319.
14. BRAND, M.D. 1990. The proton leak across the mitochondria inner membrane. Biochim. Biophys. Acta **1018:** 128–133.

15. Warburg, O., F. Wind & N.E. 1926. Uber den Stoffwechsel vonTumorim Korper. Klin. Woch. **5:** 829–832.
16. Clayton, D.A. & J. Vinorad. 1967. Circular dimer and catenate forms of mitochondrial DNA in human leukemic leucocytes. Nature **216:** 652–654.
17. Hadler, H.J., B.G. Daniel & R.D. Pratt. 1971. The induction of ATP energize mitochondrial volume changes by carcinogenic N-hydroxyl-N-acetylaminofluorenes when combined with showdomycin. A unitary hypothesis for carcinogenesis. J. Antibiotics **24:** 405–417.
18. Chiolero, R., J.P. Revelly & L. Tappy. 1997. Energy metabolism in sepsis and injury. Nutrition **13:** 45–51.
19. Takahashi, H., A.E. Kaminski & J.D. Zieske. 1996. Glucose transporter 1 expression is enhanced during comeal epithelial wound repair. Exp. Eye Res. **63:** 649–659.
20. Kreutzberg, G.W. 1996. Principles of neuronal regeneration. Acta Neurochir. (Suppl.) **66:** 103–106.
21. Carter, E.A., R.G. Tompkins, J.W. Babich, J. Correia, E.M. Bailey & A. J. Fischman. 1996. Thermal injury in rats alters glucose utilization by skin, wound, small intestine, but not by skeletal muscle. Metabolism: Clin. Exp. **45:** 1161–1167.
22. Haber, B.A., S. Chin, E. Chuang, W. Buikhuisen, A. Naji & R. Taub. 1995. High levels of glucose-6-phosphatase gene and protein expression reflect an adaptive response in porliferating liver and diabetes. J. Clin. Invest. **95:** 832–841.
23. Gamelli, R.L., H. Liu, L.K. He & C.A. Hofmann. 1994. Alterations of glucose transporter MRNA and protein levels in brain following thermal injury and sepsis in mice. Shock **1:** 395–400.
24. Dawson, J.S. 1996. Preiskel Elective Prize. The role of sugar in wound healing. Ann. Royal College of Surgeons of England **78:** 82–85.
25. Seal, D.V. & K. Middleton. 1991. Healing of cavity wounds with sugar. Lancet **338:** 571–572.
26. Trouillet, J.L., J. Chastre, J.Y. Fagon, J. Pierre, Y. Domart & C. Gibert. 1985. Use of granulated sugar in treatment of open mediastinitis after cardiac surgery. Lancet **2:** 180–184.
27. Tanner, A.G., E.R. Owen & D.V. Seal. 1988. Successful treatment of chronically infected wounds with sugar paste. Eur. J. Clin. Microbiol. Infect. Dis. **7:** 524–525.
28. Archer, H.G., S. Barnett, S. Irving, K.R. Middleton & D.V. Seal. 1990. A controlled model of moist wound healing: a comparison between semi-permeable film, antiseptics and sugar paste. J. Exp. Pathol. **71:** 155–170.
29. Pierre, E.J., R.E. Barrow, H.K. Hawkins, T.T. Nguyen, Y. Sakurai, M. Desai, R.R. Wolfe & D.N. Herndon. 1998. Effects of insulin on wound healing. J. Trauma **44:** 342–345.
30. Rosenburg, C.S. 1990. Wound healing in the patient with diabetes mellitus. Nursing Clin. N. Am. **25:** 247–261.
31. Alderson, M.R., R.J. Armitage, E. Maraskovsky, T.W. Tough, E. Roux, K. Schooley, F. Ramsdell & D.H. Lynch. 1993. Fas transduces activation signals in normal human T lymphocytes. J. Exp. Med. **178:** 2231–2235.

Regulation of Transglutaminases by Nitric Oxide

FRANCESCA BERNASSOLA, ANTONELLO ROSSI, AND GERRY MELINO[a]

Biochemistry Laboratory, IDI-IRCCS, c/o Department of Experimental Medicine and Biochemical Science, University of Rome Tor Vergata, Rome, Italy

ABSTRACT: Nitric oxide (NO) is an inorganic diffusible molecular messenger that plays several central roles in pathophysiology. NO can affect the biological activity of proteins through the direct or indirect (via intermediate S-nitrosothiols) S-nitrosylation of protein thiol groups. Transglutaminases (TGases), Ca^{++}-dependent enzymes that modify proteins by cross-linking reactions, require a cysteine residue in the active site as a prerequisite for their catalytic activity. Therefore, NO may regulate enzymatic activity of TGases and their biological effects, via S-nitrosylation of their crucial thiol groups. We here review the effects of NO on coagulation factor XIII (fXIII, or plasma TGase) and TGase 2 (or tissue transglutaminase). NO has an inhibitory effect on fXIII, thus suppressing the γ-chain cross-linking in fibrin gels, and subsequent clot formation. Tissue transglutaminase, an apoptotic effector molecule, also represents a molecular target for NO. Accordingly, the inhibition of tissue transglutaminase enzymatic activity by NO is able to prevent the induction of apoptosis.

INTRODUCTION

Nitric oxide, an inorganic free-radical gaseous molecule, is a multifunctional biologic mediator that is generated from the oxidation of the terminal guanido nitrogen atom of L-arginine by a family of NADPH-dependent enzymes, the NO synthases (NOS). Indeed, NO and related nitroso compounds have been implicated in a number of different physiopathological roles, including smooth muscle relaxation, platelet inhibition, neurotransmission, immune regulation and cytotoxicity.

Reactivity of Nitric Oxide

NO is a short-lived radical (half-life 15 seconds) and, in the presence of oxygen and water, is rapidly metabolized *in vivo* to nitrates and nitrites.[1] The chemistry of NO, however, involves interrelated redox forms ($NO\cdot$, NO^+, NO^-) and its pleiotropic biological effects are dependent on the chemical reactivity of NO. Indeed, the redox state of NO influences its preference for various target functional groups.[1] *In vivo*,

[a]Corresponding author: IDI-IRCCS, Biochemistry Lab, c/o Department of Experimental Medicine, D26/F153, University of Rome Tor Vergata, Via Tor Vergata 135, 00133 Rome, Italy. +39 6 20427299 (voice); + 39 6 20427290 (fax).
e-mail: gerry.melino@uniroma2.it

[b]ABBREVIATIONS: NO, nitric oxide; NOS, nitric oxide synthase; TGase, transglutaminase; tTG, tissue transglutaminase, fXIII, factor XIII; SNAP, S-nitroso-N-acetylpenicillamine; DTT, dithiothreitol; NMDA, N-methyl-D-aspartate.

the free radical reacts rapidly with species containing unpaired electrons such as molecular oxygen, superoxide anion and transition metal ions, potentially yielding nitrogen dioxide, dinitrogen trioxide, peroxynitrite and metal-nitrosyl complexes (FIG. 1A). The nitroxyl anion (NO^-) rapidly undergoes dimerization and dehydration, thus generating dinitrogen oxide, addition to thiols leading to sulfhydryl oxidation and reactions with metals (FIG. 1A). The nitrosonium cation (NO^+) is preferentially involved in nitrosation reactions with nucleophilic groups such as thiols, amides, carboxyls, hydroxyls, and aromatic rings (FIG. 1A).

NO has been shown to regulate the biological activity of proteins under physiological circumstances. NO can exert its effects by covalently modifying proteins through reversible reactions with available critical groups; protein modification can be achieved by NO itself or by more long-lived physiological NO-carriers such as S-nitrosothiols and metal conjugates.[1] Molecular targets for NO include Fe centers of prosthetic groups such as heme and Fe-S clusters, and nucleophilic centers such as nitrogen, oxygen and aromatic carbons.[1] Proteins can also be modified by S-nitrosylation of thiol groups, a chemical modification that may have important physiologically relevant regulatory implications. S-nitrosylation results from direct or indirect (via intermediate S-nitrosothiols) transfer of NO^+ to thiol groups of proteins.[2,3]

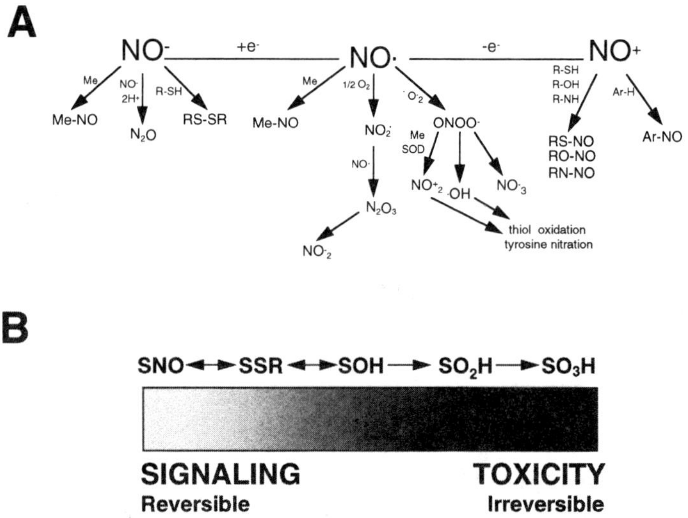

FIGURE 1. Chemical reactivity of the interrelated redox forms of NO. **A:** NO· may be converted by reaction with molecular oxygen to nitrite, react with transition metals (Me) and with superoxide to form peroxynitrite. NO^+ is involved in reactions with nucleophilic groups such as thiols and aromatic rings (Ar-H). NO^- undergoes spontaneous dimerization and addition to thiols resulting in dinitrogen oxide formation and sullfhydryl oxidation, respectively. **B:** Reversibility of the reactions of NO with thiol groups. Reversible reactions seem to be more important in physiological signaling that regulates biological responses. Irreversible reactions, such as thiol oxidation to sulfenic (SO_2H) and sulfinic (SO_3H) acid, seem to be more relevant in mediating pathological toxic effects.

S-nitrosylation is generally associated with loss of protein function, which can be achieved either by the inhibition of the redox site of cellular enzymes, or by forming hydrogen bonds and ionic interactions, which may change protein conformation. The reaction of NO with cell-surface thiols, for example, has been associated with anti-microbial actions,[4] modulation of the neuronal N-methyl-D-aspartate (NMDA) receptor channel activity,[5] and alterations of smooth muscle cell function.[6] S-nitrosylation may regulate cellular signal transduction by influencing signaling proteins such as $p21^{ras}$[7] and c-Jun N-terminal kinase 2.[8] NO reacts with thiol groups of cytosolic enzymes such as tissue transglutaminase (tTG,[9]), caspases,[9,10] glyceraldehyde-3-phosphate dehydrogenase,[11] and hemoglobin,[12,13] and it can modulate the biological activity of extracellular molecules such as coagulation factor XIII (fXIII).[14] In addition, NO affects the cellular gene transcription machinery by inhibiting the DNA-binding activity of NF-κB[15] and AP-1.[16]

The targets and mechanisms of cytotoxic NO effects are different from those involved in its physiological actions and may include irreversible protein modifications such as tyrosine nitration or thiol oxidation to sulfenic and sulfinic acid.[17,18] NO-mediated post-translational redox modification of protein thiols, such as S-nitrosylation and disulfide formation, are reversible reactions that predominantly have regulatory functions in biological systems, including cytoprotection.[18] On the contrary, irreversible modifications of cysteine residues to sulfenic and sulfinic acid are presumably associated with loss of protein function and cytotoxicity (FIG. 1B). This complex pattern of reactions may account for the diverse array of the NO-mediated biological effects. According to Stamler and Hausladen, 1998,[18] we can speculate that, while irreversible reactions might be more relevant in pathological conditions, the most important physiological role of NO is exerted through regulatory reversible reactions.

TRANSGLUTAMINASES

Transglutaminases (TGases) are a family of intracellular and extracellular Ca^{++}-and thiol-dependent enzymes that catalyze acyl-transfer reactions among polypeptide chains in which γ-carboxamide groups function as acyl donors and primary amino groups of several amines act as acceptor substrates.[19] These reactions result in post-translational modifications of proteins by establishing ε(γ-glutamyl)lysine cross-links and/or covalent incorporation of di- and polyamines into proteins. The TGase-dependent formation of stable cross-links leads to protein polymerization, which confers high resistance to mechanical breakage and chemical attack to the polypeptide involved in the linkage.[20] In mammals, a family of seven distinct TGases has been described to date, including TGase 1, a membrane-associated and cytosolic enzyme mainly expressed in epithelial tissues[21]; TGase 2 (or tTG), an ubiquitous soluble protein[22]; TGase 3, a pro-enzyme synthesized in terminally differentiating epidermal and hair keratinocytes[23]; TGase 4, secreted by the prostate, is involved in seminal coagulation[24]; TGase X, which has been recently identified in the epidermis but whose role in keratinocytes still remains largely unknown[25]; the catalytic A subunit of the blood-clotting fXIII[26]; and band 4.2, an inactive TGase-like protein which is the major constituent of the plasma membrane of red cells.[27]

The cysteine thiol active center is a prerequisite for TGase catalytic activity. Therefore, NO may regulate these enzymatic activities via S-nitrosylation of their crucial sulfhydryl groups. Indeed, TGases are complex enzymes that are post-translationally modified and their enzymatic activity can be finely modulated by different factors. For example, tTG activity is finely regulated by calcium, GTP, polyamines and NO by S-nitrosylation.[9,28] In addition to the other regulators of TGase activity, S-nitrosylation may therefore modulate the biological effects of these enzymes.

S-Nitrosylation of TGases: The Example of the Coagulation fXIII

The catalytic A subunit of the plasma fXIII, a member of the TGase family, catalyzes the cross-linking of fibrin monomers during blood coagulation; the active form of fXIII A stabilizes the blood clot, which is a quite strong structure resistant to mechanical and chemical attack.[29] In the hemostatic system, NO inhibits platelet aggregation and adhesion. Modulation of platelet aggregation by NO is mainly mediated by the activation of cytosolic guanylate cyclase resulting from the binding of NO to its iron-containing heme group.[30] Indeed, the subsequent increase in intracellular concentrations of guanosine 3′, 5′-monophosphate inhibits platelet aggregation, thus inhibiting blood clotting. Furthermore, S-nitrosylation of tissue type plasminogen activator appears to enhance its catalytic efficiency to convert plasminogen to plasmin on fibrin and thrombi, which in turn induces fibrinolysis and platelet disaggregation.[31]

We recently reported that NO is also capable of reacting with blood coagulation fXIII, thus regulating its enzymatic activity.[14] Indeed, NO-releasing compounds inhibit fXIII both *in vitro* and *in vivo* (in plasma preparations from healthy donors). The mechanism for inhibition appears to involve modification of fXIII by S-nitrosylation of a cysteine residue, as NO is able to eliminate the titratable thiol group of the enzyme. Accordingly, after S-nitrosylation, fXIII fails to efficaciously cross-link fibrin monomers. As shown in TABLE 1, the incubation of blood samples with the NO-donor S-nitroso-N-acetylpenicillamine (SNAP) completely suppresses clot formation, whereas low concentrations of the NO-donor determine its dissolution after the addition of detergents and reducing agents. Furthermore, we showed that the exposure of blood samples to an exogenous NO-generating system results in inhibition of the γ-chain cross-linking in fibrin gels.[14]

TABLE 1. Effect of SNAP on blood clot formation and lysis

	SNAP concentration (mM)				
	0	0.1	1	2	10
Clot formation	+	+	+	+	–
Clot solubilization	–	+	+	+	na

NOTES: (+), clot formation or solubilization after the addition of SDS and β-mercaptoethanol; (–), absence of clot formation or solubilization; na, not applicable.

Effect of NO on Tissue Transglutaminase

Tissue transglutaminase has been suggested to be an effector element of the apoptotic machinery whose activation leads to the irreversible assembly of a highly insoluble cross-linked intracellular protein scaffold in dying cells.[20] Protein polymerization might contribute to the ultrastructural changes occuring in cells undergoing apoptosis, including cytoplasmic and nuclear condensation;[20,32] furthermore, tTG seems to stabilize the integrity of the apoptotic cells before their clearance by phagocytosis, preventing the release of harmful intracellular components into the extracellular space and subsequent inflammation and scar formation. Transfection studies using tTG cDNA indicate that the overexpression of tTG significantly increases spontaneous apoptosis in neuroblastoma cells, whereas, reduced tTG expression results in a pronounced decrease of basal as well as retinoid-induced cell death.[32]

Tissue transglutaminase represents a molecular target for NO. Indeed, NO-releasing compounds dose-dependently reduce tTG activity both in purified enzymatic preparations and in neuroblastoma cellular extracts.[9] FIGURE 2 shows that the activity of either purified or intracellular tTG is inhibited in a dose-dependent manner by increasing concentrations of SNAP, starting from 1 μM. The inactivation of the enzyme occurs at relatively low concentrations of the nitrosothiol (half-maximal inhibition 100 μM and 250 μM for purified and cytosolic tTG, respectively). NO-donor–mediated tTG

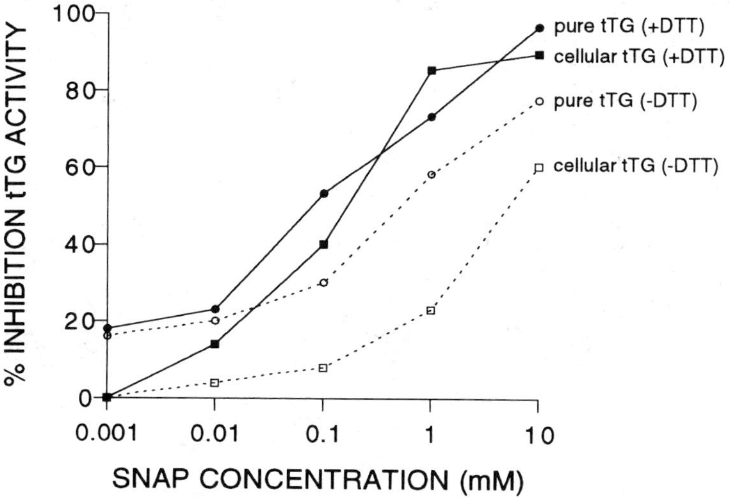

FIGURE 2. Effect of the NO-donor SNAP on tTG activity from purified enzymatic preparations and neuroblastoma cellular extracts. Increasing concentrations of SNAP were added to the enzymatic preparations or cell extracts immediately before initiating the reaction with [^{3}H]putrescine and N,N′-dimethylcasein. The incubation was carried out for 20 minutes at 37°C, in the presence or in the absence of 10 mM DTT. Data are the means of triplicate determinations of three separate experiments, SD < 10%. NO-donor concentrations result in about 1000-fold less free NO.

inhibition is prevented by NO scavengers such as oxyhemoglobin and reducing agents such as glutathione and cysteine. DL-dithiothreitol (DTT) is usually required to detect active tTG *in vitro*, as it reduces the cysteine residue of the active site. In its absence the NO concentration required to achieve the same level of enzymatic inhibition is significantly increased (FIG. 2); this indicates that NO is able to react with the thiol group of an active cysteine residue, and suggests that the inhibition of tTG activity is due to S-nitrosylation. In keeping with an inactivation due to S-nitrosylation, the titrated thiol groups of the enzyme were reduced from three to two after incubation with NO-donor compounds.[9]

Regulation of Cell Death by NO

We found a correlation between the inhibition of tTG enzymatic activity *in vivo* and the ability of exogenously supplied NO to prevent the induction of apoptosis. Cell death induced by retinoids or chemotherapeutic agents such as cisplatin in SK-N-BE(2) human neuroblastoma cells, is associated with an increase of tTG activity.[33–36] NO-generating species are able to protect neuroblastoma cells from retinoid- and drug-induced apoptosis in a dose-dependent manner.

S-nitrosylation of several critical factors may be important for cell survival. The inhibitory effect of S-nitrosylation has been demonstrated for other apoptotic effector molecules, including the cysteine protease family members (caspases).[9,10,37] Accordingly, exogenous sources of NO inhibit CD95 signaling in human leukocytes via S-nitrosylation of caspase 3-like proteases.[9,38] NO-mediated inactivation of caspases has been also reported in other apoptotic models such as tumor necrosis factor α–stimulated endothelial cell lines and rat hepatocytes.[10,37] Cytoprotective actions of NO have been described in the nervous system, where excitotoxic NMDA-mediated neuronal cell death is associated with a down-regulation of the NMDA receptor activity following its S-nitrosylation.[5]

Although NO can act through several potential toxic mechanisms eliciting necrosis as well as apoptosis, it exhibits a protective role in cell death. In addition to the other proposed mechanisms such as the inactivation of the mitochondrial respiratory chain,[39] changes in intracellular ATP levels,[40,41] the induction of heat shock proteins[42] and the inhibition of lipid peroxidation by ferrous compounds, hydrogen peroxide and reactive oxygen intermediates,[43] NO can regulate cell death by directly interfering with the apoptotic machinery.

CONCLUSIONS

TGases may be molecular targets for NO via S-nitrosylation of the active site cysteine residue that is essential for their enzymatic activity. As S-nitrosothiol groups formation in proteins represents an important cellular regulatory mechanism, the NO-mediated inactivation of TGases may have essential functional implications in several biological processes. Indeed, (i) the interaction of NO with the plasma fXIII may represent an alternative regulatory mechanism controlling blood coagulation; (ii) NO can regulate cell death by inhibiting tTG as well as other effectors of apoptosis.

The ability of NO to both induce and prevent apoptosis might be related to its redox state, concentration and exposition time, through the above described complexity of reactions (Fig. 1A, B). Therefore, cytotoxicity may result as a consequence of irreversible reactions, whereas reversible reactions may mediate transducing pathways, and thus protective actions of NO. We can speculate that the scale of the events in inhibition of cell death includes: (i) prevention of apoptosis resulting in resistance to death; (ii) inhibition of apoptosis downstream of the commitment point to die, leading to necrosis. While the inhibition of apoptosis may be physiologically relevant, NO-mediated cytotoxic actions could lead to pathological states. We would like to speculate that the anti-apoptotic signaling mechanism represents a function by far more important than the induction of cell injury.

REFERENCES

1. STAMLER, J.S., D.J. SINGEL & J. LOSCALZO. 1992. Biochemistry of nitric oxide and its redox-activated forms. Science **258:** 1898–1902.
2. STAMLER, J.S., D.I. SIMON, J.A. OSBORNE, M.E. MULLINS, O. JARAKI, T. MICHEL, D.J. SINGEL & J. LOSCALZO. 1992. S-nitrosylation of proteins with nitric oxide: synthesis and characterization of biologically active compounds. Proc. Natl. Acad. Sci. USA **89:** 444–448.
3. STAMLER, J.S. 1994. Redox signaling: nitrosylation and related target interactions of nitric oxide. Cell **78:** 931–936.
4. MORRIS, S.L., R.C. WALSH & J.N. HANSEN. 1984. Identification and characterization of some bacterial membrane sulfhydryl groups which are targets of bacteriostatic and antibiotic action. J. Biol. Chem. **259:** 13590–13594.
5. LIPTON, S.A., Y.B. CHOI, Z.H. PAN, S.Z. LEI, H.S. CHEN, N.J. SUCHER, J. LOSCALZO, D.J. SINGEL & J.S. STAMLER. 1993. A redox-based mechanism for the neuroprotective and neurodestructive effects of nitric oxide and related nitroso-compounds. Nature **364:** 626–632.
6. BATES, J.N., M.T. BAKER, R. JR. GUERRA & D.G. HARRISON. 1991. Nitric oxide generation from nitroprusside by vascular tissue. Evidence that reduction of the nitroprusside anion and cyanide loss are required. Biochem. Pharmacol. **42:** S157–165.
7. LANDER, H.M., J.S. OGISTE, S.F. PEARCE, R. LEVI & A. NOVOGRODSKY. 1995. Nitric oxide-stimulated guanine nucleotide exchange on p21 Ras. J. Biol. Chem. **270:** 7017–7020.
8. SO, H.S., R.K. PARK, M.S. KIM, S.R. LEE, B.H. JUNG, S.Y. CHUNG, C.D. JUN & H.T. CHUNG. 1998. Nitric oxide inhibits c-Jun N-terminal kinase 2 (JNK2) via S-nitrosylation. Biochem. Biophys. Res. Commun. **247:** 809–813.
9. MELINO, G., F. BERNASSOLA, R.A. KNIGHT, M.T. CORASANITI, G. NISTICÒ & A. FINAZZI-AGRÒ. 1997. S-nitrosylation regulates apoptosis. Nature **388:** 432–433.
10. DIMMELER, S., J. HAENDELER, M. NEHLS & A.M. ZEIHER. 1997. Suppression of apoptosis by nitric oxide by inhibition of ICE-like and CPP32-like proteases. J. Exp. Med. **185:** 601–608.
11. MOHR, S., J.S. STAMLER & B. BRÜNE. 1996. Posttranslational modifications of glyceraldehyde-3-phosphate dehydrogenase by S-nitrosylation and subsequent NADH attachment. J. Biol. Chem. **271:** 4209–4214.
12. JIA, L., C. BONAVENTURA, J. BONAVENTURA & J.S. STAMLER. 1996. S-nitrosohaemoglobin: a dynamic activity of blood involved in vascular control. Nature **380**: 221–226.
13. GOW, A. & J.S. STAMLER. 1998. Reactions between nitric oxide and haemoglobin under physiological conditions. Nature **391:** 169–173.

14. CATANI, M.V., F. BERNASSOLA, A. ROSSI & G. MELINO. 1998. Inhibition of clotting factor XIII activity by nitric oxide. Biochem. Biophys. Res. Commun. **249:** 275–278.

15. MATTHEWS, J.R., C.H. BOTTING, M. PANICO, H.R. MORRIS & R.T. HAY. 1996. Inhibition of NF-kappaB DNA binding by nitric oxide. Nuc. Ac. Res. **24:** 2236–2242.

16. TABUCHI, A., K. SANO, E. OH, T. TSUCHIYA & M. TSUDA. 1994. Modulation of AP-1 activity by nitric oxide (NO) in vitro: NO-mediated modulation of AP-1. FEBS Lett. **351:** 123–127.

17. BECKER, K., S.N. SAVVIDES, M. KEESE, R.H. SHIRMER & P.A. KARPLUS. 1998. Enzyme inactivation through sulfhydryl oxidation by physiologic NO-carriers. Nature Struct. Biol. **5:** 267–271.

18. STAMLER, J.S. & A. HAUSLADEN. 1998. Oxidative modifications in nitrosative stress. Nature Struct. Biol. **5:** 247–249.

19. FOLK, J.E. 1980. Transglutaminases. Annu. Rev. Biochem. **449:** 517-531.

20. FESUS, L., V. THOMAZY, F. AUTUORI, M.P. CERU'-ARGENTO, E. TARCSA & M. PIACENTINI. 1989. Hepatocytes undergoing programmed cell death become insoluble in detergent and chaotropic agents. FEBS Lett. **245:** 150-154.

21. THACHER, S.M. & R.H. RICE. 1985. Keratinocyte-specific transglutaminase of cultured human epidermal cells: relation to cross-linked envelope formation and terminal differentiation. Cell **40:** 685–695.

22. CHUNG, S-I. & J.E. FOLK. 1972. Transglutaminase from hair follicle of guinea pig. Proc. Natl. Acad. Sci. USA **69:** 303–308.

23. KIM, H-C., J. LEWIS, J.J. GORMAN, S.C. PARK, J.E. GIRARD, J.E. FOLK & S-I. CHUNG. 1990. Protransglutaminase E from guinea pig skin. Isolation and partial characterization. J. Biol. Chem. **265:** 21971–21978.

24. GRANT, F.J., D.A. TAYLOR, P.O. SHEPPARD, S.L. MATHEWES, W. LINT, E. VANAJA, P.B. BISHOP & P.J. O'HARA. 1994. Molecular cloning and characterization of a novel transglutaminase cDNA from a human prostate cDNA library. Biochem. Biophys. Res. Commun. **203:** 1117–1123.

25. AESCHLIMANN, D., M.K. KOELLER, B.L. ALLEN-HOFFMANN & D.F. MOSHER. 1998. Isolation of a cDNA encoding a novel member of the transglutaminase gene family from human keratinocytes. Detection and identification of transglutaminase gene products based on reverse transcription-polymerasechain reaction with degenerate primers. J. Biol. Chem. **273:** 3452–3460.

26. SIEBENLIST, K.R. & M.W. MOSESSON. 1996. Evidence for intramolecular cross-linked Aa.g chain heterodimers in plasma fibrinogen. Biochem. **35:** 5817–5821.

27. KORSGREN, C. & C.M. COHEN. 1986. Purification and properties of human erythrocyte band 4.2. Association with the cytoplasmic domain of band 3. J. Biol. Chem. **261:** 5536–5543.

28. MELINO, G. & M. PIACENTINI. 1998. Tissue transglutaminase in cell death: a downstream or multifunctional upstream effector? FEBS Lett. **430:** 59–63.

29. ICHINOSE, A. & E.W. DAVIE. 1988. Characterization of the gene for the A subunit of human factor XIII, a blood coagulation factor. Proc. Natl. Acad. Sci. USA **85:** 5829–5833.

30. GARBERS, D.L. 1992. Guanylyl cyclase receptors and their endocrine, paracrine, and autocrine ligands. Cell **74:** 1–4.

31. STAMLER, J.S., D.I. SIMON, O. JARAKI, J.A. OSBORNE, S. FRANCIS, M. MULLINS, D. SINGEL & J. LOSCALZO. 1992. S-nitrosylation of tissue-type plasminogen activator confers vasodilatory and antiplatelet properties on the enzyme Proc. Natl. Acad. Sci. USA **89:** 8087–8091.

32. MELINO, G., M. ANNICCHIARICO-PETRUZZELLI, L. PIREDDA, E. CANDI, V. GENTILE, P.J. DAVIS & M. PIACENTINI. 1994. Tissue transglutaminase and apoptosis: sense and antisense transfection studies with human neuroblastoma cells. Mol. Cell. Biol. **14:** 6584–6596.
33. PIACENTINI, M., M. ANNICCHIARICO-PETRUZZELLI, S. OLIVERIO, L. PIREDDA, J.L. BIEDLER & G. MELINO. 1992. Phenotype-specific "tissue" transglutaminase regulation in human neuroblastoma cells in response to retinoic acid: correlation with cell death by apoptosis. Int. J. Cancer **52:** 271–278.
34. PIACENTINI, M., L. FESUS & G. MELINO. 1993. Multiple cell cycle access to the apoptotic death programme in human neuroblastoma cells. FEBS Lett. **320:** 150–154.
35. MELINO, G., M. DRAOUI, M. PIACENTINI, L. BELLINCAMPI, F. BERNASSOLA, U. REICHERT & P. COHEN. 1997. Retinoic acid receptors α and γ mediate "tissue" transglutaminase induction in human neuroblastoma cells undergoing apoptosis. Exp. Cell Res. **255:** 55–61.
36. NAGY, L., V.A. THOMAZY, R.A. HEYMAN & P.J.A. DAVIES. 1998. Retinoid-induced apoptosis in normal and neoplastic tissue. Cell Death Differ. **5:** 11–19.
37. KIM, Y-M., R.V. TALANIAN & T.R. BILLIAR. 1997. Nitric oxide inhibits apoptosis by preventing increases in caspase-3-like activity via two distinct mechanisms. J. Biol. Chem. **272:** 31138–31148.
38. MANNICK, J.B., X.Q. MIAO & J.S. STAMLER. 1997. Nitric oxide inhibits Fas-induced apoptosis. J. Biol. Chem. **272:** 24125–24128.
39. ANKARCRONA, M., J.M. DYPBUKI, E. BONFOCO, B. ZHIVOTOVSKY, S. ORRENIUS, S.A. LIPTON & P. NICOTERA. 1995. Glutamate-induced neuronal death: a succession of necrosis or apoptosis depending on mitochondrial function. Neuron **15:** 1–13.
40. NICOTERA, P. & M. LEIST. 1997. Energy supply and the shape of death in neurons and lymphoid cells. Cell Death Differ. **4:** 435–442.
41. LEIST, M., B. SINGLE, A.F. CASTOLDI, S. KUHNLE & P. NICOTERA. 1997. Intracellular adenosine triphosphate (ATP) concentration: a switch in the decision between apoptosis and necrosis. J. Exp. Med. **185:**1481–1486.
42. KIM, Y.M., M.E. DE VERA, S.C. WATKINS & T.R. BILLIAR. 1997. Nitric oxide protects cultured rat hepatocytes from tumor necrosis factor-alpha-induced apoptosis by inducing heat shock protein 70 expression. J. Biol. Chem. **272:**1402–1411.
43. KRÖNCKE, K-D., K. FEHSEL & V. KOLB-BACHOFEN. 1997. Nitric oxide: cytotoxicity versus cytoprotection. How, why, when and where? Nitric Oxide **1:** 107–120.

The Molecular Mechanism of Programmed Cell Death in *C. elegans*

QIONG A. LIU[a] AND MICHAEL O. HENGARTNER

Cold Spring Harbor Laboratory, 1 Bungtown Road, Cold Spring Harbor, New York 11724, USA

ABSTRACT: Programmed cell death or apoptosis plays a fundamental role during animal development, metamorphosis, and tissue homeostasis. It is a genetically controlled physiological process that comprises two distinct and sequential processes: the death of cells, and their subsequent removal by engulfing cells. In the nematode *C. elegans*, genetic studies led to the discovery of 15 genes that function in programmed cell death[1] (FIG. 1). These 15 genes have been divided into four groups based on the order of their activity during the process of programmed cell death: (1) those involved in the decision making (*ces-1* and *ces-2*); (2) in the process of execution (*ced-3*, *ced-4*, *ced-9* and *egl-1*); (3) in the engulfment of dying cells by engulfing cells (*ced-1*, *ced-2*, *ced-5*, *ced-6*, *ced-7*, *ced-10*, *ced-12*); and (4) those in the degradation of cell corpses within engulfing cells (*nuc-1*). In the last five years, several genes in the genetic pathway of programmed cell death have been shown to be conserved across a wide range of species; all genes involved in the step of execution in *C. elegans* have their corresponding mammalian homologs (FIG. 2). Furthermore, emerging evidence from molecular studies of engulfment genes in several species suggests that the signaling process from apoptotic cells to engulfing cells and the subsequent engulfment process might be also conserved across species (TABLE 1).

ADVANTAGES OF USING *C. ELEGANS* TO STUDY PROGRAMMED CELL DEATH

C. elegans has proved to be one of the most valuable organisms used to understand programmed cell death. There are several reasons for this. First, the complete knowledge of anatomy and cell lineage for this organism made it possible to observe the fate of individual cells.[2,3] Second, the transparent cuticle and egg shells of animals allow each cell to be viewed in living animals under differential interference contrast (DIC) optics, and thereby allow defects in cell death and engulfment to be scored easily.[1] Third, compared to mammals, *C. elegans* is a simple organism with much less redundancy of genes, so it is an ideal organism to use to find key elements involved in programmed cell death and other biological processes.

C. elegans is the first multicellular organism to have its genome completely sequenced.[4,5] Comprehensive information provided by Genefinder and other analysis programs are stored in a central and freely distributed database (ACEDB), illustrating the location of genes and information about their splice sites, protein homologs, and EST clones.[4] Large numbers of genes have been previously identified

[a]e-mail: liuq1@yahoo.com; hengartn@cshl.org

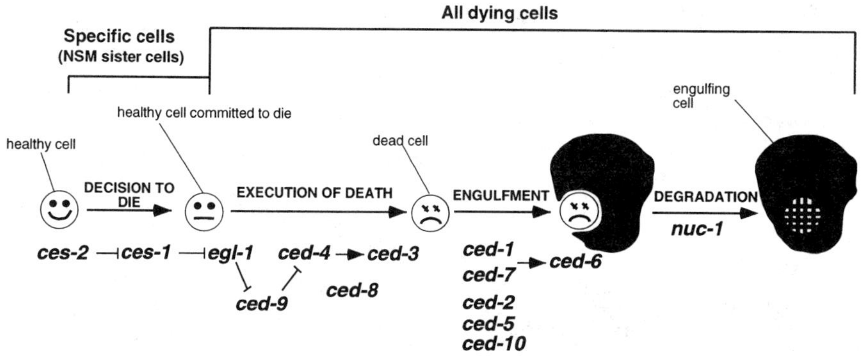

FIGURE 1. The genetic pathway for programmed cell death in *C. elegans*. In the nematode *C. elegans*, genetic studies led to the discovery of 15 genes related to programmed cell death. These 15 genes have been divided into four groups based on the order of their activity during the process of programmed cell death: *ces-1* and *ces-2* are involved in the decision making, *ced-3*, *ced-4*, *ced-8*, *ced-9*, and *egl-1* in the process of execution, ced-1, ced-2, ced-5, ced-6, ced-7, and ced-10 in the engulfinent of dying cells by engulfing cells, and *nuc-1* in the degradation of cell corpses within engulfing cells. Adapted from reference 1.

and cloned, generating excellent genetic markers for mapping newly identified genes. One consequence of this is that many of the genes previously characterized in other species have corresponding homologs in *C. elegans*. Because of the technical difficulty in evaluating many genes for their biological significance in complicated systems such as vertebrates, it has become of growing interest within the scientific community to determine the biological function of *C. elegans* homologs. In fact, several means for a reverse genetic analysis of genes have been developed; worm libraries including transposon insertion libraries,[6] and deletion libraries generated with EMS or other mutagens.[7] A recently developed technique to inactivate genes through injection of double-stranded mRNA provides an efficient way to look for the phenotypes of many genes.[8]

MOLECULES INVOLVED IN THE EXECUTION OF THE CELL DEATH

CED-3 Is a Killer

Four genes, *ced-3*, *ced-4*, *ced-9*, and *egl-1*, form the "death machinery" that is responsible for the execution of cells fated to die (FIGS. 2, 3; reviewed by Hengartner[1]; Conradt and Horvitz[9]). Genetic studies demonstrated that both *ced-3* and *ced-4* are required for the killing of cells; in *ced-3* or *ced-4* mutants, the 131 cells that normally die during the development of hermaphrodites were found alive.[10] *ced-9* antagonizes the activity of *ced-3* and *ced-4* and prevents cells from being killed. In *ced-9(gf)* animals, the 131 cells that normally die were also found alive.[12] In contrast, in *ced-9(lf)* animals, in addition to the 131 cells that normally die, many cells that normally survive were also found to die, resulting in developmental arrest and embryonic lethal-

TABLE 1. Evolutionary conservation of engulfment genes[a]

	C. elegans	*Drosophila*	*Mammals*
Receptors	CED-7	?	ABC1
	?	*Croquemort*	CD36
	?	?	lectin
	?	?	*scavenger* receptor
	?	?	CD14
	CED-1 (?)	?	?
Signal Transducers	CED-5	*Myoblast city*	DOCK180
	CED-6	Drome CED-6 ?	hCED-6 ?
Unknown	CED-2	?	?
	CED-10	?	?

[a]In *C. elegans*, six engulfment genes *ced-1*, *-6*, *-7*, *-2*, *-5*, and *-10* have been shown to be involved in the engulfment of apoptotic cells. One of the mammalian homologues of CED-7 was identified as the ABC l transporter. The homologs of CED-5 are human Dockl80 and *Drosophila Myoblast city*. CED-5 and its homologs might act as adaptor molecules in the signal transduction pathway that is required for the engulfment of apoptotic cells. CED-6 has been determined to act as an adaptor molecule in a putative tyrosine kinase pathway that is required specifically to remove apoptotic cells (see below). In mammals, several proteins including ABC 1, CD36, lectin, ventronectin receptor, scavenger receptor, and CD 14 have been suggested to act on the surface of phagocytes to recognize the apoptotic cells. The *Drosophila* homolog of CD36 has been identified as *Croquemort*.

ity.[12] Together, these genetic data demonstrated that *ced-9* is a survival factor that negatively regulates the killing activity of *ced-3* and *ced-4*.

Molecular, biochemical, and cellular biology studies on CED-3, CED-4 and CED-9 proteins implied that CED-3, rather than CED-4, is the real killer. *ced-3* encodes a protein that is similar to the mammalian interleukin-1β (IL-1β)–converting enzyme (ICE).[13] ICE belongs to a growing cysteine protease family which has at least thirteen members[14] (FIG. 2), known as caspases.

In mammals, different subsets of caspases are activated, depending on the pro-apoptotic stimulus. For example, caspases-3, -6, -7, and -8 act in the Fas/TNF-mediated death pathway and caspase-9, -3, together with Apaf- I and cytochrome c act in mitochondria-associated cell death. It is notable that in both cases, the phenotype of apoptotic cells is similar. Two groups[15,16] recently reported a link between these two groups of caspases. They found that Bid was a specific substrate for caspase-8 in the Fas-mediated cell death pathway. Normally, the full-length Bid is localized in the cytosol. The truncated form translocates to the mitochondria and releases cytochrome c, which subsequently activates caspase-9 and -3, resulting in the destruction of the cell. Bcl-2 and Bcl-XL have been found to prevent the release of cytochrome c by Bid and this prevents cell death from occurring.

Studies of the substrate specificity of *ced-3* have demonstrated that *ced-3* is more similar to mammalian caspase-3 than to mammalian caspase-1 or -2.[17] These results suggest that the CED-3/4/9 complex resembles the caspase-9/caspase-3/Apaf-1/

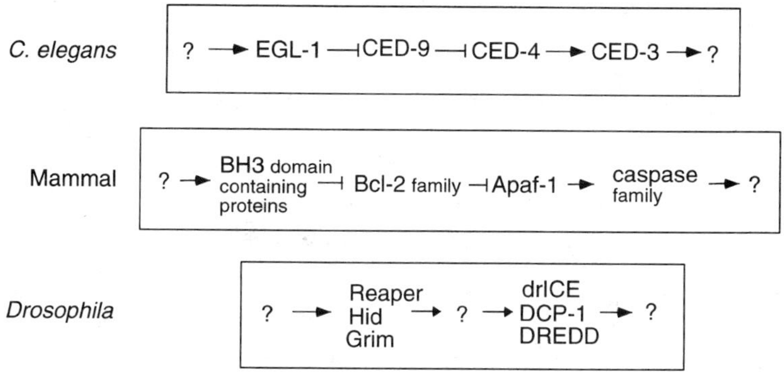

FIGURE 2. Conservation of the execution step in the genetic pathway for programmed cell death. In *C. elegans*, the execution step in the genetic pathway for programmed cell death comprises four genes, *ced-3*, *ced-4*, *ced-9*, and *egl-1*. All of these genes have their corresponding mammalian homologs: CED-9 is a member of the Bcl-2 family, CED-3 is a caspase family member, CED-4 is homologous to the Apaf- 1 protein, and EGL-1 is similar in structure to Bid and other BH3 domain only–containing proteins. In *Drosophila*, *reaper*, *hid*, and *grim* are required for the apoptotic cell death and might act upstream of the *Drosophila* caspases, drICE, DREDD, and DCP-1. The homologs of *reaper*, *hid*, and *grim* in other species have not been identified. Similarly, the homologs of EGL-1, CED-4 and CED-9 in *Drosophila* have not been found.

cytochrome c complex in mammals. However, the role of cytochrome c played in the execution of cell death has not been established in *C. elegans*. Furthermore, in *C. elegans*, no homologs of Fas or TNFR have been identified. The similarity between the roles of EGL-1 in *C. elegans* and Bid in mammals suggests that a pathway similar to Fas/TNFR-mediated cell death is likely to exist in *C. elegans*.

The difference in the morphology of apoptotic cells between *C. elegans* and mammals might reflect some difference in downstream events of apoptosis. In *C. elegans*, the phenotype of apoptotic cells includes the condensation of cytoplasm, and shrinkage of chromatin. However, the apoptotic bodies are not as apparent as those observed in mammalian apoptotic cells. Thirteen caspases in mammals have been shown to be responsible for the cleavage of many molecules. Cleavage of DFF45/ICAD by caspase-3 leads to chromatin condensation and fragmentation.[18,19] Cleavage of fodrin, gelsolin and PAK2 results in changes at the cell membrane and in the cytoskeleton.[20–22] Finally, cleavage of nuclear lamins by caspase-6 leads to nuclear membrane breakage.[23,24] In *C. elegans*, only CED- 3 and two recently characterized caspases have been identified,[25] suggesting that apoptosis in *C. elegans* might be simpler than in mammals.

ced-8 was isolated as an engulfment mutant. It was later suggested that mutations in this gene might slow down the death process.[1]

CED-4, an Adaptor Molecule that Regulates the Killing Activity of CED-3

Although the ectopic expression of *ced-4* can induce cell death, CED-4 does not engage the killing of cells directly. Sequence analysis of CED-4 showed that it contains a P-loop, an ATP binding site,[26,27] and two potential Ca^+ binding sites.[11] However, the putative Ca^+ binding sites were not required for its apoptotic activity.[11] Alternative spliced forms of CED-4 also exist; CED-4L suppresses cell death, whereas the more common CED-4S splice variant enhances cell killing.[28]

Genetic studies in which *ced-3*, *ced-4,* and *ced-9* were ectopically expressed in the six touch cells in *C. elegans*, demonstrated that *ced-4* might be regulated by *ced-9* and act upstream of *ced-3.*[29] Cell biology studies showed that CED-9 was localized mainly on the outer mitochondrial membrane, and sequestered CED-4 to the mitochondria membrane in mammalian cells.[30] Biochemical studies demonstrated that CED-9 directly interacted with CED-4.[26,31,32] Furthermore, CED-4 interacts with CED-3, together with CED-9, to form a protein complex on the mitochondrial membrane[26,33] (FIG. 3). Mutations that disrupt CED-9's ability to bind to CED-4 resulted in the activation of CED-4, and subsequently, the activation of CED-3.[32] That CED-9 regulates the activity of CED-4 was also supported by other studies; overexpression of CED-4 was found to be able to enhance the killing by CED-3, and this phenomenon can be suppressed by overexpressed CED-9 protein.[27] A recent report showed that after the release of the CED4/CED-3 complex from CED-9, CED-4 oligomerizes, which brings several copies of proCED-3 close to each other, an event that triggers the proteolytic activation of CED-3 (FIG. 3).[34]

CED-9, a Survival Factor

ced-9 encodes a 280 amino-acid protein that is similar to mammalian Bcl-2 (B cell lymphoma-2).[35] Overexpression of Bel-2 can protect cells from being killed in *C. elegans,*[35,36] suggesting that CED-9 and Bcl-2 are functional homologs. Both CED-9[30] and Bcl-2[37–39] contain a C-terminal hydrophobic tail on the carboxy termini and are localized on the mitochondria, the endoplastic reticulum, and the nuclear membrane.

Many proteins have been identified to be similar to CED-9 and Bcl-2, forming a Bcl-2 family (FIG. 2).[40-42] Structurally, Bcl-2 family proteins contain a different combination of BH domains including BH1, BH2, BH3 and BH4. Functionally, some of these proteins act to protect cells from being killed, and others promote the death process .[42] For example, Bax and Bcl-XS induce cell death, whereas CED-9, Bcl-XL and Bcl-2 prevent cells from being killed. Bik, Bid, Bad, and Harakiri contain only a BH3 domain, which is most similar to the BH3 domain in the death effectors Bak and Bax, and which can interact with Bcl-2-like proteins to induce cell death.

CED-9 and some Bcl-2 family proteins protect cells from being killed through two possible mechanisms: first, it was suggested that the three-dimentional structure of Bcl-2 or BclXL is similar to that of the membrane-inserting domains of diphtheria toxin and the pore-forming colicins A and El.[43] Thus, Bcl-2 and Bcl-XL were suggested to form pores to allow ions or small molecules across the outer mitochondrial membrane to regulate the osmotic pressure of cells.[44] Both Bcl-2 and Bcl-XL could prevent cytochrome c release from mitochondria and could save cells from being killed.[15,16,45,46] Second, in *C. elegans*, CED-9 protects cells from being killled by interacting with CED-4, which can regulate the activity of CED-3 (FIG. 3). Similarly, in mammals, Bcl-XL might regulate the activity of caspases by interacting with Apaf- 1.[47,48]

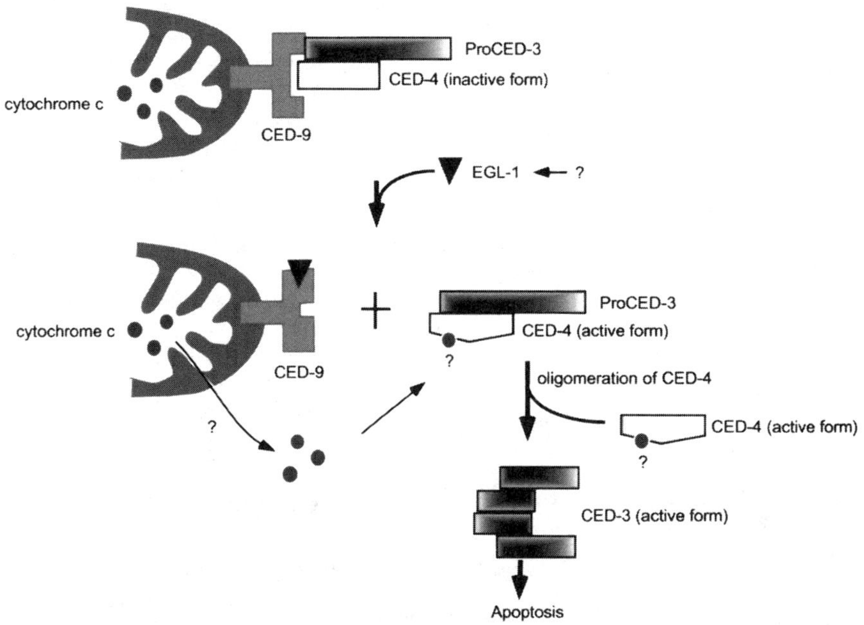

FIGURE 3. A model for the regulation of cell death in *C. elegans*. The survival factor CED-9 attaches to the outer mitochondrial membrane using its hydrophobic tail. Normally, CED-9 interacts directly with CED-4. A BH3 domain–containing protein, EGL-1, interacts with CED-9, resulting in the release of CED-3/4 complex from CED-9. The oligomerization of CED-4 brings several proCED-3 molecules into close proximity, triggering the cleavage of proCED-3 into an activated form of CED-3. The CED-3 tetramer subsequently cleaves its substrates, leading to the irreversible death of cells. Cytochrome c was shown in a mammalian system to form a complex of Apaf-1/cytochrome c/caspase-9/dATP that activates caspase-9, subsequently caspase-3. In *C. elegans*, whether cytochrome c can interact with CED-4 and is involved in triggering the oligomerization of CED-4 and the activation of CED-3 has not been proved.

EGL-1, a New Member of Death Effectors in C. elegans

One interesting cell death phenomenon is related to a pair of HSN (hermaphrodite-specific neurons). HSN innervate the vulva muscle to control the egg-laying process in the hermaphrodite; in the male these neurons undergo programmed cell death.[2,49,50] In *egl-1(gf)* (egg-laying defective) hermaphrodites, the HSNs also undergo programmed cell death in hermaphrodites. Interestingly, the loss of *egl-l* function results in survival of all cells that normally die in *C. elegans*.[9] Recently, the *egl-l* gene was cloned; it encodes a BH3-domain containing protein, which directly interacts with CED-9 (FIG. 3).[9] EGL-1 might activate programmed cell death by binding to CED-9 and inhibiting its activity, thereby, CED-4 is released from the CED-9/CED-4-containing complex.[9] Therefore, EGL-1 falls into a prodeath regulator family including the mammalian proteins Bik, Bid, Harakiri, and Bad.

Signals that Regulate the Death Effectors in C. elegans

Several genes have been found to affect the death of specific cell types. For example, in wild-type animals, the sister cells of the pharyngeal NSM neurons undergo programmed cell death. Both *ces-1* and *ces-2* (cell death specification) mutants specifically affect the fate of these two cells.[65] Gain-of-function alleles of *ces-1* and loss-of-function of *ces-2* prevent the death of the NSM sister cells.[65] In addition, the gain-of-function alleles of *ces-1* also cause the survival of the 12 sisters (another pair of pharyngeal neurons). *ces-2* has been shown to act as a negative regulator of *ces-1*, because in the double mutant of *ces-l (lf); ces-2 (lf)* double mutants, both NSM sisters and 12 sisters again died. Genetic studies also showed that both *ces-1* and *ces-2* act upstream of *ced-3*, *ced-4*, and *ced-9*.[65] These genetic studies, therefore, led to the conclusion that *ces-1* and *ces-2* are cell-type-specific death regulators. Molecular analyses have shown that *ces-2* encodes a protein with bZIP domain, suggesting that it functions as a transcription factor.[52]

APOPTOSIS DEATH TRIGGERS ENGULFMENT

Engulfment Mutants

In *C. elegans*, all cells undergoing programmed cell death are swiftly removed by engulfing cells, which are neighbors of the dying cells. Mutations in six genes, *ced-1*, *-2*, *-5*, *-6*, *-7*, and *-10* prevent dying cells from being engulfed efficiently (FIG. 1).[1] *ced-1* and *ced-2* were first isolated based on the persistence of highly retractile cell corpses past the normal wave of embryonic cell deaths.[53] *ced-5*, *-6*, *-7*, and *-10* genes were isolated by Ellis *et al.*[51] based on similar criteria. Mutation of any one of these engulfment genes affects the engulfment of all dying cells.[51]

The six engulfment genes represent two partially redundant pathways. Ellis *et al.*[51] suggested that these six genes form two partially redundant pathways: one composed of *ced-1*, *-6* and *-7*, and the other of *ced-2*, *-5*, and *-10*. They made all double and many triple mutants among the six engulfment genes, and scored for the presence of cell corpses in the pharynx. They found that the double mutants across groups had much higher numbers of cell corpses than the single mutants. Those from the same group had approximately the same number of cell corpses as the stronger single mutant. In the triple mutants, the number of cell corpses was never more than the strongest double mutants among the three. Together, these results suggest that the six engulfment genes function in two partially redundant pathways.

That the six genes can be put into two groups is supported by two additional phenotypes associated with the *ced-2*, *-5*, and *-10* group. Mutations on *ced-2*, *-5*, and *-10*, but not *ced-1*, *-6*, and *-7*, cause a distal tip cell migration defect[54] (Gumienny and Hengartner, personal communication). The distal tip cells are two cells that migrate and guide the expansion of the gonad during its development. In these mutant worms the distal tip cells either turned prematurely, or went in different directions, resulting in abnormally shaped gonads.

All engulfment genes except *ced-1* have a maternal effect, suggesting that all engulfment genes except *ced-1* might contribute to spermatogenesis, oogenesis and the early embryonic development.

Molecular Analysis of ced-5, ced-6, *and* ced-7

Molecular analysis has been reported for three engulfment genes, *ced-5,*[54] *ced-6,*[55] and *ced-7.*[56] *ced-5* encodes a 1781 amino acid protein and contains a proline- rich region. The activity of CED-5 was demonstrated to be within engulfing cells.[54,56] CED-5 was thus proposed to be a signaling molecule that regulates the rearrangement of cytoskeleton and the dynamic extension of cytoplasm to envelop apoptotic cells.

CED-5 shares a 26% identity with human DOCK1 80 and *Drosophila myoblast city* over the entire length of these proteins. Human DOCK180 was isolated on the basis of its interaction with an adaptor protein, CRK.[57] CRK has been demonstrated to transduce signals from tyrosine kinases and directly interact with cytoskeleton molecules,[58] and is involved in integrin-mediated signaling and cell movement.[59] Overexpression of DOCK 180 in 3T3 fibroblasts results in the extension of cell surfaces, and these cells adopt flat and polygonal shapes.[60] *Drosophila myoblast city* has been shown to be essential for myoblast fusion, dorsal closure, and cytoskeletal organization.[61] Interestingly, overexpression of DOCK180 in *C. elegans* did not rescue the engulfment defect of *ced-5* mutants; however, it did rescue the distal migration defect caused by *ced-5* mutation, indicating that CED-5 and DOCK180 might be functional homologs.[54]

The cloning of the *C. elegans* engulfment genes *ced-6* shed new light in the signal transduction pathway that mediates apoptosis-triggered engulfment.[55] Like *ced-5* and *ced-7*, *ced-6* is also required for removing apoptotic cells in both soma and germline. *ced-6* encodes a novel protein that contains a phosphotyrosine binding (PTB) domain at its N-terminus and several potential SH3 binding sites in its C-terminal half. *ced-6* has also been demonstrated to act within engulfing cells. Thus, CED-6 might be an adaptor molecule that acts in a novel tyrosine kinase pathway since PTB domains can interact specifically with phosphorylated tyrosine residues present within a NPXY(p) consensus sequence motif. PTB binding sites are often found within the intracellular domain of receptors on the plasma membrane.[62,63] A new subfamily of PTB domains been suggested to be involved specifically in mediating apoptosis-triggered phagocytosis[55]; these molecules are likely orthologs of *C. elegans* CED-6. Overexpression of *ced-6* can promote the engulfment of cells in the early stages of apoptosis as well as of persistent cell corpses, suggesting that the *ced-6* signal transduction pathway can be used to remove apoptotic cells regardless the stage of apoptotic death. Overexpression of *ced-6* can partially suppress the engulfment defect of both *ced-1* and *ced-7*, but not *ced-2*, *-5*, and *-10*; *ced-6* is therefore likely to act downstream of both *ced-1* and *ced-7*.

ced-7 encodes an ABC transporter and belongs to a family of over 50 transporter proteins.[64] All members in this family have a highly conserved ATP-binding cassette. Most ABC transporters utilize the energy of ATP hydrolysis to pump substrates across a membrane against a concentration gradient. Specific ABC transporters have been found to transport amino acids, sugars, inorganic ions, polysaccharides, peptides, and even proteins. Some ABC transporters import and others export; none can pump in both directions. CED-7 has been shown to be expressed in all cells and to localize mainly on the plasma 14 membrane.[54,56] Genetic mosaic analysis demonstrated that *ced-7* activity is required within both engulfing cells and dying cells.[54,56] Wu and Horvitz also proposed that CED-7 functioned by transporting some molecules that are involved in cell adhesion[56]; when adhesion is disrupted, engulfment is impaired. However, it is unclear in which direction across the plasma membrane CED-7 pumps its substrates. Since there has been no report suggesting that ABC transporters can function as receptors to transduce the extracellular signals inside cells, one or more proteins should exist on the plasma membrane to respond to the altered concentration of molecules transported by CED-7, and these proteins are likely to be receptors on the plasma membrane that interact with the unknown molecules transported by CED-7.

It has been proposed that when cells die, they present signals on their surfaces that are recognized by receptors on the plasma membrane of engulfing cells. Activated receptors subsequently transduced signals through signaling molecules to trigger the rearrangement of the cytoskeleton and dynamic extension of cytoplasm to envelop the dying cells. Although we still do not have any knowledge about the signals from dying cells and the receptors used for recognizing apoptotic cells on engulfing cells in *C. elegans*, some studies in other systems—including human and *Drosophila*—proposed that an increase in phosphatidylserine on the outer leaflet of the plasma membrane of apoptotic cells might be a universal signal from dying cells,[66,67] whereas CD36,[68–70] *Myoblast city,*[61] lectin,[71] scavenger receptors,[72] ABC transporter,[54,56,73] and CD14[74] might be involved in recognition of apoptotic cell corpses specifically.

CONCLUSION

Our molecular understanding of apoptosis is far from being complete. Over the last decade, an increasing number of molecules were discovered to be involved in the cell death process in several species. However, how most of these molecules function is not yet clear. Also, we know very little about what molecules are involved in signaling the phagocytosis of apoptotic cells and the subsequent engulfment process, and how these molecules function.

In *C. elegans*, many questions remain to be answered. We still do not know what signals are used to trigger apoptosis in both soma and germline, and whether additional factors are involved in the execution of cell death. Also, we do not know what signals are sent from dying cells to trigger the engulfment process. Many molecules listed above function as receptors in mammals to recognize apoptotic cells. Whether these molecules function *in vivo* to recognize apoptotic cells is not yet clear. Furthermore, we know very little about downstream molecules that transduce signals from receptors to effectors, cytoskeleton molecules. In *C. elegans*, *ced-7*, which encodes a ABC transporter, is the only gene that has been shown to be involved in transducing signals from dying cells to engulfing cells.[56] The mechanism of *ced-7* action has not yet been found. *Ced-5* was proposed to act as a signaling molecule,[54] and might act downstream of *ced-6*, and perhaps, of *ced-1*, and *ced-7*.[55] What molecules CED-5 might interact with is also not known. The studies on *ced-6* suggested that the signal transduction pathway that is required to remove apoptotic cells might involve tyrosine phosphorylation. The *ced-6* signal transduction pathway is likely to be conserved across species.[55]

REFERENCES

1. HENGARTNER, M. 1997. Cell death. *In C. elegans* II. D. Riddle, T. Blumenthal, B. Meyer & J. Priess, Eds. Cold Spring Harbor, NY. Cold Spring Harbor Laboratory Press. pp. 383–415.
2. SULSTON, J. E. & H.R. Horvitz. 1977. Post-embryonic cell lineages of the nematode, *Caenorhabditis elegans*. Dev. Biol. **56:** 110-156.
3. SULSTON, J.E., E. SCHIERENBERG, J.G. WHITE & J.N. THOMSON. 1983. The embryonic cell lineage of the nematode *Caenorhabditis elegans*. Dev. Biol. **100:** 64–119.
4. WATERSON, R.H., J.E. SULSTON & A.R. COULSON. 1997. The genome. *In C. elegans* II. D. Riddle, T. Blumenthal, B. Meyer & J. Priess, Eds. Cold Spring Harbor, NY. Cold Spring Harbor Laboratory Press. pp. 23–47.
5. PENNISI, E. 1998. Worming secrets from the *C. elegans* genome. Science **282:** 1972–1974.
6. PLASTERK, R.H.A. & H.G.A.M. VAN LUENEN. 1997. Transposon. *In C. elegans* II. D. Riddle, T. Blumenthal, B. Meyer & J. Priess, Eds. Cold Spring Harbor, NY. Cold Spring Harbor Laboratory Pres. pp. 97–117.
7. JOHNSEN, R. C. & D.L. BAILLIE. 1997. Mutation. *In C. elegans* II. D. Riddle, T. Blumenthal, B. Meyer & J. Priess, Eds. Cold Spring Harbor, NY. Cold Spring Harbor Laboratory Press. pp. 79–97.
8. FIRE, A., S. XU, M.K. MONTGOMERY, S.A. KOSTAS, S.E. DRIVER & C. MELLO. 1998. Potent and specific genetic interference by double-stranded RNA in *Caenorhabditis elegans*. Nature **391:** 806–811.
9. CONRADT, B. & H.R. HORVITZ. 1998. The *C. elegans* protein EGL-1 is required for programmed cell death and interacts with the Bcl-2-like protein CED-9. Cell **93:** 519–29.
10. ELLIS, H.M. & H.R. HORVITZ. 1986. Genetic control of programmed cell death in the nematode *C. elegans*. Cell **44:** 817–829.
11. YUAN, J. & H.R. HORVITZ. 1992. The *Caenorhabditis elegans* cell death gene *ced-4* encodes a novel protein and is expressed during the period of extensive programmed cell death. Development **116:** 309–320.
12. HENGARTNER, M.O., R.E. ELLIS & H.R. HORVITZ. 1992. *Caenorhabditis elegans* gene *ced-9* protects cells from programmed cell death. Nature **356:** 494–499.
13. YUAN, J., S. SHAHAM, S. LEDOUX, H.M. ELLIS & H.R. HORVITZ. 1993. The *C. elegans* cell death gene *ced-3* encodes a protein similar to mammalian interleukin-1 beta-converting enzyme. Cell **7:** 641–652.
14. THORNBERRY, N.A. & Y. LAZEBNIK.1998. Caspases: enemies within. Science **281:** 1312–1316.
15. LUO, X., I. BUDIHARDJO, H. ZOU, C. SLAUGHTER & X. WANG. 1998. Bid, a Bcl2 interacting protein, mediates cytochrome c release from mitochondria in response to activation of cell surface death receptors. Cell **94:** 481-490.
16. LI, H., H. ZHU, C.J. XU, J. YUAN. 1998. Cleavage of BID by caspase 8 mediates the mitochondrial damage in the Fas pathway of apoptosis. Cell **94:** 491–501.
17. XUE, D., S. SHAHAM & H.R. HORVITZ. 1996. The *Caenorhabditis elegans* cell-death protein CED-3 is a cysteine protease with substrate specificities similar to those of the human CPP32 protease. Genes Dev. **10:** 1073–1083.
18. LIU, X., H. ZOU, C. SLAUGHTER & X. WANG. 1997. DFF, a heterodimeric protein that functions downstream of caspase-3 to trigger DNA fragmentation during apoptosis. Cell **89:** 175–184.
19. LIU, X., P. LI, P. WIDLAK, H. ZOU, X. LUO, W.T. GARRARD & X. WANG. 1998. The 40-kDa subunit of DNA fragmentation factor induces DNA fragmentation and chromatin condensation during apoptosis. Proc. Natl. Acad. Sci. USA **95:** 8461–8466.
20. MARTIN, S.J., G.A. O'BRIEN, W.K. NISHIOKA, A.J. MCGAHON, A. MAHBOUBI, T.C. SAIDO & D.R. GREEN. 1995. Proteolysis of fodrin (non-erythroid spectrin) during apoptosis. J. Bio.l Chem. **270:** 6425-6428.

21. KOTHAKOTA, S., T. AZUMA, C. REINHARD, A. KLIPPEL, J. TANG, K. CHU, T.J. MCGARRY, M.W. KIRSCHNER, K. KOTHS, D.J. KWIATKOWSKI & L.T. WILLIAMS. 1997. Caspase-3-generated fragment of gelsolin: effector of morphological change in apoptosis. Science **278:** 294–298.
22. RUDEL, T. & G.M. BOKOCH. 1997. Membrane and morphological changes in apoptotic cells regulated by caspase-mediated activation of PAK2. Science **276:** 1571–1574.
23. LAZEBNIK, Y.A., A. TAKAHASHI, R.D. MOIR, R.D. GOLDMAN, G.G. POIRIER, S.H. KAUFMANN & W.C. EAMSHAW. 1995. Studies of the lamin proteinase reveal multiple parallel biochemical pathways during apoptotic execution. Proc. Natl. Acad. Sci. USA **92:** 9042–9046.
24. TAKAHASHI, A., E.S. ALNEMRI, Y.A. LAZEBNIK, T. FERNANDES-ALNEMRI, G. LITWACK, R.D. MOIR, G.G. POIRIER, S.H. KAUFMANN & W.C. EAMSHAW. 1996. Cleavage of lamin A by Mch2α but not CPP32: multiple interleukin-1β-converting enzyme-related proteases with distinct substrate recognition properties are active in apoptosis. Proc. Natl. Acad. Sci. USA **93:** 8395–8400.
25. SHAHAM, S. 1998. Identification of multiple *Caenorhabditis elegans* caspases and their potential roles in proteolytic cascades. J. Biol. Chem. **273:** 35109–35117.
26. CHINNAIYAN, A.M., D. CHAUDHARY, K. O'ROURKE, E.V. KOONIN & V.M. DIXIT. 1997. Role of CED-4 in the activation of CED-3. Nature **388:** 728–729.
27. SESHAGIRI, S. & L.K. MILLER. 1997. *Caenorhabditis elegans* CED-4 stimulates CED-3 processing and CED-3- induced apoptosis. Curr. Biol. **7:** 455–460.
28. SHAHAM, S. & H.R. HORVITZ. 1996. An alternatively spliced *C. elegans ced-4* RNA encodes a novel cell death inhibitor. Cell **86:** 201–208.
29. SHAHAM, S. & H.R. HORVITZ. 1996. Developing *Caenorhabditis elegans* neurons may contain both cell-death protective and killer activities. Genes Dev. **10:** 578–591.
30. WU, D., H.D. WALLEN & G. NUNEZ. 1997. Interaction and regulation of subcellular localization of CED-4 by CED-9. Science **275:** 1126–1129.
31. JAMES, C., S. GSCHMEISSNER, A. FRASER & G.I. EVAN. 1997. CED-4 induces chromatin condensation in *Schizosaccharomyces pombe* and is inhibited by direct physical association with CED-9. Curr. Biol. **7:** 246–252.
32. SPECTOR, M.S., S. DESNOYERS, D.J. HOEPPNER & M.O. HENGARTNER. 1997. Interaction between the *C. elegans* cell-death regulators CED-9 and CED-4. Nature **385:** 653–656.
33. IRMLER, M., K. HOFMANN, D. VAUX & J. TSCHOPP. 1997. Direct physical interaction between the *Caenorhabditis elegans* 'death proteins' CED-3 and CED-4. FEBS Lett. **406:** 189–190.
34. YANG, X., H.Y. CHANG & D. BALTIMORE. 1998. Essential role of CED-4 oligomerization in CED-3 activation and apoptosis. Science **281:** 1355–1357.
35. HENGARTNER, M.O. & H.R. HORVITZ. 1994. Activation of *C. elegans* cell death protein CED9 by an amino-acid substitution in a domain conserved in Bcl-2. Nature **369:** 318–320.
36. VAUX, D.L. I.L. WEISSMAN & S.K. KIM. 1992. Prevention of programmed cell death in *Caenorhabditis elegans* by human bcl-2. Science **258:** 1955–1957.
37. MONAGHAN, P., D. ROBERTSON, T.A. AMOS, M.J. DYER, D.Y. MASON & M.F. GREAVES. 1992. Utrastructural localization of bcl-2 protein. J. Histochem. Cytochem. **40:** 1819–1825.
38. AKAO, Y., Y. OTSUKI, S. KATAOKA, Y. ITO & Y. TSUJIMOTO. 1994. Multiple subcellular localization of bcl-2: detection in nuclear outer membrane, endoplasmic recticulum membrane, and mitochondrial membranes. Cancer Res. **54:** 2468–2471.
39. RIPARBELLI, M.G., G. CALLAINI, S.A. TRIPODI, M. CINTORINO, P. TOSI & R. DALLAI. 1995. Localization of the Bcl-2 protein to the outer mitochondrial membrane by electron microscopy. Exp. Cell Res. **221:** 363–369.

40. NEWTON, K. & A. STRASSER. 1998. The Bcl-2 family and cell death regulation. Curr. Opin. Genet. Dev. **8:** 68-75.
41. CHAO, D.T. & S.J. KORSMEYER. 1998. BCL-2 family: regulators of cell death. Annu. Rev. Immunol. **16:** 395–419.
42. ADAMS, J.M. & S. CORY. 1998. The Bcl-2 protein family: arbiters of cell survival. Science **281:** 1322–1326.
43. LIANG, H. & S.W. FESIK. 1997. Three-dimensional structures of proteins involved in programmed cell death. J. Mol. Biol. **274:** 291–302.
44. REED, J.C. 1997. Bcl-2 family proteins: regulators of apoptosis and chemoresistance in hematologic malignancies. Semin. Hematol. **34:** 9–19.
45. KLUCK, R.M., E. BOSSY-WETZEL, D.R. GREEN & D.D. NEWMEYER. 1997. The release of cytochrome c from mitochondria: a primary site for Bcl-2 regulation of apoptosis. Science **275:** 1132–1136.
46. YANG, J., X, LIU, K. BHALLA, C.N. KIM, A.M. IBRADO, J. CAI, T.I. PENG, D.P. JONES & X. WANG. 1997. Prevention of apoptosis by Bcl-2: release of cytochrome c from mitochondria blocked. Science **275:** 1129–1132.
47. HU, Y., M.A. BENEDICT, D. WU, N. INOHARA & G. NUNEZ. 1998. Bcl-XL interacts with Apaf-1 and inhibits Apaf-l-dependent caspase-9 activation. Proc. Natl. Acad. Sci. USA **95:** 4386–4391.
48. PAN, G., K. O'ROURKE & V.M. DIXIT. 1998. Caspase-9, Bcl-XL, and Apaf-1 form a ternary complex. J. Biol. Chem. **273:** 5841–5845.
49. TRENT, C., N. TSUNG & H.R. HORVITZ. 1983. Egg-laying defective mutants of the nematode *Caenorhabditis elegans*. Genetics **104:** 619–647.
50. DESAI, C., G. GARRIGA, S.L. MCINTIRE & H.R. HORVITZ. 1988. A genetic pathway for the development of the *Caenorhabditis elegans* HSN motor neurons. Nature **336:** 638–646.
51. ELLIS, R E., D.M. JACOBSON & H.R. HORVITZ. 1991. Genes required for the engulfment of cell corpses during programmed cell death in *Caenorhabditis elegans*. Genetics **129:** 79–94.
52. METZSTEIN, M.M., M.O. HENGARTNER, N. TSUNG, R.E. ELLIS & H.R. HORVITZ. 1996. Transcriptional regulator of programmed cell death encoded by *Caenorhabditis elegans* gene *ces-2*. Nature **382:** 545–547.
53. HEDGECOCK, E.M., J.E. SULSTON & J.N. THOMSON. 1983. Mutations affecting programmed cell deaths in the nematode *Caenorhabditis elegans*. Science ***220:*** 1277–1279.
54. WU, Y.C. & H.R. HORVITZ. 1998. *C. elegans* phagocytosis and cell-migration protein CED-5 is similar to human DOCK180. Nature **392:** 501–504.
55. LIU, Q.A. & M.O. HENGARTNER. 1998. Candidate adaptor protein CED-6 promotes the engulfment of apoptotic cells in *C. elegans*. Cell **93:** 961–972.
56. WU, Y.C. & H.R. HORVITZ. 1998. The *C. elegans* cell corpse engulfment gene *ced-7* encodes a protein similar to ABC transporters. Cell **93:** 951–960.
57. TAKAI, S., H. HASEGAWA, E. KIYOKAWA, K. YAMADA, T. KURATA & M. MATSUDA. 1996. Chromosomal mapping of the gene encoding DOCK180, a major Crk-binding protein, to lOq26.13-q26.3 by fluorescence *in situ* hybridization. Genomics **35:** 403–404.
58. ALTUN-GULTEKIN, Z.F., S. CHANDRIANI, C. BOUGERET, T. ISHIZAKI, S. NARUMIYA, P. DE GRAAF, P. VAN BERGEN EN HENEGOUWEN, H. HANAFUSA, J. WAGNER & R.B. BIRGE. 1998. Activation of Rho-dependent cell spreading and focal adhesion biogenesis by the v-Crk adaptor protein. Mol. Cell Biol. **18:** 3044–3058.
59. DUONG, L.T. & G.A. RODAN. 1998. Integrin-mediated signaling in the regulation of osteoclast adhesion and activation. Front. Biosci. **3:** d757–768.

60. Hasegawa, H., E. Kiyokawa, , S.Tanaka, , K. Nagashima, N. Gotoh, M. Shibuya, T. Kurata & M. Matsuda. 1996. DOCK180, a major CRK-binding protein, alters cell morphology upon translocation to the cell membrane. Mol. Cell Biol. **16:** 1770–1776.
61. Erickson, M.R., B.J. Galletta & S.M. Abmayr. 1997. *Drosophila myoblast city* encodes a conserved protein that is essential for myoblast fusion, dorsal closure, and cytoskeletal organization. J. Cell Biol. **138:** 589–603.
62. Kavanaugh, W.M., C.W. Turck & L.T. Williams. 1995. PTB domain binding to signaling proteins through a sequence motif containing phosphotyrosine. Science **268:** 1177–1179.
63. van der Geer, P. & T. Pawson. 1995. The PTB domain: a new protein module implicated in signal transduction. Trends Biochem. Sci. **20:** 277–280.
64. Higgins, C.F. & M.M. Gottesman. 1992. Is the multidrug transporter a flippase? Trends Biochem. Sci. **17:** 18–21.
65. Ellis, R.E. & H.R. Horvitz. 1991. Two *C. elegans* genes control the programmed deaths of specific cells in the pharynx. Development **112:** 591–603.
66. Fadok, V.A., J.S. Savill, C. Haslett, D.L. Bratton, D.E. Doherty, P.A. Campbell & P.M. Henson. 1992. Different populations of macrophages use either the vitronectin receptor or the phosphatidylserine receptor to recognize and remove apoptotic cells. J. Immunol. **149:** 4029–4035.
67. Martin, S.J., C.P. Reutelingsperger, A.J. McGahon, J.A. Rader, R.C. van Schie, D.M. LaFace & D.R. Green. 1995. Early redistribution of plasma membrane phosphatidylserine is a general feature of apoptosis regardless of the initiating stimulus: inhibition by overexpression of Bcl-2 and Abl. J Exp Med 182, 1545-56.
68. Ren, Y., R.L. Silverstein, J. Allen & J. Savill. 1995. CD36 gene transfer confers capacity for phagocytosis of cells undergoing apoptosis. J. Exp. Med. **18:** 1857–1862.
69. Savill, J., N. Hogg, Y. Ren & C. Haslett. 1992. Thrombospondin cooperates with CD36 and the vitronectin receptor in macrophage recognition of neutrophils undergoing apoptosis. J. Clin. Invest. **90:** 1513–1522.
70. Franc, N.C., J.L. Dimarcq, M. Lagueux, J. Hoffmann & R.A. Ezekowitz. 1996. Croquemort, a novel Drosophila hemocyte/macrophage receptor that recognizes apoptotic cells. Immunity **4:** 431–443.
71. Duvall, E., A.H. Wyllie & R.G. Morris. 1985. Macrophage recognition of cells undergoing programmed cell death (apoptosis). Immunology **56**: 351-358.
72. Platt, N., H. Suzuki, Y. Kurihara, T. Kodama & S. Gordon. 1996. Role for the class A macrophage scavenger receptor in the phagocytosis of apoptotic thymocytes in vitro. Proc. Natl. Acad. Sci. USA **93:** 12456–12460.
73. Luciani, M.F. & G. Chimini. 1996. The ATP binding cassette transporter ABC1, is required for the engulfment of corpses generated by apoptotic cell death. EMBO J. **15:** 226–235.
74. Devitt, A., O.D. Moffatt, C. Raykundalia, J.D. Capra, D.L. Simmons & C.D. Gregory. 1998. Human CD14 mediates recognition and phagocytosis of apoptotic cells. Nature **392:** 505–509.

Developmental Regulation of Induced and Programmed Cell Death in *Xenopus* Embryos

CARMEL HENSEY[a] AND JEAN GAUTIER[b]

Department of Genetics and Development, and Department of Dermatology, College of Physicians and Surgeons of Columbia University, 630 W est 168th Street, New York, New York 10032, USA

ABSTRACT: We have analyzed the role of cell death during early *Xenopus* development and have identified two distinct types of cell death programs during the period between fertilization and the tadpole stage. One is a maternal cell death program that is activated at the onset of gastrulation following damage to the pre-midblastula transition embryo, resulting in the death of non-viable cells. The activation of this cell death program at a specific time during development is a maternally programmed event under the control of a developmental timer set at fertilization, and does not depend on the type of stress applied, on cell cycle progression, or *de* novo protein synthesis. Subsequently, a second program corresponding to programmed cell death is initiated as part of the normal development of the embryo. Programmed cell death starts at the onset of gastrulation and we have analyzed its spatio-temporal patterns by a whole-mount *in situ* DNA end labeling technique (the TUNEL protocol).

INTRODUCTION

Apoptosis is a physiological process of cell elimination that functions as an essential mechanism in maintaining normal tissue homeostasis. Cells have a built in death or suicide program and this evolutionary conserved form of cell suicide is identifiable by its characteristic morphological and biochemical features.[1,2] This type of cell death is important in a wide range of physiological settings and can be induced by diverse stimuli. A cell death pathway can be activated as part of the normal developmental process, or as an adaptive response to external stress and perturbations, to eliminate damaged and non-viable cells. Many of the molecular components of the cell death pathway have been identified and a number are evolutionary conserved genes such as the Bcl-2 family of cell death regulators and the caspases, cysteine proteases important for dismantling the cellular structure.[3,4]

Programmed cell death refers to the naturally occurring cell death that is part of the developmental program of an organism.[5] This loss of cells can be fundamental to some developmental processes and serves many functions, such as sculpting or deleting structures, controlling cell numbers, and eliminating abnormal, misplaced, or non-functional cells.[6,7] A clear demonstration that cell death was a highly regulated component of development came from studies in the nematode *C. elegans*, where these cell deaths occur reproducibly at specific times and positions throughout

[a]Present address: Department of Pharmacology, University College Dublin, Belfield, Dublin 4, Ireland.
[b]Address for correspondence: 212-305-9586 (voice); 212-305-7391 (fax).
e-mail: jg130@columbia.edu

the cell lineage.[8] In vertebrates, the information is more sparse especially during early development, although cell death has long been recognized to occur throughout embryonic development.[9–11] Cell death has been reported as early as the blastocyst stage, and during gastrulation.[12–14] The role of programmed cell death in the formation of digits in the vertebrate limb, and in the morphogenic reorganization of organs such as heart and kidney has been clearly demonstrated.[15,16]

Neuronal death plays a major part in the development of the nervous system and is thought to be regulated by a number of different mechanisms, although studies have primarily concentrated on its function in the control of cell number caused by a limited supply of neurotrophic factors: "trophic cell death."[17,18] Neurons lacking sufficient or correct neurotrophic support undergo PCD, eliminating neurons that have not made any connections or have made incorrect ones. Thus, PCD has an important role in the establishment of functional patterns of synaptic organization and axonal pathways. PCD is also detected at earlier stages of neurogenesis when trophic support is not yet a survival determining factor, and there cell death is regulated by other, as yet poorly understood, mechanisms. In chicken, PCD is detected during the folding and closure of the neural tube. It has also been shown to have a role in sculpting cephalic neural crest cells and studies in chick have shown that regions from where neural crest cells migrate are separated by stripes where all presumptive neural crest cells undergo apoptosis.[19,20] Finally, it has been proposed that some cell deaths in the developing brain may occur through phenotypic selection based on the intrinsic properties of cells, in a manner comparable to the killing of T cells by phenotypic specific selection.[21,22]

We have characterized two types of cell death program during early development in *Xenopus*, one an induced cell death program in response to damage in the pre-MBT embryo, and the second the developmentally regulated programmed cell death. In Xenopus, early development consists of a series of rapid cell divisions driven by maternally inherited proteins and mRNAs.[23] At the mid-blastula transition (MBT),[23,24] zygotic transcription is initiated,[25] the embryonic cell cycles lacking gap phases are progressively replaced by somatic cell cycles driven by zygotic components, cell division becomes asynchronous, and cells display motile activity. A maternal cell death program induced at gastrulation following damage to the pre-MBT embryo has been described by us and others.[26–29] Here we present a detailed characterization of this program and its developmental regulation. We also present a detailed spatio-temporal analysis of the naturally occurring cell deaths during early development in *Xenopus*, using TUNEL labeling on wholemount *in situ* embryos.

METHODS

For detailed descriptions of the methods see references 27 and 30.

RESULTS

Induced Apoptosis

Induction of a Synchronous Embryonic Death at the Onset of Gastrulation

A number of toxic insults to cleavage stage *Xenopus* embryos induces embryonic death at the onset of gastrulation. Treatment of embryos with high doses of γ-radiation, or inhibitors of transcription, translation, or DNA replication, leads to the rapid death of the embryo at gastrulation, (St.10.5).[27] Irradiated embryos divide normally during the cleavage stage but die at stage 10.5, thus the rapidly cleaving Xenopus embryos do not activate a DNA damage-induced cell cycle checkpoint. Similarly, treatment with the transcription inhibitor α-amanitin lead to the death of the embryo at stage 10.5. Despite the fact that treatment with the protein synthesis inhibitor cycloheximide causes an arrest in cell division, the embryo remains intact until gastrulation at which time it rapidly dies. Embryonic death is defined as the rapid disintegration of the embryo, characterized by a white mottled aspect, and the appearance of non-adherent cells or fragments of cells, within the vitelline membrane space. Within a single embryo, all cells die rapidly and 100% of embryos die at this time following such treatments. Our results show that these morphological changes were always occurring precisely at stage 10.5, regardless of the type of insult applied to the embryos. Morphological changes reminiscent of these were reported following γ-irradiation,[26,28] and DNA replication inhibition.[29] By a number of criteria we demonstrated that dying embryos exhibit characteristic apoptotic features and these results are outlined below.

DNA Fragmentation

Analysis of genomic DNA from dying embryos showed a DNA ladder at St. 10.5 in all treated embryos (FIG. 1A; lanes 2, 4, and 6) while none was detected in control embryos (FIG. 1A, lanes 1, 3, and 5), indicating that DNA fragmentation was occurring at the time of the observed death of the embryo. Despite the fact that embryos were treated at the 1- or 2-cell stage, there were no signs of DNA fragmentation at any time prior to St. 10 (see TUNEL staining; FIG. 1B, panels 4, 7, and 10). It is remarkable that cycloheximide-arrested embryos can still undergo intra-nucleosomal DNA fragmentation despite cell cycle arrest and the limited number of cells (FIG. 1A, 6). In fact, apoptosis can be detected in a 4-cell cycloheximide-arrested embryo.[27]

TUNEL Staining

The TdT-mediated dUTP digoxigenin nick end labeling (TUNEL) staining technique demonstrates *in situ* DNA fragmentation in cells destined to die by apoptosis.[31] Whole embryos were fixed at various times from St. 9 onwards and processed for TUNEL staining. This approach allowed us to draw a spatio-temporal picture of the occurrence of cell death in embryos subjected to different treatments. FIGURE 1B shows control, (panels, 1–3), γ-irradiated, (panels, 4–6), α-amanitin-treated (panels 7–9), and cycloheximide-treated (panels 10–12), embryos that were TUNEL stained. Densely labeled nuclei are seen throughout the dying embryos (FIG. 1B, panels 6, 9, and 12), and staining was evenly distributed across the animal and vegetal halves of the embryo. TUNEL staining gradually appears at St. 10 as evidenced by the partially stained St. 10 embryos (FIG. 1B, panels 5, 8, and 11). No TUNEL staining was seen in St. 9 treated embryos (Fig. 1B, panels 4, 7, and 10), nor was there any staining in St. 9, St.10, or St. 10.5 control embryos (FIG. 1B, panels 1, 2, and 3). We could observe some slight differences in the staining of individual embryos subjected to different treatments; for example, TUNEL staining of irradiated embryos was occasionally more patchy with some variations in the size of nuclei stained compared to the evenly spaced uniformly stained nuclei in α-amanitin treated embryos. This difference may be due to some uneven cleavages in the irradiated

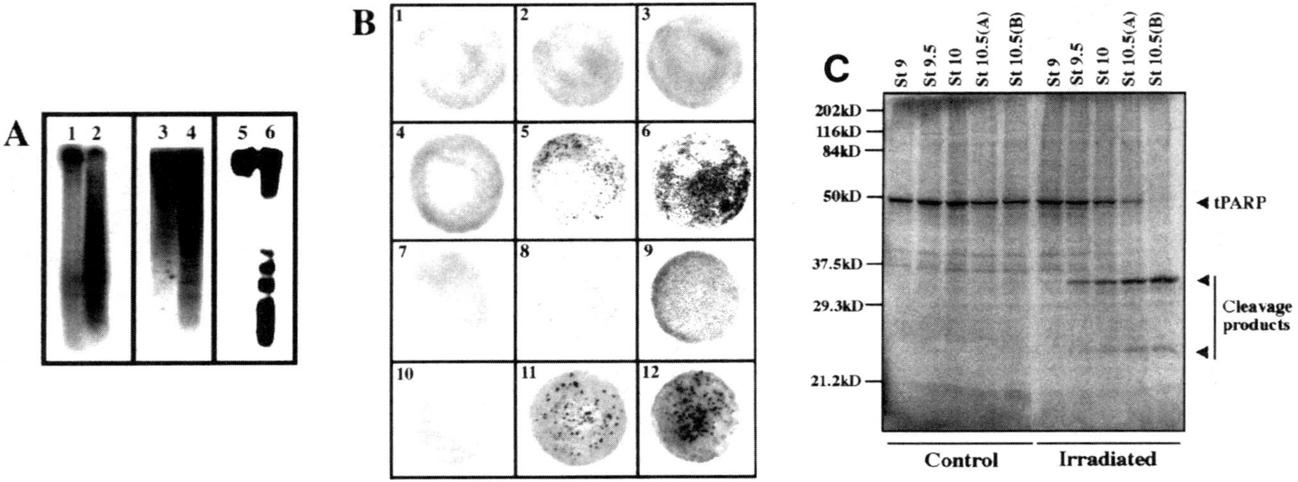

FIGURE 1. DNA fragmentation, TUNEL staining and caspase activation are evident in dying embryos. **A:** DNA fragmentation of genomic DNA from control, and treated embryos was analyzed by end-labeling, 2% agarose gel electrophoresis, and autoradiography. Irradiated embryos received 40 Gy at St. 1, α-amanitin was injected at St. 2, and embryos were treated with cycloheximide at St. 6.5, leading to an arrest in cell division at St. 7. Batches of embryos were collected at St. 10.5, or an equivalent time in cycloheximide-treated embryos and the genomic DNA extracted. Genomic DNA from control embryos was analyzed alongside treated embryos from the same batch. Amounts loaded in lanes 1, 2, 3, 4, 5 and 6, represent amounts equivalent to 0.4, 0.5, 0.5, 0.5, 1.4, and 1.4, respectively, of the genomic DNA of embryos. Lane 1, control; lane 2, irradiated; lane 3, control; lane 4, α-amanitin treated; lane 5, control; lane 6, cycloheximide treated. **B:** Embryos were fixed and TUNEL stained at the indicated stages: panels 1, 4, 7, and 10, St. 9; panels 2, 5, 8, and 11, St. 10; panels 3, 6, 9, and 12, St. 10.5. Panels 1, 2, and 3 show control embryos; panels 4, 5 and,6 show embryos irradiated at St. 1, 40 Gy; panels 7, 8, and 9, show embryos treated with α-amanitin at the 2-cell stage; and panels 10, 11 and 12, show embryos treated with cycloheximide at St. 6.5, leading to an arrest in cell division at St. 7. **C:** ^{35}S-tPARP cleavage in embryonic extracts from the indicated stages. St. 10.5-(B) represents embryos collected 20 min later than St.10.5-(A). Following incubation with tPARP samples were resolved by 12% SDS-PAGE. *Arrows* indicate the cleavage products produced by caspase activity.

embryos. The larger masses of TUNEL staining in FIGURE 1B, panel 12, represent the larger nuclei in a cell division–arrested embryo following cycloheximide treatment. The focal *in situ* staining of apoptotic nuclei, correlates with the biochemical characteristics of apoptosis, suggesting that embryonic death is mediated by activation of an apoptotic pathway between St. 9 and 10.5 in DNA damaged, α-amanitin, or cycloheximide-treated embryos.

Caspase Activity

Using a known substrate for the caspases, truncated poly ADP-ribose polymerase (tPARP),[32] we tested crude cytoplasmic extracts prepared from different embryonic stages for caspase activity (FIG. 1C). Caspase activity, as measured by the cleavage of tPARP into two fragments (indicated by arrowheads) was only evident in cell-free extracts prepared from St. 9.5–10.5-irradiated embryos, while no caspase activity was detected in extracts made at earlier stages in irradiated embryos or in control embryos. Similar PARP cleavage was observed in extracts made from hydroxyurea-treated embryos.[29] Caspase activity in an extract from St. 10.5-irradiated embryos was completely inhibited by 250 nM z-DEVD-fmk, a pseudosubstrate inhibitor of the caspase-3 subdivision of caspases (TABLE 1), thus confirming the specificity of this assay. The caspase-1 inhibitor, z-VAD-fmk, only inhibited the caspase activity at higher concentrations, 2.5 μM (TABLE 1).

Ectopic Bcl-2 Expression and the Caspase Inhibitor z-DEVD-fmk Inhibit the Onset of Cell Death

The Bcl-2 protein inhibits apoptosis in many systems.[3] We expressed human Bcl-2 protein *in vivo* in irradiated, α-amanitin- and cycloheximide-treated embryos, in an effort to modulate the apoptotic response. *In vitro* transcribed human Bcl-2 RNA was injected into one blastomere of a 2-cell embryo which had been induced to die, thus allowing the comparison of the apoptotic response in cells with or without Bcl-2, within a single embryo. TUNEL staining of irradiated embryos, at the time cell death normally occurred, shows densely stained nuclei in the half of the embryo that did not receive Bcl-2 (FIG. 2, panel 1, right half of embryo), while no staining is seen is the half of the embryo protected by Bcl-2 (FIG. 2, panel 3, left half of embryo). α-Amanitin- and cycloheximide-induced death was also inhibited following the expression of Bcl-2 in half of a treated embryo (FIG. 2, panel 2, and panel 3, left half of embryo). Microinjection of Bcl-2 RNA into embryos prior to

TABLE 1. Embryonic death is rescued by injection of the caspase inhibitors, z-DEVD-fmk and ICE-P

Caspase Inhibitor	Inhibition of tPARP Cleavage in Embryonic Extracts	Inhibition of Irradiation-induced Embryonic Death in Whole Embryos	% Rescue
z-DEVD-fmk	+ (250 nM)	+ (100 μm)	56
z-VAD-fmk	+ (2.5 μM)	– (100 μm)	0
ICE-P (IQACRG)	n/a	+ (1.00 μM)	74
p35	n/a	—	0

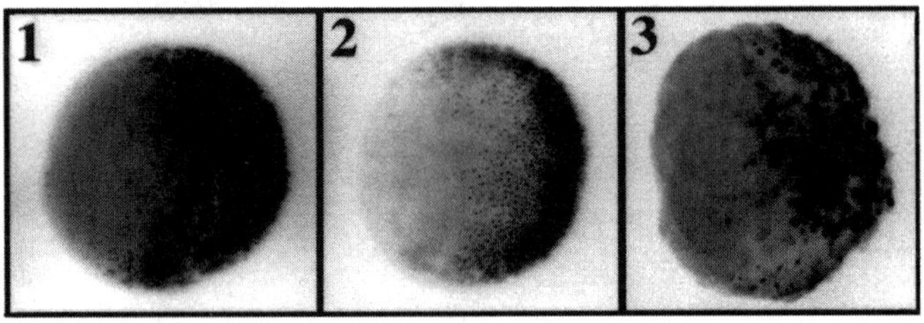

FIGURE 2. Embryonic death is rescued by ectopic expression of Bcl-2 protein. Embryos were treated with γ-irradiation at stage 1; 40 Gy (panel 1), α-aminitin at stage 2 (panel 2), and cycloheximide at stage 7 (panel 3). In each case one blastomere was injected with Bcl-2 RNA at the two-cell stage and the embryos was fixed and TUNEL stained at St. 10.5. In all treated embryos no TUNEL staining is observed in the cells arising from the blastomere injected with Bcl-2, left side of embryo, while TUNEL staining is detected throughout the half of the embryo that did not receive Bcl-2.

treatment with cycloheximide allowed for expression of Bcl-2 protein in advance of blocking protein synthesis. These results provide evidence that cell death is mediated by an apoptotic pathway which can be regulated by the Bcl-2 protein.

Embryonic death could also be rescued by injection of the caspase inhibitor z-DEVD-fmk, implicating a caspase-3 like protease in the execution of the cell death program in embryos (TABLE 1). It is not clear why the degree of rescue was not as high as for Bcl-2, but this may be due to the partial inactivation of the inhibitor over the approximate 10-h incubation period in the embryos, prior to the activation of a cell death program. The caspase inhibitor ICE-P[33] also rescued embryonic death. Equivalent concentrations of z-VAD-fmk, did not inhibit embryonic death (TABLE 1).

Embryonic Death Is Only Induced in Embryos Treated prior to the MBT

We found that there is a critical phase during which rapid, synchronous and widespread embryonic cell death could be induced in embryos by γ-irradiation, α-amanitin or cycloheximide. Only those embryos treated prior to St. 8, a time corresponding to the MBT, die synchronously at the onset of gastrulation. Embryos treated after the MBT, do not undergo this synchronous apoptosis and die later during development, following abortive gastrulation movements.[27] The precise developmental timing of this cell death program is outlined in FIGURE 4. A synchronous and rapid apoptotic program is activated in embryos damaged prior to the MBT. Traversing the MBT seems to correlate with the commitment to a cell death program thus defining a "checkpoint" in development for the elimination of damaged embryos.

Programmed Cell Death during Early Xenopus *Development*

In addition to identifying an apoptotic pathway induced at gastrulation we have also analyzed the normal programmed cell death that occurs in untreated embryos. Using the wholemount TUNEL staining technique the cell death that occurs as part of the normal developmental process was detected.

Gastrulation

The earliest TUNEL-positive embryos were detected at stage 10.5, the beginning of gastrulation. The amount and distribution of dying cells varied considerably early in gastrulation (TABLE 2). Towards the end of gastrulation, bands of staining across the dorsal region of the embryo, i.e. the future neural plate, were the characteristic feature.[30]

Neural Plate Stage

From this stage on the cell deaths we detected occurred reproducibly at specific times and places in the developing embryo, and we believe these distinct patterns of cell death represent programmed cell death (PCD).

During stage 13, cell death was primarily detected in the dorsal region with varying but distinct patterns seen in the area of the neural plate. Staining was observed in broad bands that extended from the blastopore slit towards the anterior of the embryo either medially, para-medially, or laterally. In the example shown, the stripes broadened in the anterior direction and seemed to delineate the neural plate (FIG. 3, panel 1), with only a narrow dorsal medial stripe showing no staining. Sections of stained embryos showed that PCD was mostly localized to the presumptive neuroectoderm, and in some cases staining was also detected in the mesoderm, in the region underlying the neuroectoderm (data not shown). In a few cases, staining seemed to be excluded from the neural plate and was detected entirely within the non-neural ectoderm (data not shown). Through stages 14 and 15 the TUNEL staining became localized anteriorly and appeared to be concentrated at the anterior and lateral edges

TABLE 2. Programmed cell death during *Xenopus* development

Development stage	TUNEL Staining Characteristics	% Embryos Stained
Stage 10.5–11.5	Variable amounts of cell death randomly distributed across the embryo.	67%
Stage 12.5	Majority of cell death detected dorsally in the area of the future neural plate.	67%
Stage 13	Cell death in the developing neural plate.	67%
Stage 14–15	"Horseshoe" shaped area of cell death at anterior edge of neural plate.	52%
Stage 16–17	Cell death in developing brain and sensory placodes, and in the primary sensory neurons of the developing neural tube.	64%
Stage 18–25	Cell death considerably reduced compared to earlier stages. Low levels detected in the developing brain and tailbud.	98%
Stage 26–28	Cell death in the developing brain, eyes, spinal cord, and tailbud.	100%
Stage 29–35	Cell death in the developing fore-, mid-, and hindbrain. Cell death also detected in the developing sensory organs, spinal cord, branchial arches, and tail.	99%

of the neural plate where a distinct "horseshoe shaped" pattern of staining was seen in the anterior of the embryo (FIG. 3, panel 2).

Cell death was concentrated to specific areas of the developing neural plate between stages 13 and 15. Such striking patterns of cell death during early neurogenesis have not been described previously.

Stages 16–17

A representative example of the patterns of PCD seen during the neural fold stage is shown (FIG. 3, panel 3). Unique patterns were detected in the anterior and dorsal

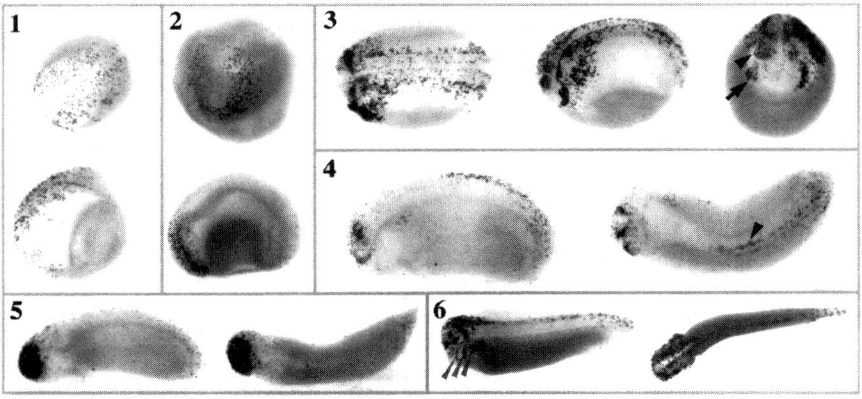

FIGURE 3. Programmed cell death as detected by whole-mount TUNEL staining. **1:** Stage 13, dorsal (*top*) and lateral (*bottom*) view. The dorsal side of the embryo is TUNEL stained with a band of staining broadening towards the anterior end of the embryo, delineating the neural plate. The pattern is representative of the staining observed following staining of 58 stage 13 embryos of which 67% were TUNEL positive. **2:** Stage 15, end on anterior view (*top*), lateral view (*bottom*), anterior is left. TUNEL staining formed a horseshoe-shaped band of staining. The embryos shown are representative of the staining observed following staining of 37 stage 14/15 embryos of which 52% were TUNEL positive. **3:** Stage 17, the neural fold stage. A dorsal, lateral, and end-on anterior view are shown (*left to right*). Strongest staining was in the anterior embryo, with patches of TUNEL staining corresponding to the optic (*arrowhead*), and olfactory (*arrow*) placodes, and developing brain. Dying cells were found in defined stripes on each side of the dorsal midline corresponding to the primary sensory neurons. A scattering of dying cells was detected throughout the neural fold. The cell death patterns are representative of the staining observed following staining of 33 embryos of which 64% were TUNEL positive. **4:** Stage 26, a lateral view (*left*) and a dorsal view (*right*) are shown. TUNEL staining is evident in the ventral forebrain and midbrain in addition to the eye vesicle. Staining was also evident in the posterior half of the spinal cord (*arrowhead*) and in the tailbud. **5:** Stage 27–28, staining in the midbrain, hindbrain, and eye vesicles. The patterns of cell death shown are representative of the cell death detected following staining of 36 embryos all of which were TUNEL positive. **6:** Stage 32. TUNEL staining in the rhombencephalon, the outer ring of the eye vesicle, the branchial arches (*arrowheads*), and the optic vesicle. Staining was also evident along the length of the spinal cord and in the tail tip. The staining is representative of the patterns seen following staining of 92 embryos of which 99% were TUNEL stained. Embryos with more than 5 TUNEL stained nuclei were considered TUNEL positive.

regions of embryos, corresponding to the developing nervous system. Apoptotic cells were found in defined stripes on each side of the dorsal midline, with the most intense staining corresponding to the primary sensory neurons, while a light scattering of staining was detected throughout the neural fold (FIG. 3, panel 3, dorsal view). The staining fanned out anteriorly, and was most intense in the future brain region (FIG. 3, panel 3, anterior view) as well as in ventral-anterior patches corresponding to the future sensory placodes (FIG. 3, panel 3, anterior view). The end on anterior view shows distinct subdivisions of staining, which correspond to the olfactory (FIG. 3, panel 3, arrow), and optic placodes (FIG. 3, panel 3, arrowhead). A degree of left-right asymmetry is also evident in the staining (FIG. 3, panel 3), which in some cases was so striking that only one half of the embryo was stained (data not shown). This is the earliest stage at which asymmetries were evident in the cell death patterns, and such staining occurred randomly on either the right, or the left side of the embryo.

During this stage cell death continued to be localized to the developing nervous system, particularly the developing brain and sensory placodes, in addition to being detected in the population of primary sensory neurons in the developing neural tube.

Stages 18–25

At the late neural stage (stages 18–20) the degree of TUNEL staining was greatly reduced compared to previous stages (TABLE 2). This may be due to the fact that neural induction has already taken place, and remodeling of the developing CNS has not yet begun. Between stages 21 and 25 the majority of cell death was detected as a diffuse staining pattern in the head possibly corresponding to migrating neural crest cells.[30]

Stages 26–28

During the tailbud stage programmed cell death was detected throughout the developing brain, in the eyes, spinal cord, and developing tail (FIG. 3, panels 4 and 5). PCD was detected as two bilaterally symmetrical regions in the forebrain and midbrain (FIG. 3, panels 4 and 5). Two bilaterally symmetrical stripes of dying cells were detected dorsally in some embryos (FIG. 3, panel 4, arrowhead) and sectioning revealed that this pattern was due to cell death in the spinal cord. The other major site of programmed cell death was the developing tailbud (FIG. 3, panel 4). In a number of cases cell death was detected in the optic vesicle (FIG. 3, panels 4 and 5).

Stages 29–35

During these stages PCD continued to take place primarily in the central nervous system, and was also detected in the sensory organs, branchial arches, and elongating tail. PCD was most frequently observed in the midbrain, with the most extensive PCD being detected in the ventral forebrain, midbrain and hindbrain (FIG. 3, panel 6). This is the period in development when the brain acquires most of its form and structure,[34] and PCD is likely playing a role in the morphogensis of the brain at this period. Additionally, cell death was frequently observed in the spinal cord, branchial arches, tail, and the eye and optic vesicles (FIG. 3, panel 6).

DISCUSSION

Two cell death programs were identified during early *Xenopus* development, one is a an induced cell death program which is only activated following damage to pre-MBT embryos whereas the second is the normal programmed cell death that occurs during development. Our analysis of cell death showed that a classical cell death pathway exists in both cases. Dying embryos exhibited characteristic apoptotic features such as DNA fragmentation, caspase activation, and regulation by Bcl-2. These features also correlated with TUNEL staining, a method we used to map the normally occurring programmed cell death in whole embryos.

The relative timing of these cell death programs with respect to development is outlined in FIGURE 4. The developmental regulation of the two programs is quite different. The induced cell death is a maternally encoded event, the fact that the pathway can still be activated following the inhibition of protein synthesis throughout the cleavage stage indicates that the apoptotic machinery must be present in the early embryo but maintained in an inactive state. Traversing the MBT appears to activate a checkpoint in development for the elimination of damaged embryos. *Xenopus* embryos display several developmental programs under the control of maternal gene expression the timing of which is tightly regulated.[23,24,35]

Interestingly, we never detected either programmed or induced cell death prior to gastrulation (FIG. 4). This could be due to either the intrinsic inability of the embryo to activate a cell death pathway prior to stage 10.5, or to an excess of inhibitory component of the cell death pathway in early embryos. This maternally inherited inhibitor could be degraded using a similar mechanism as the one described for cyclin degradation between the MBT and the onset of gastrulation,[29,35] thus leading to the onset of cell death in cases where the embryo is damaged. The fact that genotoxic insults after the MBT do not lead to the activation of a cell death program at stage 10.5 suggests a new zygotic program blocking further cell death is synthesized in the short window of time between stage 8 and stage 9. If transcription of translation of these factors is prevented apoptosis would be activated.

On the other hand PCD is a zygotic program that only occurs following the activation of zygotic transcription. The cell death that was detected in the presumptive and developing nervous system, exhibited distinct, reproducible patterns which occurred in a stereotypic manner at specific times in the development. This therefore represents the normally occurring programmed cell death (PCD) that is an essential part of embryonic development.

During neural induction and early neurulation the spatio-temporal distribution of TUNEL stained nuclei roughly coincides with the area of the neural plate and the varying patterns we detect within this region suggest that PCD may have an important role during the process of neural induction and patterning of the plate area (FIG. 3, A and B). At the neural fold stage the distinct stripes of cells along the dorsal side of the embryo (FIG. 3C), are coincident with stripes of primary neurons seen with the neural specific marker N-tubulin.[36] At this time cell death also occurs in the developing placodes (FIG. 3C). As development progressed cell death was localized to specific brain regions and the developing sensory organs (Fig. 3, D, E, F, and G). We frequently observed cell death in the spinal cord (FIG. 3D).

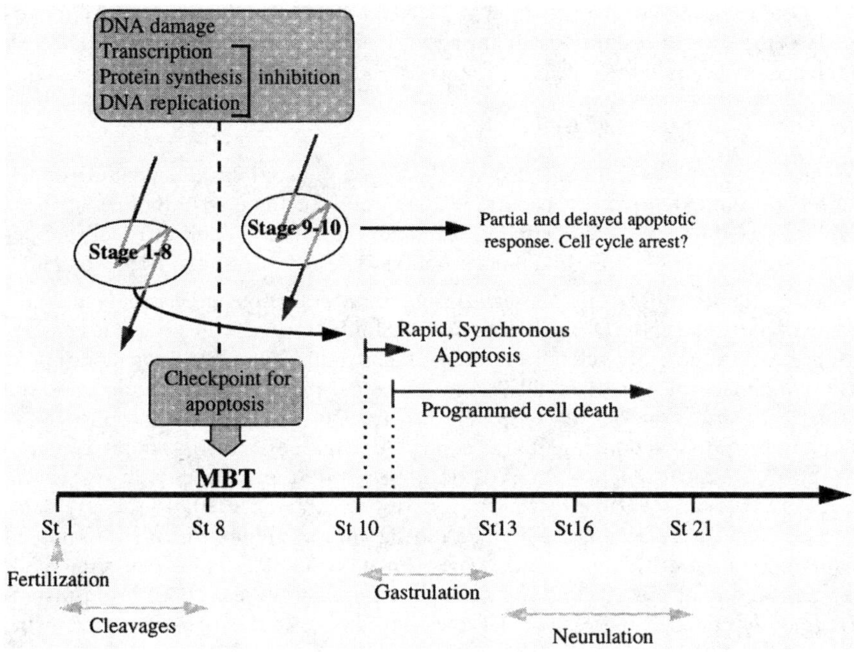

FIGURE 4. Apoptosis and programmed cell death during early development in *Xenopus laevis*. The arrow is a linear representation of the early developmental stages of *Xenopus laevis*. γ-Irradiation-induced DNA damage, or treatment of embryos with inhibitors of transcription, translation, or replication, between the time of fertilization and the MBT, elicits a rapid and synchronous apoptotic program throughout the cells of the embryo at a time corresponding to stage 10–10.5. Similar treatments following the MBT do not induce embryonic death at this time and thus a developmental checkpoint has been defined at the MBT. Naturally occurring programmed cell death commences during gastrulation at stage 10.5.

It is striking that during gastrulation and neurulation no cell death was detected in a fraction of embryos. The rapid clearance of dead cells from the embryo may be one reason for such an effect. Cells that undergo cell death are usually degraded rapidly, and a clearance time of one hour or less has been observed.[5] Thus embryos showing no staining may be passing through similar waves of cell death, but due to the rapid clearance of dead cells it was not detected. Analysis of clearance times in vertebrate embryos has proved difficult, and *Xenopus* may provide a useful system to estimate these clearance times using vital fluorescent dyes to detect dying cells *in vivo*.

During gastrulation the absence of cell death in some embryos could also be explained if the function of cell death is the elimination of damaged cells. However, during neurulation, where the patterns of cell death detected are highly reproducible, the lack of staining in some embryos cannot be explained as an absence of damaged cells. Other reasons for the presence of TUNEL negative embryos might be a variability in the staging of embryos, or variations in the timing of cell death with respect to the age of the embryo.

Regulation of Programmed Cell Death

There are probably different mechanisms regulating cell death depending on the stage in development when it occurs. We can only speculate as to the mechanisms involved based on what is already known about the causes of cell death during development in other systems, and the regulation of cell death pathways in general.

Cell death during gastrulation may be due to the detection and subsequent elimination of damaged cells. During the period of rapid cleavage which precedes gastrulation there are no cell cycle checkpoints which might allow for the generation of cells with incomplete sets of DNA, chromosome abnormalities, or chromosome breaks.[37] Gastrulation follows the midblastula transition (MBT), and the onset of zygotic transcription, at which point the cell cycle lengthens and checkpoints are activated.[23,38,39] Since *Xenopus* embryos do not activate a cell death pathway before stage 10.5, it is conceivable that during gastrulation, at stage 10.5–11.5, any damaged cells that have accumulated since fertilization are removed. In addition, cell death at this time could also be the consequence of the mechanical stress that cells are subjected to at this stage. The fact that we never observe cell death in the noninvaginating endoderm, a region that does not undergo such extensive movements during gastrulation, supports this hypothesis. Mechanical stress-induced apoptosis has been reported.[40]

Studies of cell death in the developing nervous system have focused on the role of neurotrophic factors in regulating neuronal survival. However, we observe cell death in the developing nervous system at stages in development prior to axonogenesis or the establishment of synaptic connections, therefore this cell death cannot be regulated by target derived signals. Regulation of survival at this stage is poorly understood and multiple mechanisms may be involved in the regulation of cell survival at this time. It has been suggested that cell death may reflect cell specific phenotype recognition mechanisms.[22] Additionally, neuronal death might be integrated with cell cycle regulation and differentiation.[41] During the later stages, target-derived trophic support could be playing a role in regulating neuronal numbers.

We can only speculate as to the genes involved in regulating these cell death pathways. It is interesting to note that some of the growth factors implicated in these early neuronal patterning events, such as the bone morphogenetic protein 4 (BMP-4), have also been shown to play a role in PCD during development.[42] Interestingly, there is an area of overlap between the area of cell death in the developing nervous system and the BMP-4 expression pattern in *Xenopus*,[43] reinforcing the idea that this gene may play a role in the execution of this morphogenetic cell death program.

Another gene whose expression pattern overlaps in a striking fashion with the cell death pattern observed at the neural fold stage is Pax-6 (FIG. 3B).[44] Pax-6 expression follows eye development, and is also expressed in cells fated to form part of the brain, suggesting a role for cell death in the developing eye placode and brain. Analysis of Pax-6 mutant mice revealed defects in PCD during development, suggesting a role for Pax-6 in the regulation of PCD.[45]

Functions of Cell Death during Xenopus *Development*

The fact that the maternal cell death program can be induced following a wide variety of treatments demonstrates its fundamental importance in developing embryos. The lack of cell cycle checkpoints in the cleaving embryo does allow for unchecked cell cycles and therefore could be detrimental to the survival of the organism, however a system whereby an apoptotic response is favored over cell cycle arrest would allow for such uncontrolled cell divisions during early development. This strategy could therefore palliate the lack of cell cycle checkpoints in the early embryo and provides a mechanism to eliminate non-viable cells during early development. We predict that this developmental strategy will be conserved in other organisms with similar developmental programs such as amphibians, fishes, and insects, the eggs of which all commence development as large cells that go through a series of rapid reductive divisions.[37,46,47]

Nothing is known about the function of programmed cell death during the earliest stages of neural induction and neural plate formation. Cell death could be utilized to phenotypically select cells within a population that did not receive the proper signals, therefore strengthening boundaries between presumptive territories. During early vertebrate development the cells of the ectoderm choose between two possible fates, neural and epidermal[48] and cell death could provide a means of early phenotypic selection to "sharpen" boundaries between neural and non-neural tissues. It is possible that ectodermal cells receiving intermediate levels of neuralizing factors could be considered "phenotypically ambiguous" and be eliminated. The high levels and reproducible patterns of cell death that were detected during neural induction, neurogenesis and CNS development, strongly suggests that PCD may play an important role during early embryogenesis in *Xenopus*. Among the vertebrate model systems *Xenopus* will be very useful for studying the role of PCD in early development, particularly during early neurogenesis.

REFERENCES

1. WYLLIE, A.H. 1980. Glucocorticoid-induced thymocyte apoptosis is associated with endogenous endonuclease activation. Nature **284:** 555–556.
2. KERR, J.F., G.C. GOBE, C.M. WINTERFORD & B.V. HARMON. 1995. Anatomical methods in cell death. Methods Cell Biol. **46:** 1–27.
3. YANG, E. & S.J. KORSMEYER. 1996. Molecular thanatopsis: a discourse on the BCL2 family and cell death. Blood **88:** 386–401.
4. NICHOLSON, D.W. & N.A. THORNBERRY. 1997. Caspases: killer proteases. Trends Biochem. Sci. **22:** 299–306.
5. JACOBSON, M.D., M. WEIL & M.C. RAFF. 1997. Programmed cell death in animal development. Cell **88:** 347–354.
6. ELLIS, R.E., J.Y. YUAN & H.R. HORVITZ. 1991. Mechanisms and functions of cell death. Annu. Rev. Cell Biol. **7:** 663–698.
7. SANDERS, E.J. & M.A. WRIDE. 1995. Programmed cell death in development. Int. Rev. Cytol. **163:** 105–173.
8. HENGARTNER, M.O. & H.R. HORVITZ. 1994. Programmed cell death in *Caenorhabditis elegans*. Curr. Opin. Genet. Dev. **4:** 581–586.
9. GLUCKSMANN, A. 1951. Cell deaths in normal vertebrate ontogeny. Biol. Rev. **26:** 59–86.
10. CLARKE, P.G. & S. CLARKE. 1996. Nineteenth century research on naturally occurring cell death and related phenomena. Anat. Embryol. **193:** 81–99.

11. LOCKSHIN, R.A. 1997. The early modern period in cell death. Cell Death Differentiation **4:** 347–351.
12. COUCOUVANIS, E. & G.R. MARTIN. 1995. Signals for death and survival: a two-step mechanism for cavitation in the vertebrate embryo. Cell **83:** 279–287.
13. IMOH, H. 1986. Cell death during normal gastrulation in the newt, *Cynops pyrrhogaster*. Cell Differ. **19:** 35–42.
14. SANDERS, E.J., P.H. TORKKELI & A.S. FRENCH. 1997. Patterns of cell death during gastrulation in chick and mouse embryos. Anat. Embryol. (Berl.) **195:** 147–154.
15. COLES, H.S., J.F. BURNE & M.C. RAFF. 1993. Large-scale normal cell death in the developing rat kidney and its reduction by epidermal growth factor. Development **118:** 777–784.
16. PEXIEDER, T. 1975. Cell death in the morphogenesis and teratogenesis of the heart. Adv. Anat. Embryol. Cell Biol. **51:** 3–99.
17. BUREK, M.J. & R.W. OPPENHEIM. 1996. Programmed cell death in the developing nervous system. Brain Pathol. **6:** 427–446.
18. OPPENHEIM, R.W. 1991. Cell death during development of the nervous system. Annu. Rev. Neurosci. **14:** 453–501.
19. GRAHAM, A., I. HEYMAN & A. LUMSDEN. 1993. Even-numbered rhombomeres control the apoptotic elimination of neural crest cells from odd-numbered rhombomeres in the chick hindbrain. Development **119:** 233–245.
20. JEFFS, P., K. JAQUES & M. OSMOND. 1992. Cell death in cranial neural crest development. Anat. Embryol. (Berl.) **185:** 583–588.
21. BLASCHKE, A.J., K. STALEY & J. CHUN. 1996. Widespread programmed cell death in proliferative and postmitotic regions of the fetal cerebral cortex. Development **122:** 1165–1174.
22. VOYVODIC, J.T. 1996. Cell death in cortical development: How much? Why? So what? Neuron **16:** 693–696.
23. NEWPORT, J. & M. KIRSCHNER. 1982. A major developmental transition in early Xenopus embryos: I. characterization and timing of cellular changes at the midblastula stage. Cell **30:** 675–686.
24. SIGNORET, J., & J. LEFRESNE. 1971. Contribution a l'etude de la segmentation de l'oeuf d'axolotl. I. Definition de la transition blastuleenne. Ann. Embryol. Morphogen. **4:** 113–123.
25. NEWPORT, J. & M. KIRSCHNER. 1982. A major developmental transition in early *Xenopus* embryos: II. Control of the onset of transcription. Cell **30:** 687–696.
26. ANDERSON, J.A., A.L. LEWELLYN & J.L. MALLER. 1997. Ionizing radiation induces apoptosis and elevates cyclin A1-Cdk2 activity before but not after the midblastula transition in *Xenopus*. Mol. Biol. Cell **8:** 1195–1206.
27. HENSEY, C. & J. GAUTIER. 1997. A developmental timer that regulates apoptosis at the onset of gastrulation. Mech. Dev. **69:** 183–195.
28. SIBLE, J.C., J.A. ANDERSON, A.L. LEWELLYN & J.L. MALLER. 1997. Zygotic transcription is required to block a maternal program of apoptosis in *Xenopus* embryos. Dev. Biol. **189:** 335–346.
29. STACK, J.H. & J.W. NEWPORT. 1997. Developmentally regulated activation of apoptosis early in Xenopus gastrulation results in cyclin A degradation during interphase of the cell cycle. Development **124:** 3185–3195.
30. HENSEY, C. & J. GAUTIER. 1998. Programmed cell death during Xenopus development: a spatio-temporal analysis. Dev. Biol. **203:** 36–48.
31. GAVRIELI, Y., Y. SHERMAN & S.A. BEN-SASSON. 1992. Identification of programmed cell death in situ via specific labeling of nuclear DNA fragmentation. J. Cell Biol. **119:** 493–501.

32. Stefanis, L., D.S. Park, C.Y. Yan, S.E. Farinelli, C.M. Troy, M.L. Shelanski & L.A. Greene. 1996. Induction of CPP32-like activity in PC12 cells by withdrawal of trophic support. Dissociation from apoptosis. J. Biol. Chem. **271:** 30663–30671.
33. Troy, C.M., L. Stefanis, A. Prochiantz, L.A. Greene & M.L. Shelanski. 1996. The contrasting roles of ICE family proteases and interleukin-1beta in apoptosis induced by trophic factor withdrawal and by copper/zinc superoxide dismutase down-regulation. Proc. Natl. Acad. Sci. USA **93:** 5635–5640.
34. Nieuwkoop, P.D. & J. Faber. 1967. Normal Table of *Xenopus laevis* (Daudin). North Holland Publishing Company, Amsterdam.
35. Howe, J.A. & J.W. Newport. 1996. A developmental timer regulates degradation of cyclin E1 at the midblastula transition during Xenopus embryogenesis. Proc. Natl. Acad. Sci. USA **93:** 2060–2064.
36. Oschwald, R., K. Richter & H. Grunz. 1991. Localization of a nervous system-specific class II beta-tubulin gene in *Xenopus laevis* embryos by whole-mount in situ hybridization. Int. J. Dev. Biol. **35:** 399–405.
37. Kimelman, D., M. Kirschner & T. Scherson. 1987. The events of the midblastula transition in *Xenopus* are regulated by changes in the cell cycle. Cell **48:** 399–407.
38. Graham, C.F. & R.W. Morgan. 1966. Changes in the cell cycle during early amphibian development. Dev. Biol. **14:** 439–460.
39. Frederick, D.L. & M.T. Andrews. 1994. Cell cycle remodeling requires cell-cell interactions in developing *Xenopus* embryos. J. Exp. Zool. **270:** 410–416.
40. Cheng, W., L. Baosheng, J. Kajstura *et al.* 1995. Stretch-induced programmed cell death. J. Clin. Invest. **96:** 2247–2259.
41. Ross, M.E. 1996. Cell division and the nervous system: regulating the cycle from neural differentiation to death. Trends Neurosci. **19:** 62–68.
42. Ganan, Y., D. Macias, M. Duterque-Coquillaud, M.A. Ros & J.M. Hurle. 1996. Role of TGF beta s and BMPs as signals controlling the position of the digits and the areas of interdigital cell death in the developing chick limb autopod. Development **122:** 2349–2357.
43. Hemmati-Brivanlou, A. & G.H. Thomsen. 1995. Ventral mesodermal patterning in *Xenopus* embryos: expression patterns and activities of BMP-2 and BMP-4. Dev. Genet. **17:** 78–89.
44. Hirsch, N. & W.A. Harris. 1997. Xenopus Pax-6 and retinal development. J. Neurobiol. **32:** 45–61.
45. Grindley, J.C., D.R. Davidson & R.E. Hill. 1995. The role of Pax-6 in eye and nasal development. Development **121:** 1433–1442.
46. Raff, J.W. & D.M. Glover. 1988. Nuclear and cytoplasmic mitotic cycles continue in Drosophila embryos in which DNA synthesis is inhibited with aphidicolin. J. Cell Biol. **107:** 2009–2019.
47. Kane, D.A. & C.B. Kimmel. 1993. The zebrafish midblastula transition. Development **119:** 447–456.
48. Hemmati-Brivanlou, A. & D. Melton. 1997. Vertebrate neural induction. Annu. Rev. Neurosci. **20:** 43–60.

Bone Morphogenetic Proteins Regulate Interdigital Cell Death in the Avian Embryo

RAMÓN MERINO,[a] YOLANDA GAÑÁN,[b] DOMINGO MACIAS,[b] JOAQUÍN RODRÍGUEZ-LEÓN,[b] AND JUAN M. HURLE[a,c]

[a]*Departamento de Anatomía y Biología Celular, Facultad de Medicina, Universidad de Cantabria, Santander, 39011, Spain*

[b]*Departamento de Clencias Morfológicas y Biología Animal y Celular, Universidad de Extremadura, Badajoz, 06071, Spain*

ABSTRACT: The embryonic limb bud provides an excellent model for analyzing the mechanisms that regulate programmed cell death during development. At the time of digit formation in the developing autopod, the undifferentiated distal mesodermal cells may undergo or chondrogenic differentiation or apoptosis depending whether they are incorporated into the future digital rays or into the interdigital spaces. Both chondrogenesis or apoptosis are induced by local BMPS. However, whereas the chondrogenic-promoting activity of BMPs appears to be regulated through the BMPR- 1b receptor, the mechanism by which the BMPs execute the death program remains unknown. The BMP proapoptotic activity requires the expression of members of the *msx* family of closely related homeobox-containing genes and is finally mediated by caspase activation, but the nature of the caspase(s) directly responsible for the cell death is also unknown. Finally, other growth factors present in the developing autopod at the stages of digit formation such as members of the FGF and TGFβ families modulate the ability of BMPs to induce cell death or chondrogenesis.

INTRODUCTION

The maintenance of an appropriate balance between cell proliferation, differentiation and cell death plays an essential role in embryonic development and tissue homeostasis.[1] During animal development the formation of multiple organs involves the elimination by programmed cell death (PCD) of a large fraction of cells or even all the cellular components of an organ rudiment. PCD is involved in the control of a variety of processes during development. Thus, cell death participates in the regulation of the number of cells in the different tissues, contributes to sculpturing the shape of organs, is essential in the elimination of structures that are needed during specific stages of development or in one sex but not in the other and, finally, is responsible for the elimination of abnormal, non-functional or potentially harmful cells.[1] Multiple studies have clearly demonstrated that the cell death mechanism is

Address for correspondence: Dr. Juan M. Hurle, Departamento de Anatomía y Biología Celular, Facultad de Medicina, Universidad de Cantabria, C/ Cardenal Herrera Oria s/n, Santander, 39011, Spain. 34-942-201903 (fax).
e-mail: hurlej@galeno.medi.unican.es

genetically regulated and several regulatory components of the PCD pathway have been identified in various living organisms.[2,3] Remarkably, the death machinery is highly conserved through evolution and similar molecules regulate PCD in the nematode *Caenorhabditis elegans* or in mammals.[2,3] This review is focused in the analysis of the control of PCD during the development of the vertebrate limb.

The early developing limb consists of a relatively simple structure that appears as a growing bud on the lateral surface of the embryonic body. The limb bud is composed by a homogeneous core of mesenchymal cells covered by a layer of ectoderm. The ectoderm in the distal margin of the bud adopts a thickened shape and differentiates in an specialized region termed the Apical Ectodermal Ridge (AER) that is responsible for the proximo-distal growth of the limb.[4,5] Members of the Fibroblast

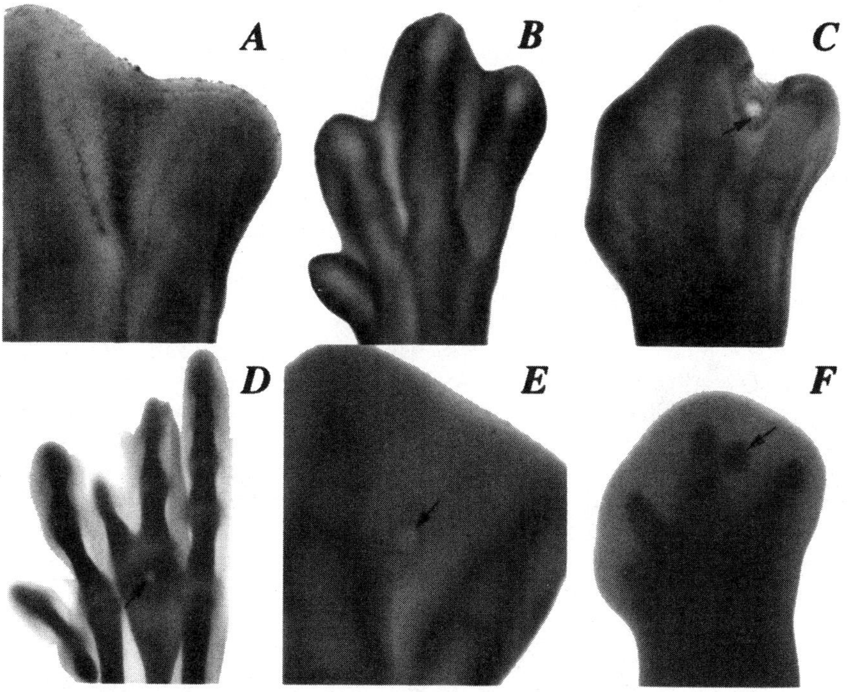

FIGURE 1. A: Embryonic chick leg bud stained with neutral red showing the presence of apoptotic cell death in the third interdigital space. **B:** Expression of *bmp-7* in the autopod at stage 33. **C:** Induction of interdigital cell death 20 h after implantation of a bead (*arrow*) incubated in BMP-7 in the third interdigital space. Note the absence of cell death (neutral red) in the adjacent interdigit at this stage of development (stage 30). **D:** Administration of a BMP-4 bead (*arrow*) at the tip of digit III at stage 28 promotes intense chondrogenesis at digit bifurcation. The autopod has been stained for cartilage with alcian green four days after implantation of the bead. **E:** Implantation of a Noggin bead (*arrow*) in the third interdigital space completely blocks physiological cell death (compare with **A**). **F:** Induction of an ectopic interdigital domain of *bmpR-lb* expression 15 h after implantation of a bead incubated in TGFβ1 (*arrow*) at stage 28.

Growth Factor family (FGF) are produced by the AER and promote the proliferation of the undifferentiated mesenchymal cells located under the AER in a region designated the progress zone (PZ).[4] Once the PZ cells are displaced proximally from the AER, they initiate differentiation, forming the cartilaginous primordia of the limb skeleton. However, an important fraction of mesenchymal cells that are not incorporated in the cartilage primordia undergo extensive PCD in precise areas of the developing limb.[6,7] Ultrastructural and biochemical studies indicate that the cell death observed in these cells has the characteristics of apoptosis.[8]

In the early chick limb bud, there are three areas of massive apoptosis that have been termed the anterior necrotic zone (ANZ), the posterior necrotic zone (PNZ) and the opaque patch (OP).[9] The ANZ and PNZ are characteristic of the limb bud of avian embryos and seem to play a role in the reduction of the antero-posterior axis of the limb, limiting the amount of mesenchyme available for digit formations.[10,11] The OP is located between the two skeletal pieces of the zeugopod and is observed in both the fore- and hind-limb buds of most avian as well as mammalian species.[9] Later in limb development, during the period of digit formation in the autopod, the cells located in the interdigital regions also undergo PCD. These areas of cell death have been named the interdigital necrotic zones (INZ) and are characteristic of all amniote species. The temporal and spatial distribution of INZs is closely related to the morphogenesis of digits and in species with free digits such as the chick, mouse or human, the INZs extend through the entire interdigital mesoderm (FIG. 1A).[6] However, in species with webbed digits such as the duck,[12] or in avian species having free digits with membranous lobulations along the digital margin, such as the moorhen or the coot,[13] the INZs are restricted to the distal or central part of the interdigital region, respectively. In addition to these major areas of mesodermal cell death, PCD is also observed in the AER controlling the extension of this structure,[14] during the establishment of the developing joints[15] and in the formation of the axon pathways.[16] In the following sections of this review, we will summarize our knowledge on the mechanisms that trigger and regulate PCD during limb development, and particularly in the apoptotic cell death of the interdigital undifferentiated mesoderm.

BONE MORPHOGENETIC PROTEINS (BMPS) ARE THE MEDIATORS OF THE INTERDIGITAL APOPTOSIS

BMPs constitute a large family of secreted factors belonging to the Transforming Growth Factor-β (TGFβ) superfamily.[17] These proteins were first identified from bone extracts by their capacity to promote ectopic chondrogenesis and ossification.[18] However, in addition to their role in bone formation, BMPs have been now identified as key signals for multiple developmental processes.[19] Like other members of the TGFβ superfamily, BMPs exert their biological function by interactions with cell surface receptors consisting in heterodimers of type I and type II receptors with intracellular serine/threonine kinase domains. Ligand binding of BMPs to both receptors results in phosphorylation of the type I receptor by the type II receptor. Then, the type I receptor is responsible for the transduction of the signal by phosphorylating members of the Smad family.[20]

Multiple members of the BMP family, including *bmp-2*, *bmp-4*, *bmp-6*, *bmp-7*, and *gdf-5*, are expressed in a regulated pattern during embryonic limb development.[21–28] In this regard, *bmp-2*, *bmp-4* and *bmp-7* are expressed in the undifferentiated mesenchyme of the PZ, AER and interdigital spaces[21–25] (Fig. 1B). *Bmp-2* and *bmp-7* also exhibit an intense region of expression in the joint-forming regions and in the perichondrium of the developing cartilages, respectively.[24] In contrast to the above mentioned *bmps*, *gdf-5* expression is restricted to the regions of digital joint formation and transiently at the tip of each digit coincidentally with the formation of the last phalanxes.[27,28] Similarly, *bmp-6* expression is restricted to the developing cartilages.[26]

In an attempt to clarify the role of BMPs during limb development, recombinant human BMP proteins have been administered into the autopod using beads incubated in the different factors. These studies indicate that BMPs are able to promote intense apoptosis in the interdigital regions but also chondrogenesis at the tip of the digits[24,29] (FIG. 1 C and D). Cell death following BMP treatment is also detected in the AER but not in the dorsal or ventral ectoderm.[24,30] The involvement of BMPs in cell death and chondrogenesis during limb development has been further demonstrated in experiments using antagonists of these factors or after retrovirus-induced overexpression of dominant negative BMP-receptors. Several inhibitors of BMPs have been described. We and others have recently characterized the expression and function of one of them, Noggin,[31] during limb development. Noggin is a natural antagonist of BMP that exerts its function by binding BMPs and preventing their interaction with specific BMP-receptors. Noggin is expressed in the autopod at the level of the digital rays. Implantation of beads soaked in recombinant human Noggin in the interdigital regions completely inhibits the apoptotic cell death in the undifferentiated mesoderm (FIG. 1E).[32] Similarly, administration of Noggin at the tip of digits blocks the physiological chondrogenesis and induces digit truncations.[32] Similar conclusions have been obtained in experiments of retrovirus-mediated overexpression of *noggin* in the developing limb, in mice with a targeted disruption of the *noggin* gene and after overexpression of dominant negative BMPR-1 receptors (see below).[33–31]

One interesting observation is that the ability of BMPs to induce cell death or chondrogenesis is different between the members of this family expressed in the autopod. Thus, whereas BMP-2, BMP-4 and BMP-7 promote both chondrogenesis and cell death[24,29] GDF-5 lacks significant cell death activity but induces intense chondrogenesis at the tip of the digits. These results, first, are indicative for the existence of functional redundancy between the BMPs in the autopod that guaranteed the appropriate development of the limb in the absence of some of these factors and, second, demonstrate the presence of a functional specialization within the BMP-family.

As discussed above, the BMP signaling is transduced through heterodimers of type I and type II serine-threonine kinase BMP receptors.[19]Two type I and one type II BMP-receptors have been identified invertebrates.[37–39] To try to understand the mechanism by which BMPs may promote either chondrogenesis or apoptosis in the undifferentiated mesodermal cells leaving the PZ, we and others have analyzed the expression and function of the different type I BMP-receptors during the stages of digit formation and interdigital cell death in the chick autopod. In the stages of digit formation, *bmpR-1a* gene is expressed at low levels in the limb undifferentiated mesenchyme.[32,39] Later in development, the expression of *bmpR-1a* become restricted to the prehypertrophic cartilage of the developing digits, suggesting a role for this receptor in chondrocyte differentiation.[40] Consistent with this interpretation, overexpression of a constitutively active BMPR-la construct in the developing digits blocks chondrocyte differentiation at the prehypertrophic stage.[40]

In contrast to *bmpR-1a*, *bmpR-1b* is expressed at high levels in the condensing mesenchyme of the digital rays from the onset of digit formation.[32,36,40] This restricted pattern of expression is compatible with a role for this receptor in the chondrogenic function of BMPS. Accordingly, administration of BMPs at the tip of the digits, in the regions where *bmpR-1b* is normally expressed, promotes an expansion in the area of expression of this receptor that mimics the shape of the future enlarged cartilage that is formed.[32] In addition, overexpression of a constitutively active BMPR-1b or a dominant negative BMPR-lb receptor in limb buds using retroviral vectors promotes or inhibits chondrogenesis, respectively.[36,40] Finally, the formation of extra digits following interdigital administration of TGFβ is preceded by the induction of an ectopic domain of *bmpR-1b* gene (FIG. 1F), which adopts a characteristically elongated shape reminiscent of the distal phalanx of a digit.[32]

Whereas the above results clearly indicate that the chondrogenic-promoting activity of BMPs is regulated through the BMPR-lb receptor, the mechanisms by which these factors induce cell death remain elusive. In this regard, retrovirus-induced overexpression of a constitutively active BMPR-1b receptor causes massive cell death or cartilage overgrowth depending on the stage at which the limb is infected, suggesting that the BMPR-lb receptor is also involved in the transduction of the BMP-mediated cell death activity.[40] However, neither *bmpR-1a* nor *bmpR-1b* genes are expressed in the interdigital mesenchyme in a pattern compatible with a role for these receptors in apoptosis, and the induction of the *bmpR-1b* gene in the interdigital region after the administration of TGFβ, is followed by the inhibition of interdigital cell death and the formation of extra digits.[32] Further support of this interpretation is that GDF-5, in spite of its ability to interact with high affinity with the BMPR-lb receptor,[41,42] fails to promote interdigital cell death.[28] Together, these data suggest that the BMP-induced cell death activity may be transduced by other BMP-receptors. The nature of these BMP-receptors remains unknown, but an attractive hypothesis is that they may contain in their intracellular region the characteristic cell death domain present in other proapoptotic receptors as in the case of the neurotrophins.[43]

FGFS ARE SURVIVAL SIGNALS PRODUCED BY THE AER THAT SENSITIZE THE INTERDIGITAL MESENCHYME TO CELL DEATH THROUGH THE REGULATION OF *MSX* GENES

The survival of the undifferentiated cells in the PZ, despite the presence of high levels of BMPs in this location, suggests the existence of local factors that regulate the cell death–promoting activity of BMPS. Several members of the FGF family (FGF-2, FGF-4, FGF-8, and FGF-9) are produced by the AER at different periods of development and promote the proliferation of mesodemal cells located in the

PZ.[4,44–46] The proliferation of these cells accounts for the proximo-distal outgrowth of the limb and surgical removal of the AER results in limb truncations.[47] Conversely, implantation of beads bearing FGFs in limbs in which the AER was previously removed can sustain normal limb outgrowth.[44,45] There is strong experimental evidence that FGFs antagonize the BMP function.[48] Implantation of beads bearing FGF-2 or FGF-4 in the interdigital region blocks physiological and BMP-induced cell death, resulting in the formation of limbs with webbed digits.[49] In addition, implantation of FGF-beads at the tip of digits delays normal chondrogenesis, a function also mediated by BMPs.[32,49] The mechanisms by which FGFs inhibit BMP function are beginning to be clarified. Thus, implantation of FGF-beads at the tip of digits downregulates the physiological digital expression of the *bmpR-1b* gene.[32] Similarly, FGF blocks the induction of *bmpR-1b* and the formation of ectopic extra digits that follows the interdigital administration of TGFβ[32] On the basis of these results, it is tempting to speculate that the anti-apoptotic activity of FGFs may be also related to the negative regulation in the expression of the hypothetical pro-apoptotic BMP-receptor discussed above.

In the same way that the FGFs produced by the AER regulate the function of BMPS, the activity of the AER is controlled negatively by BMPS. The antagonistic effect of BMPs on FGF function is best illustrated at the later stages of limb development during the formation of the digits. At these stages, fgf-8 is the member of the FGF-family present in the AER (FIG. 2A), and its normal expression disappears first at the level of the interdigital regions concomitantly with the initiation of cell death in this area and some hours later also disappears from the tip of the digits in association with the establishment of the last phalanx of each digit.[30] During these periods BMPs are highly expressed in the PZ and implantation of BMP-beads in close association with the AER promotes the downregulation of *fgf-8* expression from the AER and the rapid degeneration the AER (FIG. 2B), suggesting that BMPs may be also involved in the disappearance of the AER once the limb bud reaches the appropriate proximo-distal size.[30]

At the same time that FGFs inhibit interdigital apoptosis, these factors sensitize the undifferentiated mesoderm of the interdigital spaces to the pro-apoptotic effect of BMPS. Studies in the duck limb bud demonstrate that despite the high levels of expression of *bmps* in the interdigits[25] these cells are very resistant to the apoptotic effect of BMPS.[30] In contrast to the chick leg bud, exogenous administration of either BMP-4 or BMP-7 into the third interdigital space of the duck leg promotes a small area of cell death restricted to the most distal part of the interdigit (FIG. 2C), one area that normally disappears by apoptosis.[30] Similarly, the interdigital space of the duck leg is very resistant to the chondrogenic effect of TGFβ and again only the distal region of the interdigit retains this ability.[30] However, the physiological and BMP-induced interdigital cell death in the duck are greatly increased when the interdigit has been previously treated with FGFs (FIG. 2D) indicating that FGFs act as anti-apoptotic signals during the period that they are present but at the same time render the cells sensitive to apoptotic stimuli once they disappear.[30]

The attenuated response of the duck interdigits to the effects of BMPs and TGFβs correlates with the expression of *msx* genes,[30] a family of closely related homeobox-containing genes homologous to the *msh* gene in drosophila. Both *msx-1* and *msx-2* genes show a restricted pattern of expression in the duck limb compared to the expression observed in the chick. At the stages of digit formation, the expression of *msx-1* and *msx-2* is limited, in both the leg and wing bud, to the most distal mesenchyme of the autopod in close correlation with the pattern of cell death observed in the foot and wing of the duck embryo[30] (FIG. 2E). Interestingly, ectopic application of FGFs in the interdigit expands the area of *msx-2* expression prior to the appearance of cell death (FIG. 2F), demonstrating the existence of a close correlation between cell death susceptibility and *msx-2* gene expression. Since *msx* genes function as repressors of gene transcription, thus maintaining the limb mesoderm in an

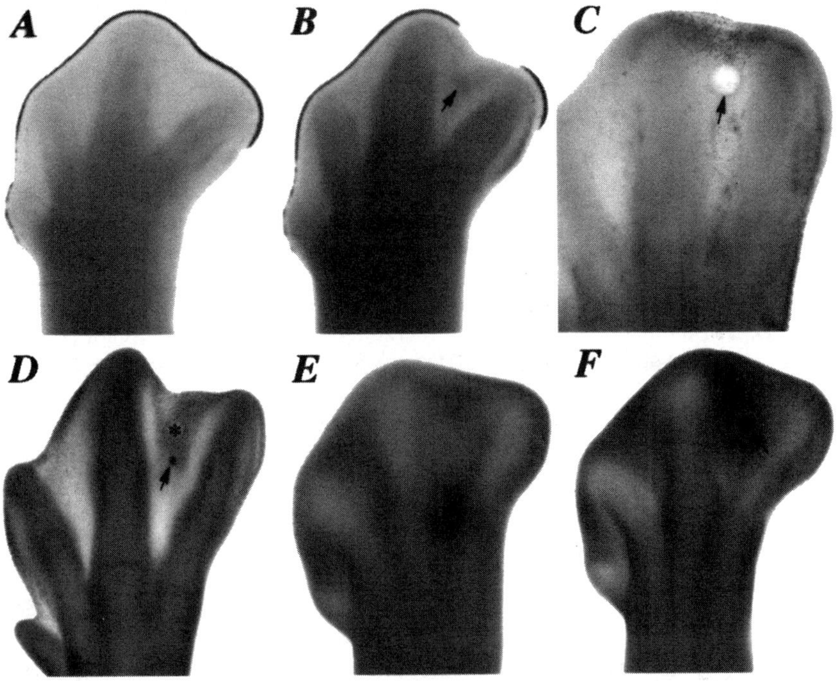

FIGURE 2. A: Expression of *fgf-8* in the AER of embryonic chick autopod at stage 30. **B:** Downregulation of *fgf-8* expression 10 h after implantation of BMP-2 (*arrow*) in the third interdigital space at stage 29. **C:** Induction of apoptotic cell death in the duck autopod after implantation of BMP-4 (*arrow*) in the third interdigital. Note that the area of cell death-induced by BMP-4 in the duck autopod is restricted to the distal margin of the interdigit (compare with the enlarged area of cell death induced by BMPs in the chick; FIG. 1C). **D:** Administration of FGF-2 (*arrow*) in the third interdigital space of a duck autopod 24 h prior to implantation of a bead incubated in BMP-7 (*) promotes an enlarged area of BMP-induced cell death (compare with **C**). **E:** Expression of *msx-2* gene in the duck autopod. Note that the domain of expression of *msx-2* is restricted to the distal margin of the interdigit and correlates with the area of cell death induced by BMP treatment (compare with **C**). **F:** Expression of *msx-2* in the interdigit of a duck autopod 24 h after administration of FGF-2 (*arrow*). Note that the treatment with FGF-2 enlarges the territory of expression of *msx-2* which shows again a correlation with the area of BMP-induced cell death (compare with **D**).

undifferentiated state, it is possible that the presence of interdigital membranes in species with webbed digits, such as the duck, is the result of a premature differentiation of the undifferentiated mesenchyme into connective tissue resistant to the action of both BMPs and TGFβs.

CASPASE ACTIVATION IS INVOLVED IN THE INTERDIGITAL CELL DEATH

Studies in both vertebrate and invertebrate species have clearly demonstrated that the apoptotic cell death is executed through the activation of the caspase/Ced-3 family of cysteine-proteases.[52,53] At the present time, at least 12 members of this family have been identified in mammals. These proteases are synthesized as proenzymes composed of an amino-terminal prodomain of variable length and a large and small subunits. The prodomain of the caspases serves as a protein dimerization domain that regulates their activation.The procaspases have little catalytic activity and their activation requires proteolytic processing.This activation that can be achieved either by previously activated capases of by the formation of procapase complexes results in the generation of a proteolytic cascade that serves to transmit and amplify the death signal. Not all the identified caspases play an active role in the induction of apoptosis *in vivo*. Caspase-1, -4, -5 and - 11are involved in inflammatory responses rather than in apoptosis.[52,53] Among the proapoptotic caspases, one group—caspases-3, -8, and -9, with long prodomains—participate in the caspase activation cascade. The second group, caspases-3, -6, and -7, have short prodomains and are downstream mediators.[52,53] Recent findings indicate that the apoptotic cell death is mediated by caspase activation. Treatment of chicken or mouse limb buds with the broad-based caspase inhibitor zVAD-fmk or with more selective caspase inhibitors, Ac-YVAD-CHO or Ac-YVAD-CMK, significantly reduces the number of interdigital apoptotic cells.[54,55] However, the nature of the caspase/s activated during this developmental cell death is not clear. Thus, whereas Ac-YVAD-CHO, or Ac-YVAD-CMK, which are strong inhibitors of caspase-1 and -4, but weak or ineffective inhibitors of caspase-3 and -7, inhibit interdigital apoptosis,[54] both the wing and leg buds develop normally in mice deficient in caspase-1.[56,57] More interestingly, mice deficient in either caspase-2,[58] caspase-3,[59]or caspase-8,[60] which have a demonstrated role in the execution of the apoptotic program *in vivo*, also show a normal limb phenotype. Finally, although the inactivation of the *apaf-1* gene delays the onset of interdigital cell death,[61,62] the regression of the interdigital webbing is complete in animals deficient in caspase-9,[63,64] the cysteine-protease activated by Apaf-1. The absence of an abnormal cell death pattern in the limbs of these caspase-mutant mice may be indicative of the existence of functional redundancy between the different caspases involved this process. The analysis of mice with combined deficiencies in caspases may help to elucidate this important question.

OTHER POTENTIAL SIGNALS INVOLVED IN THE REGULATION OF INTERDIGITAL CELL DEATH

In addition to the signals described above, several factors are expressed in the autopod in a pattern compatible with a role in the regulation of PCD during limb development either in the mouse or chicken. Among the gene products expressed in the undifferentiated intcrdigital mesenchyme concomitantly with cell-death are *c-fos,*[65] c-rel,[66] *cdk-5*[67] and *pax1,*[68] growth factors as TNFα-like proteins,[69] Insulin-like growth factors (IGF) and IGF-binding proteins,[70] tissue proteins, plasminogen activator,[71] tissue transglutaminase, and surface molecules as clusterin[72,73] and cadherin-6.[73]Although, participation in the regulation of cell death in different experimental systems has been reported for most if not all of these molecules, there is no direct evidence for their involvement in cell death during limb development. Similarly, although the interdigital undifferentiated mesenchyme expresses high levels of retinoic acid receptor-β (RAR-β) and cellular retinol-binding protein (CRBP),[74,75] and exogenous administration of retinoic acid enhances cell death in the developing limb,[76] the pattern of cell death is normal in the limbs from mice deficient for several elements of the retinoic acid signaling pathway including RAR-β or cellular retinoic acid binding proteins (CRABP-1 and CRAPB-2).[76,80]

Several spontaneous or induced gene mutations that promote alterations in the pattern of cell death during embryonic limb development have been described. These include the Talpid mutation in the chick[10] and Dominant hemimelia,[81] Hemimelia extratoe,[82] Fused toes,[83] and Hammertoe mutations[76] in the mouse. With the exception of a gene, *ft1*, present in the large chromosome deletion of the Fused toes mutation,[83] none of the gene(s) responsible for these mutations have been isolated. Since the Fused toes mutation is secondary to a deletion of 300 Kb of the mouse chromosome 8 induced by insertional mutagenesis by a transgene, the specific participation of the *ft1* gene in mutation awaits also clarification.

ACKNOWLEDGMENTS

This work was supported by grants from the DGICYT to J.M.H. (PM95-0090) and to Y.G. (PM96-0020). R.M. is the recipient of a grant from the Spanish "Ministerio de Educación y Cultura" (Reincorporación de doctores y tecnólogos).

REFERENCES

1. JACOBSON, M.D., M. WEIL & M.C. RAFF. 1997. Programmed cell death in animal development. Cell **88:** 347–354.
2. GOLSTEIN, P. 1997. Controlling cell death. Science **275:** 1081–1082.
3. WHITE, E. 1996. Life, death, and the pursuit of apoptosis. Genes Dev. **10:** 1–15.
4. MARTIN, G. 1998. The roles of FGFs in the early development of vertebrate limbs. Genes Dev. **12:** 1571–1586.
5. SUMMERBELL, D., J.H. LEWIS & L. WOLPER. 1973. Positional information in chick limb morphogenesis. Nature 244: 492-495.
6. HURLE, J.M., M.A. ROS, V. GARCIA-MARTINEZ, D. MACIAS & Y. GAÑAN. 1995. Cell death in the embryonic developing limb. Scan. Microsc. **9:** 519–534.
7. HURLE, J.M., M.A. ROS, V. CLIMENT & V. GARCIA-MARTINEZ. 1996. Morphology and significance of programmed cell death in the developing limb bud of the vertebrate embryo. Micros. Res. Tech. **34:** 236–246.
8. GARCIA-MARTINEZ, V., D, MACIAS, Y. GARAN, *et al.* 1993. Internucleosomal DNA fragmentation and programmed cell death (apoptosis) in the interdigital tissue of the embryonic chick leg bud. J. Cell. Sci. **106:** 201–208.

9. HINCHLIFFE, J.R. 1982. Cell death in vertebrate limb morphogenesis. Progress in Anatomy. **2:** 1–19. Cambridge University Press. Cambridge.
10. HINCHLIFFE, J.R. & D.E. EDE. 1967. Limb development in the polydactylous talpid mutant of the fowl. J. Embryol. Exp. Morphol. **17:** 385–404.
11. HINCHLIFFE, J.R. & D.E. EDE. 1993. Cell death and development of limb forms and skeletal pattern in normal and wingless (ws) chick embryos. J. Embryol. Exp. Morphol. **30:** 753–772.
12. SAUNDERS, J.W. & J.F. FALLON. 1967. Cell death in morphogenesis. *In* Major Problems in Developmental Biology. 289–314. Academic Press, New York.
13. HURLE, J.M. & V. CLIMENT. 1987. The regression of the interdigital tissue in rallidae avian embryos (*Fulika atra* and *Gallinula chloropus*). Arch. Biol. **98:** 299–316.
14. TODT, W.L. & J.F. FALLON. 1986. Development of the apical ectodermal ridge in the chick leg bud and a comparation with the wing bud. Anat. Rec. **215:** 288–304.
15. MORI, C., N. NAKAMURA, S. KIMURA, *et al.* 1995. Programmed cell death in the interdigital tisue of the fetal mouse limb is apoptosis with DNA fragmentation. Anat. Rec. **242:** 103–110.
16. TOSNEY, K.W., S. SCHROETER & J. POKRZYWINSKI. 1988. Cell death delineates axon pathways in the hindlimb and does so independently of the neurite outgrowth. Dev. Biol. **130:** 558–572.
17. HOGAN, B.L.M. 1996. Bone morphogenetic proteins: multifunctional regulators of vertebrate development. Genes Dev. **10:** 1580–1594.
18. WOZNEY, J.M., V. ROSEN, A.J. CELESTE, *et al.* 1988. Novel regulators of bone formation: molecular clones and activities. Science **242:** 1528–1534.
19. MASSAGUE, J. 1996. TGFP signaling: receptors, transducers, and Mad proteins. Cell **85:** 947–950.
20. WHITMAN, M. 1998. Smads and early developmental signaling by the TGFβ superfamily. Genes Dev. **12:** 2445–2462.
21. FRANCIS, P.H., M.K. RICHARDSON, P.M. BRICKELL & C. TICKLE. 1994. Bone morphogenetic proteins and a signaling pathway that controls patterning in the developing chick limb. Development **120:** 209–218.
22. FRANCIS-WEST, P.H., K.E. ROBERTSON, D.E. EDE, *et al.* 1995. Expression of genes encoding bone morphogenetic proteins and sonic hedgehog in Talpid (*ta3*) limb buds: their relationship in the signaling cascade involved in limb patterning. Dev. Dyn. **203:** 187–197.
23. LYONS, K.M., HOGAN, B.L.M. & E.J. ROBERTSON. 1995. Colocalization of BMP7 and BMP2 RNAs suggest that these factors cooperatively mediate tissue interactions during murine development. Mech. Dev. **50:** 71–83.
24. MACIAS, D., Y. GAÑAN, T.K. SAMPATH, *et al.* 1997. Role of BMP-2 and OP-1 (BMP-7) in programmed cell death and skeletogenesis during chick limb development. Development **124:** 1109–1117.
25. LAUFER, E.S., H. ZOU, O.E. OROZCO & L. NISWANDER. 1997. BMP expression in duck interdigital webbing: a reanalysis. Science **78:** 305.
26. VORTKAMP, A., K. LEE, B. LANSKE, *et al.* 1996. Regulation of rate of cartilage differentiation by Indian Hedgehog and PTH-related protein. Science **273:** 613–622.
27. STORM, E. & D.M. KINGSLEY. 1996. Joint patterning defects caused by single and double mutations in members of the bone morphogenetic protein (BMP) family. Development **122:** 3969–3979.
28. MERINO, R., D. MACIAS, Y. GAÑAN, *et al.* 1999. Expression and function of *Gdf-5* during digit skeletogenesis in the embryonic chick leg bud. Dev. Biol. **206:** 33–45.
29. GAÑAN, Y., D. MACIAS, M. DUTERQUE-COQUILLAUD, M.A. ROS & J. M. HURLE. 1996. Role of TGFP and BMPs as signals controlling the position of the digits and the areas of cell death in the developing chick autopod. Development **122:** 2349–2357.

30. GAÑAN, Y., D. MACIAS, R.D. BASCO, R. MERINO & J.M. HURLE. 1998. Morphological diversity of the avian foot is related with the pattern of *msx* gene expression in the developing autopod. Dev. Biol. **196:** 33–41.
31. ZIMMERMAN, L.B., J.M. DE JESUS-ESCOBAR & R.M. HARLAND. 1996. The Speman organizer signal noggin binds and inactivates bone morphogenetic protein 4. Cell **86:** 599-606.
32. MERINO, R., Y. GAÑAN, D. MACIAS, *et al.* 1998. Morphogenesis of digits in the avian limb is controlled by FGFs, TGFβs, and Noggin through BMP signaling. Dev. Biol. **200:** 35-45.
33. CAPDEVILA, J. & R.L. JOHNSON. 1998. Endogenous and ectopic expression of noggin suggests a conserved mechanism for regulation of BMP function during limb and somite pattern. Dev. Biol. **197:** 205–217.
34. BRUNET, L.J., J.A. MCMAHON, A.P. MCMAHON & R.M. HARLAND. 1998. Noggin, cartilage morphogenesis, and joint formation in the mammalian skeleton. Science **280:** 1455–1457.
35. ZOU, H. & L. NISWANDER. 1996. Requirement for BMP signaling in interdigital apoptosis and scale formation. Science **272:** 738–741.
36. KAWAKAMI, Y., T. ISHIKAWA, M. SHIMABARA, *et al.* 1996. BMP signaling during bone pattern determination in the developing limb. Development **122:** 3557–3566.
37. KOENIG, B.B., J.S. COOK, D.H. WOLSING, *et al.* 1994. Characterization and cloning of a receptor for BMP-2 and BMP-4 from NIH 3T3 cells. Mol. Cell Biol. **14:** 5961–5974.
38. TEN DIJKE, P., H. ICHIJO, P. FRANZEN, *et al.* 1993. Activin receptor-like kinases: a novel subclass of cell-surface receptors with predicted serine/threonine kinase activity. Oncogene **8:** 2879–2887.
39. NOHNO, T., T. ISHIKAWA, T. SAITO, *et al.* 1995. Identification of a human type 11 receptor for Bone Morphogenetic Protein-4 that forms differential heteromeric complexes with Bone Morphogenetic Protein type I receptors. J. Biol. Chem. **270:** 22522–22526.
40. ZOU, H., R. WIESER, J. MASSAGUE & L. NISWANDER. 1997. Distinct roles of type I bone morphogenetic protein receptors in the formation and differentiation of cartilage. Genes Dev. **11:** 2191–2203.
41. NISHITOH, H., H. ICHIJO, M. KIMURA, *et al.* 1996. Identification of type I and Type 11 serine/threonine kinase receptors for growth/differentiation factor-5. J. Biol. Chem. **271:** 21345–21352.
42. ERLACHER, L., J. MACCARTNEY, E. PIEK, *et al.* 1998. Cartilage-derived morphogenetic proteins and osteogenic protein-1 differentially regulate osteogenesis. J. Bone Min. Res. **13:** 384–392.
43. DECHANT, G. & Y.A. BARDE. 1997. Signaling through the neurotrophin receptor p75 (NTR). Cuff. Opin. Neurobiol. **73:** 413–418.
44. NISWANDER, L., C. TICKLE, A. VOGEL, I. BOOTH & G.R. MARTIN. 1993. FGF-4 replaces the apical ectodermal ridge and directs outgrowth and patterning of the limb. Cell **75:** 579–587.
45. FALLON, J.F., A. LOPEZ, M.A. ROS, *et al.* 1994. FGF-2, apical ridge growth signal for chick limb development. Science **264:** 104–107.
46. VOGEL, A., C. RODRIGUEZ & J.C. IZPISUA-BELMONTE. 1996. Involvement of FGF-8 in iniciation, outgrowth and patterning of the vertebrate limb. Development **122:** 1737–1750.
47. ROWE, D.A., J.M. CAIRNS & J.F. FALLON. 1982. Spatial and temporal patterns of cell death in limb bud mesoderm after apical ectodermal ridge removal. Dev. Biol. **93:** 83–91.

48. NISWANDER, L. & G. MARTIN. 1993. FGF-4 and BMP-2 have opposite effects on limb outgrowth. Nature **361:** 68–71.

49. MACIAS, D., Y. GAÑAN, M.A. ROS & J.M. HURLE. 1996. In vivo inhibition of programmed cell death by local administration of FGF-2 and FGF-4 in the interdigital areas of the embryonic leg bud. Anat. Embryol. **193:** 533–541.

50. FERRARI, D., R.A. KOSHER & C.N. DEALY. 1994. Limb mesenchymal cells inhibited from undergoing cartilage differentiation by a tumor promoting phorbol ester maintaining expression of the homeobox-containing gene *MSX1* and fail to exhibit gap junctional communication. Biochem. Biophys. Res. Commun. **205:** 429–434.

51. CATRON, K.M., H. WANG, G. HU, M.M. SHEN & C. ABATE-SHEN. 1996. Comparison of MSX-1 and MSX-2 suggests a molecular basis for functional redundancy. Mech. Dev. **55:** 185–199.

52. CRYNS, V. & J. YUAN. 1998. Proteases to die for. Genes Dev. **12:** 1551–1570.

53. THORNBERRY, N.A. & Y. LAZEBNIK. 1998. Caspases: enemies within. Science **281:** 1312–1316.

54. MILLIGAN, C.E., D. PREVETTE, H. YAGINURA, *et al.* 1995. Peptide inhibitors of ICE protease family arrest programmed cell death of motoneurons in vivo and in vitro. Neuron **15:** 385–393.

55. JACOBSON, M.D., M. WEEL. & M.C. RAFF. 1996. Role of Ced-3/ICE-family proteases in staurosponne-induced programmed cell death. J. Cell Biol. **133:** 1041–1051.

56. LI, P., H. ALLEN, S. BENERJEE, *et al.* 1995. Mice deficient in IL-1β-converting enzyme are defective in production of mature IL-1β and resistant to endotoxic shock. Cell **80:** 401–411.

57. KUIDA, K., G. LIPPKE, G. KU, *et al.* 1995. Altered cytokine export and apoptosis in mice deficient in interleukin-1β converting enzyme. Science **267:** 2000–2002.

58. BERGERON, L., G.I. PEREZ, G. MACDONALD, *et al.* 1998. Defects in regulation of apoptosis in caspase-2-deficient mice. Genes Dev. **12:** 1304–1314.

59. KUIDA, K., T.S. ZHENG, S. NA, *et al.* 1996. Decreased apoptosis in the brain and premature lethality in CPP32-deficient mice. Nature **384**: 368–372.

60. VARFOLOMEEV, E.E., M. SCHUCHMANN, V. LURIA, *et al.* 1998. Targeted disruption of the mouse caspase 8 gene ablates cell death induction by the TNF receptors, Fas/Apol, and DR3 and is lethal prenatally. Immunity **9:** 267–276.

61. CECCONI, F., G. ALVAREZ-BOLADO, B.I. MEYER, K.A. ROTH & P. GRUSS. 1998. Apaf 1 (CED-4 homolog) regulates programmed cell death in mammalian development. Cell **94:** 727–737.

62. YOSHIDA, H., Y-Y. KONG, R. YOSHIDA, *et al.* 1998. Apaf 1 is required for mitochondrial pathways of apoptosis and brain development. Cell **94:** 739–750.

63. KUIDA, K., T.F. HAYDAR, C-Y. KUAN, *et al.* 1998. Reduced apoptosis and cytochrome-c–mediated caspase activation in mice lacking caspase 9. Cell **94:** 325–337.

64. HAKEM, R., A. HAKEM, G.S. DUNCAN, *et al.* 1998. Differential requirement for caspase 9 in apoptotic pathways in vivo. Cell **94:** 339–352.

65. YANO, H., A. OHTSURU, M. ITO, T. FUJII & S. YAMASHITA. 1996. Involvement of *c-Fos* proto-oncogene during palatal fusion and interdigital space formation in the rat. Develop. Growth Differ. **38:** 351–357.

66. ABBADIE, C., N. KABRUN, F. BOUALI, *et al.* 1993. High levels of *c-rel* expression are associated with programmed cell death in the developing avian embryo and in bone marrow cells in vitro. Cell **75:** 899–912.

67. ZHANG, Q., H.S. AHUJA, Z. ZAKERI & J. WOLGEMUTH. 1997. Cyclin-dependent kinase 5 is associated with apoptotic cell death during development and tissue remodeling. Dev. Biol. **183:** 222–233.

68. TIMMONS, P.M., J. WAI-LIN, P.W.J. RIGBY & R. BALLING. 1994. Expression and function of *Pax1* during development of the pectoral girdle. Development **120:** 2773–2785.
69. WRIDE, M.A., P.H. LAPCHAK & E.J. SANDERS. 1994. Distribution of TNFα-like proteins correlates with some regions of programmed cell death in the chick embryo. Int. J. Dev. Biol. **38:** 673–682.
70. VAN KLEFFENS, M., C. GROFFEN, R.R. ROSATO, *et al.* 1998. mRNA expression patterns of the IGF system during mouse limb bud development, determined by whole mount in situ hybridization. Mol. Cell Endocrinol. **138:** 151–161.
71. CARROLL, P.M., S.E. TSIRKA, W.G. RICHARDS, M.A. FROHMAN & S. STRICKLAND. 1994. The mouse tissue plasminogen activator gene 5′flanking region directs appropriate expression in development and a seizure-enhanced response in the CNS. Development **120:** 3173–3183.
72. MOALLEM, S.A. & B.F. HALES. 1996. Transglutaminase and clusterin induction during normal and abnormal limb development in the mouse. Biol. Reprod. **55:** 281– 290.
73. INOUE, T., O. CHISAKA, H. MATSUNAMI & M. TAKEICHI. 1997. Cadherin-6 expression transiently delineates specific rhombomeres, other neural tube subdivisions, and neural crest subpopulations in mouse embryos. Dev. Biol. **183:** 183–194.
74. DOLLE, P., E. RUBERTE, P. KASTNER, *et al.* 1989. Differential expression of genes encoding α, β and γ retinoic acid receptors and CRABP in the developing limbs of the mouse. Nature **342:** 702–705.
75. DOLLE, P., E. RUBERTE, P. LEROY, G. MORRISS-KAY & P. CHAMBON. 1990. Retinoic acid receptors and cellular retinoid binding proteins. I. A systematic study of their differential pattern of transcription during mouse organogenesis. Development **110:** 1133–1151.
76. AHUJA, H.S., W. JAMES & Z. ZAKERI. 1997. Rescue of the limb deformity in Hammertoe mutant mice by retinoic acid-induced cell death. Dev. Dyn. **208:** 466–481.
77. MENDELSOHN, C., M. MARK, P. DOLLE, *et al.* 1994. Retinoic acid receptor P2 (RARβ2) null mutant mice appear normal. Dev. Biol. **166:** 246–258.
78. LUO, J., P. PASCERI, R.A. CONLON, J. ROSSANT & V. GIGUERE. 1995. Mice lacking all isoforms of retinoic acid receptor β develop normally and are susceptible to the teratogenic effects of retinoic acid. Mech. Dev. **53:** 61–71.
79. LAMPRON, C., C. ROCHETTE-EGLY, P. GORRY, *et al.* 1995. Mice deficient in cellular retinoic acid binding protein II (CRABPII) or in both CRABPI and CRABPII are essentially normal. Development **121:** 539–548.
80. FAWCETT, D., P. PASCERI, R. FRASER, *et al.* 1995. Postaxial polydactyly in forelimbs of *CRABP-II* mutant mice. Development **121:** 671–679.
81. ROOZE, M.A. 1977. The effects of the *Dh* gene on limb morphogenesis in the mouse. *In* Morphogenesis and Malformations of the Limb. D. Bergsma & W. Lenz, Eds. :69–95. Alan R. Liss, Inc. New York.
82. KNUDSEN, T.B. & D.M. KOCHHAR. 1981. The role of morphogenetic cell death during abnormal limb-bud outgrowth in mice heterozygous for the dominant mutation Hemimelia-extra toe (*Hm*^{x}). J. Embryol. Exp. Morphol. **65:** 289–307.
83. LESCHE, R., A. PEERZ, F. VAN DER HOEVEN & U. RÜTHER. 1997. *Ft1*, a novel gene related to ubiquitin-conjugating enzymes, is deleted in the Fused toes mouse mutation. Mamm. Genome **8:** 879–883.

Oxidative Damage, Bleomycin, and Gamma Radiation Induce Different Types of DNA Strand Breaks in Normal Lymphocytes and Thymocytes

A Comet Assay Study

LUIS BENÍTEZ-BRIBIESCA[a] AND PATRICIA SÁNCHEZ-SUÁREZ

Oncological Research Unit, Oncology Hospital, National Medical Center, IMSS, 06725 Mexico, DF, Mexico

ABSTRACT: Most anticancer treatments such as chemo- and radiotherapy induce DNA damage and apoptosis in normal cells. The aim of this study was to assess the induction of single and double DNA strand breaks (ssb and dsb, respectively) and apoptosis in normal human lymphocytes and rat thymocytes subjected to the action of H_2O_2, bleomycin and ionizing radiation. Normal human peripheral thymocytes and young rat thymocytes were subjected to the following treatments: a) H_2O_2; b) bleomycin, and c) γ-radiation, all with various doses. DNA strand breaks were studied with the alkaline and neutral comet assay for detection of ssb and dsb. Apoptosis was quantified morphologically and with DNA agarose gel electrophoresis. After H_2O_2 treatment, a dose-dependent increase of ssb was observed. Bleomycin treatment produced a moderate increase of ssb at lower concentrations and a striking increase of dsb at higher concentrations that coincided with the presence of apoptosis and DNA ladders. Gamma radiation initially induced the formation of ssb, and after three hours an increase of dsb in a dose-dependent manner. Apoptosis and DNA laddering appeared only 3 hours post-irradiation. The biomonitoring of DNA damage inflicted by antineoplastic agents can be easily performed with the comet assay and could be useful to monitor and modulate chemo- and radiotherapeutic regimes in cancer patients.

INTRODUCTION

Most antineoplastic treatments currently in use, such as chemotherapy and ionizing radiation, are known to induce various types of DNA damage including strand scissions.[1] On the other hand, these treatments are also known to induce apoptosis which, in its final stage, leads to a systematic destruction of the genome by different endonucleases.[2] The formation of distinct DNA fragments of 180 bp has been the biochemical hallmark of apoptosis that yields the classical DNA ladder pattern in agarose gel electrophoresis.[3] However, other types of DNA fragmentation have been

[a]Address for correspondence: Luis Benítez-Bribiesca, Oncological Research Unit, Oncology Hospital, National Medical Center, Av. Cuauhtémoc 330, 06725 México, DF, Mexico; 525-578-61-74 (voice/fax).

e-mail: luisbenbri@mexis.com

recently reported to occur during apoptosis owing to an ordered series of stages commencing with the production of DNA fragments of 300 Kb, which are then degraded into fragments of 50 Kb[4,5] of both double and single DNA strand breaks.[6] The 50 Kb fragments are further degraded, in some but not all cells, into small fragments (10–40 Kb) and release the small oligonucleosome fragments that are recognized as the characteristic DNA ladder in conventional agarose gels.[7,8] Therefore, endonucleolytic DNA fragmentation is frequently used as the sole criterion for the detection of apoptosis.[9] Nevertheless, no clear evidence has been presented so far that DNA fragmentation due to DNA-damaging agents plays a primary and causative role in apoptotic cell death; fragmentation appears more likely to be a consequence than a cause. Furthermore, recent reports have described that cell death with the key morphological features of apoptosis can be induced in the absence of detectable DNA fragmentation,[10,11] and that DNA fragmentation does not necessarily induce programmed cell death. Thus, DNA fragmentation during antineoplastic therapy can be due either to the direct effect of therapy on chromatin or to the indirect induction of apoptosis.

Apoptosis can be initiated by various pathological and physiological inducers such as hormones, toxins, chemotherapeutic drugs, carcinogens, ionizing radiation, and others.[12–15] Radio- and chemotherapy for treatment of malignant tumors cause interphase death by apoptosis, both in normal cells such as lymphocytes,[16–18] and in neoplastic cells, and DNA is considered to be the primary target for cell killing by these treatments[1,2]; nonetheless, DNA fragmentation can occur without the induction of apoptotic cell death. For instance the cytotoxic activity of the anticancer agent bleomycin (BLM) is usually believed to be correlated with its ability to introduce heterogenic DNA strand breaks.[19] Several molecular mechanisms have been proposed to explain the ability of BLM to induce single-and double-stranded breaks (ssb and dsb) and also to liberate free bases.[20–22] Single-strand breaks (ssb) result in double- strand breaks (dsb) that occur in close proximity and that are produced only at higher concentrations of bleomycin.[19] Recent studies describe the ability of BLM to associate with DNA and form a complex with ferrous ions.[24] Therefore, the genotoxic actions of BLM and ionizing radiation are quite similar, both producing DNA strand breaks in living cells.[24]

Radiation produces a broad spectrum of lesions including several chromosome aberrations, DNA ssb and dsb, interstrand- and protein-crosslinking, and damage to the DNA bases and sugars. While several of these lesions may be involved in cell killing by radiation, unrepaired dsb but not ssb are likely to lead to cell death.[25] Furthermore radiation and antineoplastic drugs stimulate a substantial production of oxygen–free radicals and it has been demonstrated that a major class of DNA damage is caused by reactive oxygen species leading to oxidized bases and to strand breaks.[26,27] Ames has suggested that significant oxidative damage occurs *in vivo* as a result of endogenous free radical attack and that it may contribute to the etiology of cancer.[28] Hence oxidative damage is thought to be a common pathway not only for most antitumor therapies but also for mutagenesis and carcinogenesis.[26,29] Malignant cells, however, exhibit a decreased ability to undergo apoptosis.[30] In fact, resistance to antineoplastic treatments results mainly from the selection of malignant cell clones incapable of activating the complex cell death mechanism. Their resistance is due principally to mutations or deletions in the genes that express proteins involved in this process, such as Bcl-2 and p53.[2,31] However, those clones overexpressing bcl-2 and exposed to DNA damaging agents can exhibit DNA strand breaks and a large number of mutations without inducing apoptosis. So far, many studies have suggested in many ways that DNA fragmentation could be considered the initial response to programmed cell death after chemo- or radiotherapy despite, the fact that no distinction has been clearly made among the different roles that ssb and dsb might play in triggering apoptosis.[32] Therefore, the measurement of DNA damage and the induction of apoptosis in cells subjected to the action of chemotherapeutic agents or ionizing radiation could be indicative of the different types of response to these treatments. In fact, many studies *in vitro* have demonstrated the effect of different chemotherapeutic agents and gamma radiation on DNA in different cell lines.[25] But, since anticancer treatments are not selective and affect tumor and normal cells as well, measurement of DNA damage and apoptosis induction in normal cells such as circulating lymphocytes could serve as suitable indicators for the modulation of antineoplastic therapy.

Several methods for DNA-strand-break analysis are known but require rather time-consuming and complicated procedures.[33–35] Single-cell gel electrophoresis (SCGE), or the "comet assay" introduced by Östling and Johanson,[36] has proven to be both sensitive and rapid in the detection of DNA-strand breaks in individual cells. The alkaline comet assay has been used for several years as an alternative method for the quantitation of oxidative damage to DNA.[26,37] Recently its application has been extended to the study of mutagenesis[38] and additionally, for the study of apoptosis.[39,40] Olive *et al.*[41,42] have shown that the microgel electrophoresis technique can be adapted for the detection of double-strand breaks under nondenaturing conditions. The advantage of studying DNA fragmentation in single cells and differentiating ssb from dsb is that it allows the assessment of the heterogeneity of this phenomenon within a cell population independently of the presence or absence of programmed cell death in the same population.[36,40]

The aim of this study was to assess DNA damage and apoptosis in human lymphocytes and rat thymocytes subjected in vitro to the action of two of the most commonly used anticancer agents, radiation and bleomycin, and to compare them with oxidative damage using a modified single cell gel electrophoresis method that distinguishes single- from double-DNA strand breaks.

MATERIALS AND METHODS

Chemicals

Bleomycin (BLM) 3 mg/ml was from Bristol; hydrogen peroxide (H_2O_2); sodium dodecyl sulfate (SDS), Trypan Blue, Triton X-100, Histopaque 1.077 g/ml and acridine orange (AO) were from Sigma, and RPMI, fetal bovine serum, streptomycin and penicillin were from Flow.

Cell Preparation

Peripheral blood lymphocytes were obtained from freshly obtained blood by means of the Boyum method[43] from healthy non-smoker male donors of 20 to

37 years of age. Cells were resuspended in PBS buffer and cultured at 2×10^6 cells/ml in RPMI 1640 culture medium supplemented with 10% bovine calf serum and penicillin (50 units/ml) and streptomycin (50 μg/ml) at 37°C in a humidified 5% CO_2/95% air atmosphere until used. Thymocytes were obtained from the thymus glands of male Sprague-Dawley rats 4–6-week-old (150–200 g) by mechanical disruption. Thymocytes were maintained at a density of 5×10^6 cells/ml in PBS, and were placed in 15 ml plastic conical tubes with PBS until subjected to radiation.

DNA-damaging Agents

Oxidative Damage

Lymphocytes, (10^5 cells/ml) were treated with H_2O_2 at concentrations ranging from 10 to 500 μM in PBS at 4°C for 5 minutes in order to avoid DNA repair. Cells were then washed and resuspended with PBS.

γ-Radiation

Thymocytes were subjected to ionizing radiation with ^{60}Co-gamma source (1.25 MEV) at a dose rate from 0.1 to 10 Gy/min while submerged in water at ambient temperature. After irradiation, the cell suspension was kept at 4°C to stop DNA repair. An aliquot of this cell suspension was processed immediatly as an initial sample (time zero). Cells were maintained in culture as described above; subsequently, 30-, 60-, 180-min, and 24-h samples of 5×10^6 cells were removed from the medium and processed.

Bleomycin

Bleomycin (BLM) was dissolved in PBS to give a stock solution of 6 mg/ml. To obtain final concentrations ranging from 75 to 1000 μg/ml an appropriate amount of the stock solution was added to and mixed with the cell culture medium. Incubation with BLM was carried out for 4 h at 37°C. Cultures were washed free of BLM and resuspended with PBS prior to the assays.

The Comet Assay

The alkali comet assay was slightly modified from the original from Collins *et al.*[26] as follows: 100 μl of 1% normal melting agarose (Sigma Type I) was added at 40°C to a 100 μl cold cell suspension of 10^4 cells at 4°C in PBS. The content was pipetted onto a fully frosted microscope slide positioned on an ice-cold surface and allowed to gel for 10 min. The slide was immersed in ice-cold lysis solution. The alkali assay lysis solution was composed of 2.5 M NaCl, 100 mM EDTA, 10mM Tris (1N NaOH was added to adjust pH to 10) plus 1% Triton X-100 immediately before use. Slides were kept in lysis at 4°C for 1 h and then placed for 20 min in a solution containing 0.3 N NaOH, 1mM EDTA, at pH 10 to allow DNA unwinding. Electrophoresis was performed at 35 V for 20 min. Slides were washed in neutralization buffer (0.4M Tris, pH 7.5) for 10 min postfixed in 1:1 (v/v) alcohol-acetone and then stained with 10 μg/ml acridine orange (AO) solution in Veronal-Acetate buffer pH 4.5. The neutral non-denaturating comet lysis solution was used as suggested by Olive *et al.*,[41] and consisted of 30 mM EDTA, 0.5% sodium-dodecylsulfate; pH was adjusted to 8.3 for 1h. Following lysis, slides were thoroughly rinsed with 90 mM boric acid, 2mM EDTA pH 8.3 with incubation for 1 h. Slides were electrophoresed at 75 V for 20 min, washed in PBS, fixed in 1:1 (v/v) alcohol-acetone and stained with the same AO solution. Observation was carried out under an epifluorescence microscope (Olympus AX70). The number of comets among 200 cells per slide were counted and expressed as a percentage. Comets in the alkali assay reveal DNA single strand breaks (ssb) and those in the neutral non-denaturating assay indicate DNA double strand breaks (dsb).

DNA Agarose Gel Electrophoresis

DNA was isolated from 10^6 cells by the method of Guntincich *et al.*[44] DNA was then pelletted by centrifugation at 13 000 × *g* for 10 min, washed with 1 volume of 70% ethanol, and dried under vacuum. DNA was resuspended with 10 mM Tris-HCl, 1 mM EDTA (pH 8.0) and supplemented with loading buffer (50% glycerol-0.05% bromophenol blue-0.05% xylene cyanol). Electrophoresis was performed for 2 h at 70 V in 1.5% agarose (Sigma Type I). Slab gels were submerged in TAE buffer (40 mM acetic acid, 1mM EDTA, pH 8). The gel was stained with 10 μg/ml AO during 10 min and washed three times with TAE buffer. DNA was visualized under a transilluminator with UV light for photography.

Analysis of Nuclear Morphology and Cell Viability

Cells were stained with AO and analyzed by fluorescent microscopy. The apoptotic morphology was scored according to the criteria suggested by Wyllie *et al.* and counted per 200 cells.[31] Cell viability was assessed using the Trypan Blue exclusion assay.

RESULTS

SCGE and DNA Electrophoresis

The comet assay performed on a single agarose layer and stained with acridine orange proved to be simple and sensitive, rendering well-defined bright green tails that could be counted easily. The neutral and alkali assays allowed a clear differentiation of the type of DNA breaks, because, with a short oxidative challenge, most comets were seen in the alkali range (ssb), as previously reported for DNA oxidative damage.[26,45] In contrast, the neutral assay produced few comets under the same conditions. In our experience AO staining gave a brighter green tail than the somewhat less-bright-red color obtained with ethidium bromide. The morphology of comets was similar in the alkaline and neutral assays (FIG. 1A). DNA agarose gel electrophoresis rendered the well-known patterns of normal non-fragmented DNA, the smeared and laddering images depending on the type of DNA-damage present (FIG. 1B).

Oxidative Challenge

Treatment of cells with H_2O_2 was purposefully maintained brief (5 min) with no recovery time to avoid repair and induction of apoptosis. This experiment was per-

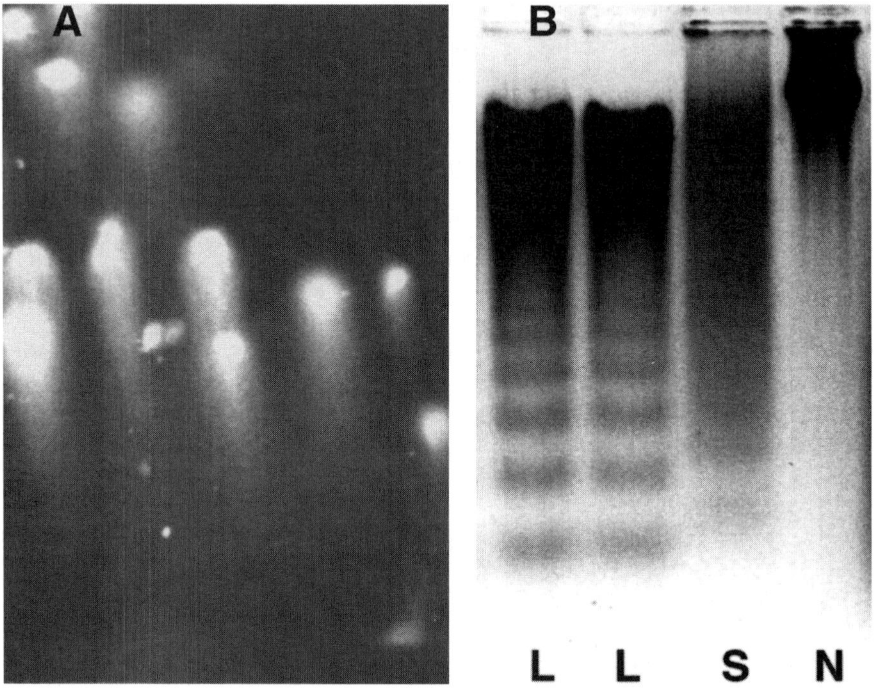

FIGURE 1. **A:** Typical comets in the SCGE stained with acridine orange. **B:** Typical DNA agarose gel electrophoresis patterns, (N) non-migrating normal DNA pattern, (S) smeared pattern, (L) ladder pattern (see corresponding graphs for variation in each experiment).

formed to serve as a comparison with the other two with bleomycin and γ-radiation, as abundant literature points to the fact that the action of these two treatments is mediated via generation of reactive oxygen species. Furthermore, the comet assay has been used in the alkali range to study oxidative DNA damage, but surprisingly little information is available concerning DNA oxidative damage with the neutral non-denaturating comet assay.[46] Untreated lymphocyte population contained 8 ± 2.0% cells with ssb, and 2 ± 1.1% with dsb. After treatment with H_2O_2 ssb increased in a dose-dependent manner, reaching 96 ± 4.0% at 500 μM. Only a minimum increase in dsb was observed, reaching only 19 ± 3.1% at 500 μM H_2O_2. No change in viability as defined by trypan blue was observed with concentrations up to 100 μM of H_2O_2, but at higher concentrations viability decreased to 51 ± 5.5% and cellular morphology revealed necrotic changes but no apoptosis. DNA gel electrophoresis showed a slightly smeared pattern at only the two highest concentrations (200 and 500 μM), but no ladder was produced (FIG. 2). In this experiment, it was clear that ssb and dsb could be quantitated separately in the absence of apoptosis with the comet assay, and that the main damage was that of ssb, even at toxic levels.

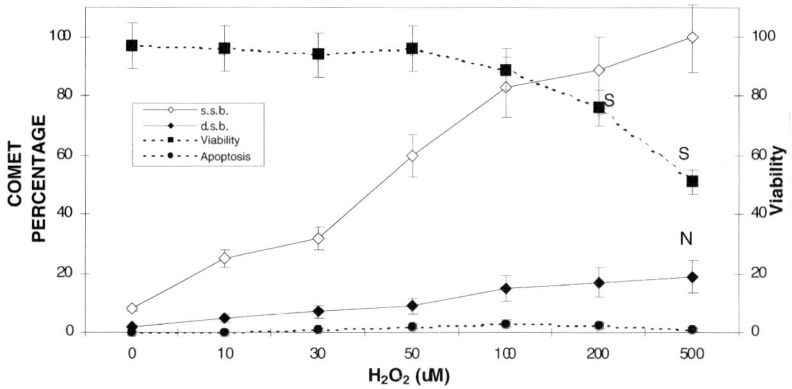

FIGURE 2. Treatment of human peripheral lymphocytes with increasing concentrations of H_2O_2 for 5 min. The number of comets at pH 10 (DNAss) increased in a dose-dependent manner up to 96 ± 4.0% at the highest dose. Comets at pH 8.3 (DNA ds) showed a moderate increase only at doses higher than 50 μM, reaching only 19 ± 3.1% at the highest concentration. Viability diminished progressively above 100 μM, and DNA agarose gel electrophoresis showed a smeared pattern only at 200 and 500 μM (s). Only occasional apoptotic figures were found throughout. (Each point is the mean of 10 determinations ± SEM).

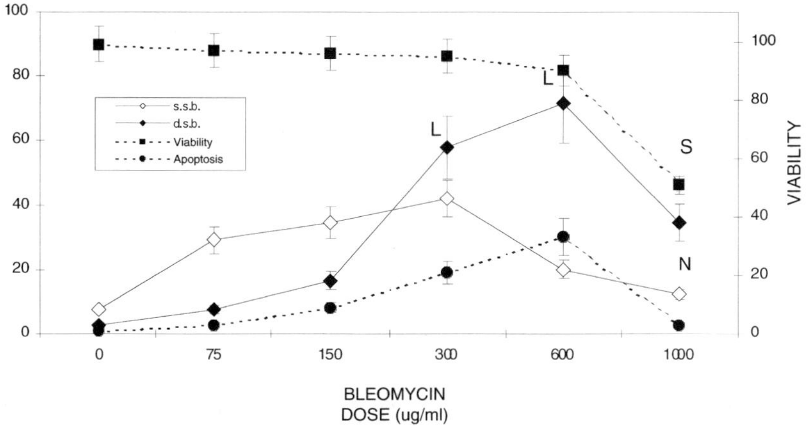

FIGURE 3. Treatment of human peripheral lymphocytes with increasing concentration of bleomycin (see METHODS). At between 75 and 300 μg/ml, the number of comets at pH 10 (DNAss) increased progressively to 42 ± 4.8%, but decreased to 12.2% at 1,000 μg/ml. The number of comets at pH 8.3 showed an opposite trend, increasing significantly at 150 μg/ml and reaching its maximum value at 600 μg/ml (58 ± 7.5%). A striking drop in viability was seen at 600 and 1000 μg/ml. DNA agarose gel electrophoresis showed clear ladder formation (L) at 300 and 600 μg/ml, but a smeared (S) pattern at the highest concentration. Apoptotic figures increased in a dose- dependent manner up to 600 μg/ml and were substituted by necrosis (N) at 1,000 μg/ml. (Each point is the mean of 5 determinations ± SEM).

Bleomycin Treatment

With the lowest bleomycin concentrations (75 and 150 μg/ml) there was a moderate increase in ssb (29 ± 4.5 and 34.5 ± 5.5%, respectively), and only a slight increase in dsb (7.5 ± 2.1% and 16.5 ± 3.2% respectively); at these concentrations, cell viability was not affected and no change in the DNA electrophoresis pattern occurred. From 300 to 600 μg/ml there was a striking increase of dsb (58 ± 7.5% and 71.5 ± 11.0%, respectively), while the ssb decreased (42 ± 4.8% and 20 ± 5.8%, respectively). Cell viability was slightly lower; DNA electrophoresis showed a ladder pattern, and about 33 ± 5.2% of cells had apoptotic features. In contrast, with the highest concentration of bleomycin (1 mg/ml) ssb and dsb breaks diminished (34.5 ± 5.2% and 12.2 ± 2.2%, respectively); viability was decreased to 49 ± 10.3% and DNA electrophoresis showed a smeared pattern. Cell morphology revealed necrotic changes in approximately 60% of the cells. These cells appeared swollen or burst but rendered no DNA tail in the comet assay (FIG. 3). It should be noted that increase of ssb does not coincide with apoptosis, with DNA laddering, or with decreased viability as opposed to the increase of dsb, which, at their highest levels always coincide with DNA laddering and apoptotic morphology.

Ionizing Radiation

Immediately after irradiation (Time 0), the number of ssb and dsb increased progressively in relation to the amount of Gys applied. There was, however, a twofold predominance of ssb. No change was noted in cell viability, morphology or DNA electrophoresis. (FIG. 4a). This suggests that γ-radiation can immediately induce severe DNA damage, mainly ssb that do not affect cell viability and that cannot be detected in DNA electrophoresis, Thirty minutes after irradiation, there was a moderate decrease of DNA strand breaks, principally ssb, probably due to DNA repair. Cell viability decreased progressively with high radiation doses and a smeared DNA electrophoresis pattern appeared only at 10 Gy 60 min after irradiation. No morpho-

FIGURE 4. Rat thymocytes subjected to various doses of ionizing radiation and post-cultured at different times: **a:** At zero time after irradiation, a dose-dependent increase in ssb was seen, almost twofold higher than dsb. Viability and DNA agarose gel electrophoresis remained without significant variation. Apoptosis did not vary from the basal figures; **b:** Thirty min after irradiation, the ratio of ssb and dsb was equalized at lower radiation doses, although ssb maintained higher values (almost twofold) with larger radiation doses; **c:** Sixty min after radiation, ssb and dsb had equivalent values, although clearly lower than those found at zero time. Cell viability was already lower (less than 80%) and DNA gel electrophoresis showed a smeared pattern (S) exclusively at the highest dose. No increase in the amount of apoptosis was noted; **d:** Three h after radiation, the number of dsb increased in a dose-dependent manner while the ssb remained along the same figures 60 min after radiation. Cell viability dropped further, and DNA agarose gel electrophoresis showed faint ladders (fL) at 0.87, 1.75, and 2.5 Gy, and clear ladders (L) at 5 and 10 Gy. Apoptotic figures increased moderately in relation to dose; **e:** Twenty-four h after radiation, ssb increased moderately, while dsb reached its maximum values in relation to dose. Cell viability was severely decreased and DNA agarose gel electrophoresis showed a smeared pattern (S) beginning already at 0.1 and 0.21 Gy, a faint ladder (fL) at 0.43 Gy, and well-defined ladders (L) at all remaining doses. The highest number of apoptotic figures occurred at this time. (Each point is the mean of four determinations ± SEM).

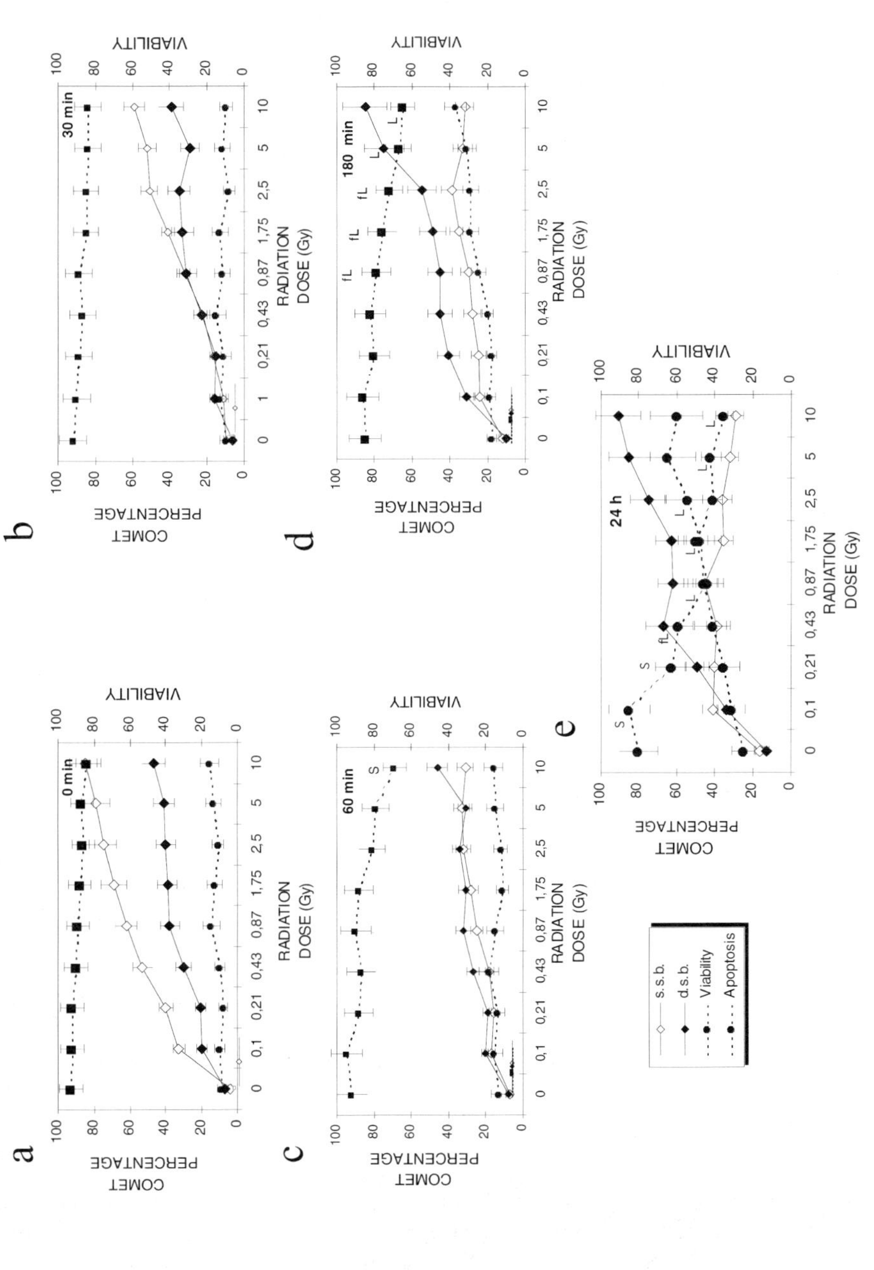
a
b
c
d
e
0 min
30 min
60 min
180 min
24 h
COMET PERCENTAGE
VIABILITY
RADIATION DOSE (Gy)
0 0,1 0,21 0,43 0,87 1,75 2,5 5 10
s.s.b.
d.s.b.
Viability
Apoptosis

logical change indicating apoptosis was detected. (FIG. 4b,c). Three hours after irradiation, there was a clear shift in the relation of ssb to dsb also related to dosage. Above 0.21 Gys, there was a greater amount of dsb almost threefold at higher doses and viability was then reduced to 64 ± 11.2%. The DNA electrophoresis pattern showed well-formed ladders at 5 and 10 Gy, and cell morphology showed abundant apoptotic cells (FIG. 4d). Twenty four hours after irradiation, there was a marked dose-dependent increase of dsb up to 91 ± 6.6% at 10 Gy; in contrast ssb increased only slightly. At this time 59 ± 12.2% apoptotic cells were found and DNA electrophoresis showed ladders from 0.43 Gy on to the highest dose. Viability was also decreased in a dose-dependent manner down to 35 ± 12.0% at the highest radiation dose (See FIG. 4e). Untreated thymocytes cultured for 24 h as controls maintained a high viability index (79 ± 8.3%), ssb 16 ± 6.1%, dsb 13 ± 1.4%, and apoptotic cells 21 ± 5.1%.

DISCUSSION

Our results show that all agents used in this study were capable of inducing DNA strand breaks. There were, however, striking differences in the type of breaks produced by these damaging agents. The type of DNA strand breaks produced after H_2O_2 treatment were predominantly ssb. Bleomycin treatment induced a heterogenous mixture of DNA ssb and dsb, depending on dose, and γ-radiation produced also different types of strand breaks depending on recovery time and dose applied. This suggests that the kinetics of DNA scissions differ with each treatment and merit separate discussion.

H_2O_2

Our results are concordant with previous studies performed with different methodologies that demonstrate that in DNA oxidative damage, the spectrum of oxidation products formed includes various types of strand breaks, base-less sugars and oxidized bases, principally 7,8-dihydro-8-oxo-guanine.[47] Collins *et al.* have shown dose-dependent strand breakage immediately after H_2O_2 treatment using SCGE, although the authors did not distinguish between ssb and dsb.[26] Other authors have shown that oxidative damage in cell lines and thymocytes treated with H_2O_2 and other oxidative agents can induce apoptosis, which could result in double DNA strand breaks and the formation of a ladder in DNA agarose gel electrophoresis if sufficient time is allowed for this process to occur. These studies differ from ours in that the detection of DNA strand breaks was studied only after apoptotic cell death had occurred; therefore, it is not clear whether H_2O_2 treatment or the apoptotic process was responsible for the DNA damage observed.[48,49] In experiments not shown in this report, a 2-h challenge of peripheral lymphocytes with 100 μM H_2O_2 produced 79% ssb, but also 44% dsb. At this time, a few apoptotic figures appeared, although no DNA ladder pattern was observed. Marini *et al.*[10] demonstrated that H_2O_2 was an effective inducer of nuclear fragmentation at levels as low as 10 μM, and that these effects increased as a function of the oxidant concentration. The authors did not find the formation of oligonucleosomal DNA fragments, but, using pulse field gel electrophoresis, they were able to demonstrate the formation of large-scale 50 Kb double stranded DNA fragments. In their experiments, treatment with H_2O_2 ranged from 1 h to 6 h and, in some, a post-incubation time of 3 h was permitted. Under those conditions, the DNA damage observed could not be attributed only to the direct oxidative action of H_2O_2 but rather, additionally to the occurrence of apoptosis as a consequence of the oxidative insult. In our experiments, the time after H_2O_2 treatment was too short to expect apoptotic changes and ladder formation. In fact, our purpose was to ascertain the direct effect of oxidative damage on cellular DNA and to see how efficiently it could be detected with the comet assay. Our results demonstrate that H_2O_2 treatment is capable of inducing ssb and dsb before any detectable change of apoptosis takes place. Forrest *et al.*[49] were able to induce apoptosis in thymocytes subjected to H_2O_2 treatment and Kane *et al.*[50] and Hockenbery *et al.*[51] have proposed that apoptosis may be mediated by oxidative pathways. Additionally, these groups noted that many apoptosis inducing agents, such as ionizing radiation and chemotherapeutic agents have in common the generation of excess reactive oxygen species. On the other hand, Bump *et al.*[52] have shown that treatment of various cell lines with H_2O_2 at concentrations similar to those generated after radiation did not induce apoptosis. In other experiments, Meyer *et al.*[53] were unable to demonstrate that apoptosis could be mediated via oxygen-generated free radical formation, although Sanfer *et al.*[54] showed that reactive oxygen species can be part of the triggering mechanism of programmed cell death. These different opinions could be explained by assuming that sensitivity to the induction of apoptosis could vary under different conditions and in cell lines. Martin and Cotler[55] observed that when HL60 were treated with 15 μM H_2O_2, apoptosis appeared after 4 h, reaching its peak at 8 h. In the treatment of those cells with 400 μM H_2O_2 (known to be toxic) no apoptosis occurred and viability decreased below 20% at 2 h. It is, however, important that our results demonstrate a direct effect of H_2O_2 on normal lymphocyte DNA prior to any activation of an apoptotic program. Nevertheless, the generation of reactive oxygen species seems to be a common pathway both for radio- and chemotherapy, and the extent of DNA damage induced by these molecules in normal cells is known to be mutagenic, carcinogenic and may contribute to the aging process.[26,27] It is, therefore, of paramount importance that these types of damage be assessed in patients subjected to chemo- or radiotherapy to estimate the possible mutagenic effect induced in normal cells, such as peripheral lymphocytes. Our results show that ssb induced by oxidative damage can be detected in single cells and could be used to monitor the effect of antineoplastic treatments independent of the induction of apoptotic cell death.

Bleomycin

Bleomycin cytotoxicity results from its interaction with DNA, binding both with double- and single-stranded DNA, producing single- and double- strand scissions. It is known that dsb result from a number of ssb occurring in close proximity, and that they are produced at higher concentrations of bleomycin.[22] Several molecular mechanisms have been proposed to explain the ability of BLM to induce single- and double-stranded breaks, and also to liberate free bases. BLM forms complexes with DNA and ferrous ions and, through the action of ferrous oxidase activity, highly reactive oxygen species are liberated in close proximity to DNA.[56] Therefore the genotoxic actions are assumed to be similar, producing DNA strand breaks by means of free radical attack.[57] These characteristics make this substance a suitable drug against a number of neoplasms.

Our studies show that, at low BLM concentrations ssb predominate, but at above 150 μg/ml, the number of dsb increases manyfold, while ssb show only a moderate increase. This was further confirmed by DNA agarose gel electrophoresis, because, at 300 and 600 μM concentration, when the highest number of comets in the neutral assay were found, a DNA ladder was evident. At the highest concentrations used (1 mg/ml) there was a marked drop of the amount of dsb coinciding with a diminution of viability to about 50%, and a smeared pattern in DNA electrophoresis that is quite characteristic of necrosis. Apoptotic morphology was evident in the higher concentrations, except at the highest, where necrotic cells were predominant. It follows that bleomycin induces double DNA strand breaks and apoptosis only at concentrations ranging from 150 to 600 μg/ml, and necrosis at higher doses. Using S.C.G.E., Östling and Johanson[58] showed a heterogeneous formation of comets after treating Chinese hamster ovary cells with 75 to 30 μg/ml of BLM, but no distinction between ssb and dsb was attempted. Item and Burkatt[59] also found a marked heterogeneous induction of DNA strand breaks beginning at concentrations as low as 10 μg/ml and 5 min post-incubation, that appears to be quite different from the dose-dependent uniform production of the DNA strand breaks occurring after ionizing radiation. No information is, however, provided in regard to the type of DNA strand breaks produced. Sausville *et al.*[60] have shown that bleomycin at 10 μg/ml concentration had little effect on DNA during a 15-min incubation, whereas, at a drug concentration of 50 μg/ml, breakage of DNA occurs. These studies were not performed in intact cells, and no mention of the type of DNA damage was made. Antioxidants do not inhibit the formation of strand breaks by BLM (data to be published elsewhere). This phenomenon appears different from that observed after oxidative damage.

Radiation

Ionizing radiation–induced DNA damage has been studied extensively by means of different methods that allow the distinction of various types of DNA strand breaks. Östling and Johnson[36] were the first to point out the advantages of microelectrophoresis to study DNA strand breaks induced by ionizing radiation in mammalian cells. The refinements of the technique introduced by Olive *et al.*[40,41] demonstrated the possibility of detecting ssb and dsb using the alkaline or the neutral assay, respectively. In a different study, Olive *et al.*,[40] described a dose-dependent increase in apoptosis in TK-6 cells measured 24 h after irradiation by means of the alkali and neutral lysis comet assay, assuming that comets might indicate apoptosis. The authors were able to detect apoptosis as early as 2 h after exposure to doses as low as 0.5 Gy. Our results agree with those of Olive *et al.* in that there is a dose-dependent increase in the number of comets after irradiation, but differ on several other points. The first difference is that the number of apoptotic cells does not correspond to the number of comets in both the neutral and alkali assays, suggesting that the presence of ssb and dsb do not necessarily mean an early stage of apoptotic cell death. The second difference is that the type of DNA strand breaks changes with post-irradiation (recovery) time. In the early stages (0 to 30 min) ssb outnumber the dsb and, after 60 min, dsb predominate, becoming almost 80% at 10Gy after 24 h (compare panels a to e in FIG. 4). The shift in the ratio of single vs. double DNA breaks can be explained in part by assuming that ionizing radiation produces a sudden accumulation of reactive oxygen species resulting in single-stranded DNA breaks. After one-h recovery time some of this damage is repaired while the programmed cell death mechanism is being activated. This could explain why at this time the number of ssb decreased and some dose-dependent increase of dsb was seen (FIG. 4c), but still no apoptosis or loss of cell viability was evident. In fact, apoptotic figures began to appear at 3 h post-irradiation, when a marked increase of dsb took place. It is also well established that ionizing radiation induces DNA dsb, and that this lesion is critical for the induction of cell death.[1] The dose-response for DNA dsb could vary markedly between cell lines, and could be an important cause of differences in radiosensitivity.[1,61] The appearance of dsb soon after irradiation is probably due to a direct radiation effect, but those occurring after 3 h could be best explained by the action of endonucleases as a consequence of the activation of the program for apoptotic cell death. Liu *et al.*[62] observed a dose-dependent increase in apoptosis in mice thymocytes when subjected to doses above 0.5 Gy, but not with doses between 0.2 Gy and 0.5 Gy. They found that low radiation doses induce over expression of Bcl-2 protein 12 to 24 h post-irradiation but, at higher doses up to 4 Gys, this phenomenon was not observed. Over-expression of this antiapoptotic protein with low-dose radiation challenge could explain the relatively lower level of apoptosis and dsb found with similar doses in our study (FIG. 4d).

Woudstra *et al.*[63] did not find an altered ratio of ssb to dsb in three human tumor cell lines with different radiosensitivities and subjected to the same amount of ionizing radiation. They cannot explain the differences of radiosensitivity by means of the amount or type of DNA breakage. Variations in damage induction may also be related to distinct variation in chromatin structure among cells lines.[64] These specific chromatin factors not only influence radiosensitivity, but also DNA migration in the comet assay.

Our studies show the kinetics of radiation-induced DNA strand breaks in a dose- and time-dependent scale. Between 0 and 30 min post-irradiation, ssb predominate; at 60 min, a marked reduction of ssb and a moderate increase of dsb at the highest dose occur and between three and twenty-four hours, dsb predominate. The dose-dependent increase of apoptosis takes place after 3 hours post irradiation but does not correlate quantitatively with the number of cells exhibiting dsb. Thus, the presence of DNA strand breaks cannot be taken as an indication of apoptosis, although most cells with severely damaged chromatin and numerous DNA strand breaks certainly end up eventually with apoptotic death.

In summary, this study shows the importance and feasibility of measuring DNA strand breaks in normal cells induced by the most commonly used antineoplastic treatments. It has long been believed that most anti-cancer therapies induce tumor cell necrosis. It is, instead, becoming clear that these treatments induce apoptotic cell death, and that most of these treatments are DNA damaging agents. These agents at low levels of exposure might mimic tumor promotion and stimulate cell proliferation,[1] however, only marginally high levels can induce apoptosis.[65] Cells in which bcl-2 is over-expressed or cells with defective or deleted p53 will be resistant to apoptosis and more prone to acquire new mutations. Therefore current treatments against malignancies, such as radiation or chemotherapeutic drugs, may ultimately result in the selection of more malignant clones. The exposure to different doses to DNA-damaging agents could have opposite effects. Because the probability of efficient, error-free repair decreases with extensive DNA damage, removal of heavily compromised cells is essential to prevent a potential proliferation of mutated clones. The assessment of DNA strand breaks could serve as a reliable indicator of the extent of DNA damage inflicted by any given anti-cancer treatment. The comet assay provides a reliable and sensitive method for estimating of DNA damage, and can be used for assessing the effect of antineoplastic treatments in both malignant and normal cells. The biomonitoring of DNA damage in normal cells is of particular importance, because chemo- and radiotherapy are not cancer-specific. It is known that immune cells such as lymphocytes and thymocytes are quite sensitive to these treatments, and that, in many cases, immune depletion appears during the course of anti-cancer treatments. The assessment of DNA damage and apoptosis in normal cells such as peripheral lymphocytes during treatment could be of help for modulating chemo- or radiotherapy.

ACKNOWLEDGMENTS

This work was supported by grant No. 264100-5-25781M from CONACYT, México.

REFERENCES

1. Harms, R.M., P. Nicotera & I.R. Radford. 1996. Radiation induced apoptosis. Mutation Res. **366:** 171–179.
2. Benítez-Bribiesca, L. 1998. Assessment of apoptosis in tumor growth: importance in clinical oncology and cancer therapy. *In* When Cells Die. R.A. Lockshin, Z. Zakeri & J.L. Tilly, Eds. : 453–482. Wiley-Liss, Inc. New York.
3. Wyllie, H.A., J.F. Kerr & A.R. Currie. 1980. Cell death: the significance of apoptosis. Int. Rev. Cytol. **68:** 251–325.
4. Oberhammer, F., J.W. Wilson, C. Dive, *et al.* 1993. Apoptotic death in epithelial cells: cleavage of DNA to 300 and/or 50 kb fragments prior to or in the absence of internucleosomal fragmentation. EMBO J. **12:** 3679–3684.
5. Brown, G.D., X.M. Sun & G.M. Cohen. 1993. Dexamethasone-induced apoptosis involves cleavage of DNA to large fragments prior to internucleosomal fragmentation. J. Biol. Chem. **268:** 3037–3039.
6. Peitsch, C.M., C. Muller & J. Tschopp. 1993. DNA fragmentation during apoptosis in caused by frequent single-strand cuts. Nucleic Acids Res. **21:** 4206–4209.
7. Walker, R.P., L. Kokileva, J. LeBlanc, *et al.* 1993. Detection of the initial stages of DNA fragmentation in apoptosis. Biotechniques **15:** 1032–1040.
8. Bortner, C.D, N.B.E. Oldenbur & J.A. Cidlowski. 1995. The role of DNA fragmentation in apoptosis. Trends Cell Biol. **5:** 21–26.
9. Walker, R.P., M. Valerie, L. Boleslaw, *et al.* 1994. Endonuclease activities associated with high molecular weight and internucleosomal DNA fragmentation in apoptosis. Exp. Cell. Res. **213:** 100–106.
10. Marini, M., D. Musiani, P. Sestili, *et al.* 1996. Apoptosis of human lymphocytes in the absence or presence of internucleosomal DNA cleavage. Biochem. Biophys. Res. Commun. **229:** 910–915.

11. COHEN, M.G., X.M. SUN, R.T. SNOWDEN, *et al.* 1992. Key morphological features of apoptosis may occur in the absence of internucleosomal DNA fragmentation. Biochem. J. **286:** 331–334.
12. SCHWARTZMAN, A.R. & J.A. CIDLOWSKI. 1993. Apoptosis: the biochemistry and molecular biology of programmed cell death. Endocrinol. Rev. **14**: 133–151.
13. KAUTMANN, H.S. 1989. Induction of endonucleolytic DNA cleavage in human acute myelogenous leukemia cells by etoposide, camptothecin, and other cytotoxic anticancer drugs: a cautionary note. Cancer Res. **49:** 5870–5878.
14. NIETO, M.A. & A.R. LÓPEZ. 1989. IL-2 protects T lymphocytes from glucocorticoid-induced DNA fragmentation and cell death. J. Immunol. **143:** 4166–4170.
15. IKUSHIMA, T., H. ARITOMI & J. MORISITA. 1996. Radioadaptive response: efficient repair of radiation-induced DNA damage in adapted cells. Mutation Res. **358:** 193–198.
16. SUN, M.X., S.D. DINSDALE, M.G. ORMEROD, *et al.* 1994. Changes in nuclear chromatin precede internuclesomal DNA cleavage in the induction of apoptosis by etoposide. Biochem. Pharmacol. **47:** 187–195.
17. CATHCHPOOLE, R.D. & B.W. STEWART. 1995. Formation of apoptotic bodies is associated with internucleosomal DNA fragmentation during drug-induced apoptosis. Exp. Cell. Res. **216:** 169–177.
18. COLIN, F.A., J.E. LOWE, S.A. HARCOURT, *et al.* 1993. Hypersensitivity of human lymphocytes to UV-B and solar irradiation. Cancer Res. **53:** 609–614.
19. DE VITA V.T., S. HELLMAN & S.A. ROSENBERG. 1997. Cancer Principles and Practice of Oncology. Lippincott-Raven Pub. Philadelphia-New York, pp.495–497.
20. GREENE, J.J., C.W. DIEFFENBACH & P.O. TS'O. 1978. Bleomycin causes release of nucleosomes from chromatin and chromosomes. Nature **271:** 83–84.
21. MUBASHER, E.D. & T.J. JORGENSEN. 1995. Deletions at short direct repeats and base substitutions are characteristic mutations for bleomycin-induced double- and single-strand breaks respectively, in a human shuttle vector system. Nucleic Acids Res. **23:** 3224–3230.
22. LAZO, J.S., J.M. SEBTI & J.H. SCHELLENS. 1996. Bleomycin. *In* Cancer Chemotherapy and Biological Response Modifiers Annual. H.M. Pinedo, D.L. Longo & B.A. Chabner, Eds. : 39–47. Elsevier. New York.
23. ABSALON, J.M., J.W. KOZARICH & J. STUBBE. 1995. Sequence-specific double-strand cleavage of DNA by Fe-Bleomycin. 1. The detection of sequence-specific double-strand breaks using hairpin oligonucleotides. Biochemistry **34:** 2065–2075.
24. CHATTERJEE, A., M. J. RAMAN & B. MOHAPATRA. 1989. Potentiation of bleomycin-induced chromosome aberrations by the radioprotector reduced glutathione. Mutation Res. **214:** 207–213.
25. WOUDSTRA, C.E., J.F. BRUNSTING, J.M. ROESINK, *et al.* 1996. Radiation induced DNA damage and damage repair in three human tumour cell lines. Mutation Res. **362:** 51–59.
26. COLLINS, R.A., S.J. DUTHIE & V.L. DOBSON. 1993. Direct enzymic detection of endogenous oxidative base damage in human lymphocyte DNA. Carcinogenesis **14:** 1733–1735.
27. COLLINS, R.A., M. DUSINSKÁ, C.M. GEDIK, *et al.* 1996. Oxidative damage to DNA: Do we have reliable biomarker? Environ. Health Perspect. **104:** 465–469.
28. AMES, B.N. 1983. Dietary carcinogens and anticarcinogens. Science **221:** 1256–1264.
29. WAGNER, J R., C. HU & N.B. AMES. 1992. Endogenous oxidative damage of deoxycytidine in DNA. Proc. Natl. Acad. Sci. USA **89:** 3380–3384.
30. HOFFMAN, B. & D.A. LIEBERMANN. 1994. Molecular controls of apoptosis differentiations/growth arrest primary response genes, proto-oncogens and tumor suppressor genes as positive and negative modulators. Oncogene **9:** 1807–1812.

31. KERR, F.J. 1994. Apoptosis. Its significance in cancer and cancer therapy. Cancer **73:** 2013–2026.
32. WOJCIK, A. 1996. Analysis of DNA damage recovery processes in the adaptive response to ionizing radiation in human lymphocytes. Mutagenesis **11:** 291–297.
33. KHON, K.W. 1991. Principles and practice of DNA filter elution. Pharmacol. Ther. **49:** 55–77
34. HALLIWELL, B. 1992. The measurement of oxidative damage to DNA by HPLC and GC/MS techniques. Free Radical Res. Commun. **16:** 75–87.
35. HENGSTLER, G.J., J. FUCHS & B. TANNER. 1997. Analysis of DNA single-strand breaks in human venous blood: A technique which does not require isolation of white blood cells. Enviro. Mol. Mutagenesis **29:** 58–62.
36. ÖSTLING, O. & K.J. JOHANSON. 1984. Microelectrophoresis study of radiation-induced DNA damages in individual mammalian cells. Biochem. Biophys. Res. Commun. **123:** 291–298.
37. ANDERSON, D., T.W. YU & B.J. PHILLIPS. 1994. The effect of various antioxidants and other modifying agents on oxygen-radical-generated in human lymphocytes in the COMET assay. Mutation Res. **307:** 261–271.
38. SPEIT, G., S. HANELT, R. HELBIG, *et al.* 1996. Detection of DNA effects in human cells with the comet assay and their relevance for mutagenesis. Toxicol. Lett. **88:** 91–98.
39. FAIRBAIRN, W.D., K.G. CARNAHAN, R.N. THWAITS, *et al.* 1994. Detection of apoptosis induced DNA cleavage in scrapie-infected sheep brain. FEMS Microb. Lett. **115:** 341–346.
40. OLIVE, L.P., G. FRAZER & J. BANÁTH. 1993. Radiation-induced apoptosis measured in TK6 human B lymphoblast cells using the comet assay. Radiat. Res. **136:** 130–136.
41. OLIVE, L.P., D. WLODEK & J.P. BANÁTH. 1991. DNA double-strand breaks measured in individual cells subjected to gel electrophoresis. Cancer Res. **51:** 4671–4676.
42. OLIVE, L. P. & J.P. BANÁTH. 1995. Sizing highly fragmented DNA in individual apoptotic cells using the comet assay and a DNA crosslinking agent. Exp. Cell. Res. **221:** 19–26.
43. BOYUM, A. 1964. Separation of white blood cells. Nature **204:** 793.
44. GUSTINCICH, S. 1991. A fast method for high-quality genomic DNA extraction from whole human blood. Biotechniques **11:** 256–260.
45. O'NEILL, L.K., W.D. FAIRBAIRN & D.M. STANDING. 1993. Analysis of single-cell gel electrophoresis using laser-scanning microscopy. Mutation Res. **319:** 129–134.
46. MCKELVEY, V.J., M.H. GREEN, P. SCHMEZER, *et al.* 1993. The single cell gel electrophoresis assay (comet assay): a European review. Mutation Res. **288:** 47–63.
47. SHIGENAGA, K.M., E.N. ABOUJAOUDE, Q. CHEN & B.N. AMES. 1994. Assays of oxidative DNA damage biomarkers 8-oxo-2′-deoxyguanosine and 8-Oxoguanine in nuclear DNA and biological fluids by high-performance liquid choromatography with electrochemical detection. Methods Enzymol. **234:** 16–32.
48. OLIVE L.P., J.P. BANÁTH & R.E. DURAND. 1990. Heterogeneity in radiation-induced DNA damage and repair in tumor and normal cells measured using the "Comet" assay. Radiation Res. **122:** 86–94
49. FORREST V.J., Y.H. KANG & D.E. MACCLAIN. 1994. Oxidative stress-induced apoptosis prevented by Troxol. Free Radical Biol. Med. **16:** 675–684.
50. KANE, S.A., A. NATRAJAN & S.M. HECHT. 1994. On the role of the bithiazole moiety in sequence-selective DNA cleavage by Fe-bleomycin. J. Biol. Chem. **269:** 10899–10904.
51. HOCKENBERY, D.M., G. NUNEZ, C. MILLIMAN, *et al.* 1990. Bcl-2 is an inner mitochondrial membrane protein that blocks programmed cell death. Nature **348:** 334–336.
52. BUMP E.A., S.J. BRAUNHUT & S.T. PALAYOOR. 1994. Novel concepts in modification of radiation sensitivity. Int. J. Radiat. Oncol. Biol. Phys. **29:** 249–253.

53. MEYER, M., R. SCHRECK & P.A. BAEUERLE. 1993. H_2O_2 and antioxidants have opposite effects on activation of NK.KB and AP-1 in intact cells: AP-1 as secondary antioxidant responsive factor. EMBO J. **12:** 2005–2015.
54. SANFER, H., G.B, BULKLEY & J.L. CAMERON. 1984. The role of oxygen-derived free radicals in the pathogenesis of acute pancreatites. Ann. Surg. **200:** 405–410.
55. MARTIN J.S. & T.J. COTLER. 1994. Apoptosis of human leukemia: induction, morphology, and molecular mechanism. *In* Apoptosis II: The Molecular Basis of Apoptosis in Disease. L.D. Tomei & F.O. Cope, Eds.: 185–229. Cold Spring Harbor Lab. Press. New York.
56. CASPARY, W.J., D.A. LANZO & C. NIZAK. 1982. Effect of deoxiribonucleic acid on the production of reduced oxygen by bleomycin and iron. Biochemistry **21:** 334–338.
57. IQBAL, Z.M., K.W. KOHN, R.A. EWING, *et al.* 1976. Single strand scission and repair of DNA in mammalian cells by bleomycin. Cancer Res. **36:** 3834–3837.
58. ÖSTLING, O. & K.J. JOHANSON. 1987. Bleomycin, in contrast to gamma irradiation, induces extreme variation of DNA strand breakage from cell to cell. Int. J. Radiat. Biol. **52:** 683–691.
59. ITEM, A.M. & W. BURKATT. 1995. Resolution of DNA damage at the single-cell level shows largely different actions of x-ray and bleomycin. J. Histochem. Cytochem. **43:** 229–235.
60. SAUVILLE, A.E., E.W. STEIN, J. PEISACH, *et al.* 1987. Properties and products of the degradation of DNA by bleomycin and iron (II). Biochemistry **17:** 2746–2755.
61. RADFORD, I.R. & T.A MURPHY. 1994. Radiation response of mouse lymphoid and myeloid cell lines. Part III. Different signals can lead to apoptosis and may influence sensitivity to killing by DNA double-strand breakage. Int. J. Radiat. Biol. **65:** 229–239.
62. LIU, S.Z., V.C ZHANG, Y. MU, *et al.* 1996. Thymocyte apoptosis in response to low-dose radiation. Mutation Res. **358:** 185–191.
63. WOUDSTRA, E.C., F.J. BRUNSTING, M.J. ROESINK, *et al.* Radiation induced DNA damage and damage repair in three human tumour cell lines. Mutation Res. **362:** 51–59.
64. OLIVE, P.L., J. HILTON & R.E. DURAND. 1986. DNA conformation of chinese hamster V79 cells and sensitivity to ionizing radiation. Radiation Res. **107:** 115–124.
65. DYPBUKT, J.M., M. ANKARCRONA, M. BURKITT, *et al.* Different prooxidant levels stimulate growth, trigger apoptosis, or produce necrosis of insulin-secreting RINm5F cells. The role of intracellular polyamines. J. Biol. Chem. **269:** 30553–30560.

Tumor Cells Utilize Multiple Pathways to Down-modulate Apoptosis

Lessons from a Mouse Model of Islet Cell Carcinogenesis

JEFFREY H. HAGER AND DOUGLAS HANAHAN[a]

Department of Biochemistry and Biophysics and The Hormone Research Institute, University of California at San Francisco, San Francisco, California 94143-0534, USA

ABSTRACT: Apoptosis, the process of programmed cell death, plays a critical role in many normal and pathological (disease) processes.[1] In normal tissues, apoptosis functions in the homeostatic maintenance of proper tissue and organ size by eliminating aged cells to offset the birth of new cells that arise by mitosis. In disease, apoptosis can affect the pathological process is two disparate ways. There are diseases that have too much apoptosis such as autoimmune diabetes and Alzheimer's, or those that have too little apoptosis, such as cancer. This review will focus on the latter and, more specifically, detail and summarize some important lessons learned about apoptosis and cancer from studying a transgenic mouse model of islet cell carcinoma, RIP-Tag, as outlined below.

INTRODUCTION: THE RIP-TAG MODEL OF MULTISTAGE TUMORIGENESIS

The RIP1-Tag transgene directs the expression of SV40 T-antigen (Tag) to β-cells of the endocrine pancreas within multi-focal nodules called the islands of Langerhans.[2] T-antigen is a potent oncoprotein that exerts its oncogenic effects by binding to and inactivating the proteins of the tumor suppressor genes p53 and Rb.[3] The RIP-Tag tumor phenotype is fully penetrant, with all animals developing tumors by 10–12 weeks of age, and each tumor progressing through a similar histological pathway. The multi-focal nature of this organ has allowed for an accurate identification, quantification and physical isolation of the distinct stages of this pathway (FIG. 1). In short, Tag expression commences in the developing pancreas at embryonic day 9 and is maintained in all ~400 islets for the life of the animal.[4] Interestingly, the expression of this potent oncogene is without apparent effect until 4–5 weeks of age, when the relatively quiescent β-cells of individual islets begin proliferating.[5] Although this stage is routinely referred to as hyperplastic (islet), in fact, these islets exhibit hallmarks of carcinoma *in situ* (CIS), as evidenced by an increase in the nuclear/cytoplasmic ratio and cell density. From this hyperplastic state emerge angiogenic islets (7–9 weeks) wherein the developing tumor nodule activates the growth of new

[a]Address for correspondence: University of California, Hormone Research Institute, 1090 HSW, 513 Parnassus Avenue, San Francisco, CA 94143-0534, USA. 415-476-8277 (voice); 415-71-3612 (fax).
e-mail: dh@biochem.ucsf.edu

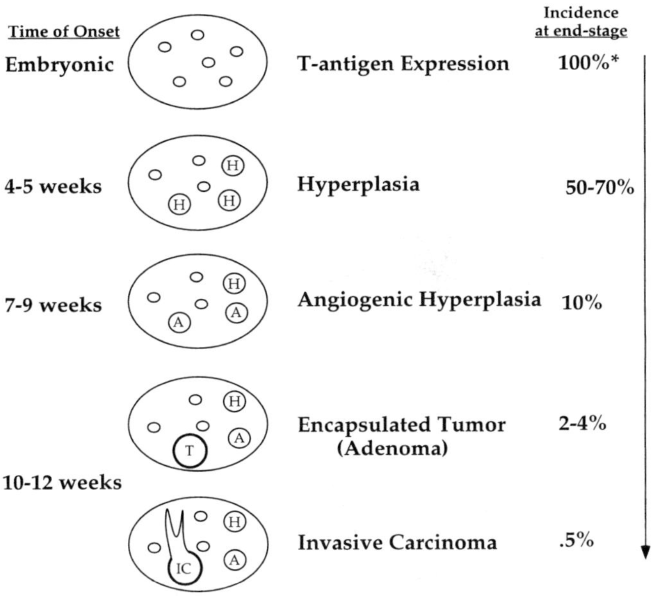

FIGURE 1. A schematic representation of RIPTag tumorigenesis. T-antigen expression commences during embryogenesis but is without apparent effect until 4–5 weeks when sporadic islets become hyperplastic (H); over time 50–70% become hyperplastic. At 7–9 weeks, angiogenic islets (A) appear. From these angiogenic islets emerge encapsulated tumors (10–12 weeks; T) (adenomas) of which a subset develop into invasive carcinoma (IC). * Incidence at all stages.

blood vessels (neovascularization), a process critical to the growth of a tumor.[6] From these angiogenic islets arise well encapsulated tumors (adenoma; 10–12 weeks) which then, at low frequency, invade into the surrounding exocrine pancreas (invasive carcinoma).[7] Although T-antigen is expressed in all β-cells of all islets, only 50–70% become hyperplastic, 10% angiogenic, 2–4% encapsulated tumor and less than 1% invasive carcinoma. What these statistics clearly suggest is that while necessary, T-antigen expression itself is not sufficient to produce a tumor. By inference, other relatively rare genetic and epigenetic changes ("hits") are required for tumor development. From multiple avenues of investigation it has become increasingly evident that one target of these "hits" are genes that control apoptosis. The down-modulation of apoptosis is a rate-limiting step in the development of RIP-Tag tumors, and perhaps most types of cancer.

PATTERN OF APOPTOSIS IN RIP-TAG TUMORIGENESIS

Apoptotic labeling using the TUNEL procedure on the different histological stages reveals that apoptosis undergoes a dramatic modulation in the RIP-Tag tumor pathway (FIG. 2A, [8]). The apoptotic rate in oncogene expressing but pre-

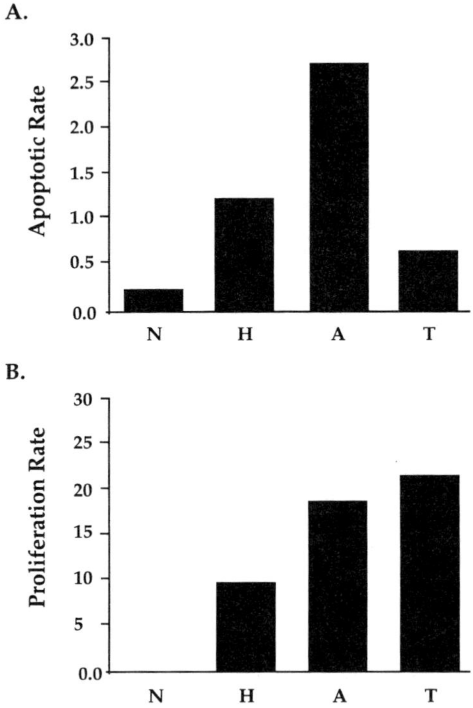

FIGURE 2. The apoptotic and proliferation indices of RIP-Tag tumorigenesis pathway.[8] **A:** Apoptotic cells were visualized by TUNEL labeling and **B:** proliferating cells visualized by BrdU labeling. Both indices are expressed as the percentage of total cells that are positive for TUNEL or BrdU staining. N, normal, Tag^{+} islets; H, hyperplastic islets; A, angiogenic islets; T, large, end-stage tumor.

hyperplastic, quiescent islets is comparable to that of non-transgenic controls. This indicates that oncogene expression is not sufficient to induce apoptosis. Rather, apoptosis is activated in response to the oncogene driven proliferation in hyperplastic islets. The apoptotic index peaks during the angiogenic islet stage and then falls in two successive steps in small and large tumors. Thus, the rise in apoptotic rate is coincident with an increase in the rate of cell proliferation. However, the cancer cell finds a way to subvert this apoptotic response while maintaining hyperproliferation, a step we infer is critical to expansive tumor growth. As depicted in FIGURE 2B, the proliferative rate is approximately the same in the angiogenic islet stage as it is in the tumor stage, yet there is 100–1000× increase in size of the nodule, indicating that the drop in the apoptotic rate is the critical step to tumor expansion. The molecular control of this down-modulation of apoptosis is a continued focus of the laboratory and below we detail our current understanding and ongoing efforts.

APOPTOSIS IS p53 INDEPENDENT IN RIP-TAG TUMORIGENESIS

p53 is a key mediator of apoptosis in response to aberrant proliferative signals, such as those derived from oncogene expression.[9,10] It was therefore surprising that a robust apoptotic response was observed in the early stages of this tumor pathway, since p53 should be rendered non-functional through its interaction with T-antigen.

However, it had been shown that not all p53 protein is complexed with T-antigen.[11] To determine if the observed apoptosis is a function of free, unbound p53, and, therefore, p53 dependent, the apoptotic rate was monitored in the absence of p53 by generating RIP-Tag; $p53^-/p53^-$ mice.[8] The apoptotic rate in $p53^-/p53^-$ hyperplastic and angiogenic islets was similar to that observed in their $p53^+/p53^+$ counterparts, indicating that these cells are functionally p53 negative, and that the substantial apoptosis observed early in the pathway is p53 independent. This robust, p53-independent apoptosis may reflect a process of tissue homeostasis, the integration of proliferation, life span and apoptotic elimination of β-cells so as to maintain islet size, a process that is distinct from the "damage sensor" role of p53.[12] Examination of the size and apoptotic rate of islets of $p53^-/^-$ mice as they age would directly test the possibility that p53-independent apoptosis plays a role in normal, homeostatic β-cell maintenance.

ACTIVATION OF IGF-II: A GROWTH FACTOR THAT SUPPRESSES APOPTOSIS

In an expression screen of ~ thirty growth factors and their receptors, insulin-like growth factor-2 (IGF-2) was shown to be up-regulated in tumors in comparison to normal islets.[13,14] A subsequent *in situ* analysis indicated that IGF-2 expression was below the level of detection in normal, Tag-expressing islets, was focally activated in hyperplastic islets, and was uniformly expressed in angiogenic islets and tumors. The timing of activation suggested that IGF-2 might function in the switch from the quiescent to proliferative state. This growth factor is known to have both proliferative and anti-apoptotic effects depending on the cellular context.[15] The functional significance of IGF-2 activation was tested genetically by crossing the RIP-Tag mice to IGF-2 KO mice.[13,14] Surprisingly, lack of IGF-2 did not affect the mitotic rate in any of the distinct stages but resulted in a reduction in the size and number of tumors, associated with an increase in the apoptotic rate. In addition to an effect on apoptotic rate, the IGF-2 null tumors that did arise had a less malignant morphology, with less densely packed cells having a more "normal" nuclear to cytoplasmic ratio than that observed in $^+/^+$ controls. It is not clear whether this effect on morphology is the direct result of IGF-2 signaling or reflects the (indirect) outgrowth of more malignant cells that die in the absence of IGF-2. There are two known receptors for IGF-2, the IGF-1R and the IGF-2R.[15–19] Current data suggest that the signaling receptor for IGF-2 is the IGF-I receptor and that the IGF-2R acts as a non-signaling, competitive sink.[20,21] Recent work has shown that the IGF-1R, at least in part, suppresses apoptosis by signaling through PI3K and the serine/threonine kinase AKT/PKB.[22,23] Activated AKT/PKB in turn phosphorylates the intracellular transducer, BAD, which modulates the activity of the apoptosis suppressors bcl-2 and bcl-x_L.[23,24] In addition, it has been shown recently that AKT/PKB can directly phosphorylate caspase-9, which inhibits its pro-apoptotic protease activity.[25] This suggests that IGF-1R signaling may suppress apoptosis by regulating the activity of multiple molecules. Both IGF-1R and IGF-2R are uniformly expressed throughout the pathway, albeit at relatively low levels, suggesting that IGF-2 function may be limited by its receptor levels (ref. 14 and J.H. & D.H., unpublished results).

EXPRESSION OF BCL-2 FAMILY MEMBERS: BCL-X_L EXPRESSION IS UP-REGULATED IN TUMORS

At the center of a cell's decision to live or die in response to an apoptotic signal is the bcl-2 family of apoptotic regulators.[9,26] The bcl-2 family members fall into two broad groups; those that are anti-apoptotic, of which bcl-2 and bcl-x_L are the prototypical members, and those that are pro-apoptotic, such as bax and bak. It has been shown through a variety of cell culture studies that the ratio of the pro-apoptotic and anti-apoptotic proteins will, to some degree, determine if the cell is likely to die upon encountering an apoptotic signal. One way in which a cancer cell could escape death is by modulating the expression of the bcl-2 family members—either by down-regulating the expression of a pro-apoptotic gene or by up-regulating an anti-apopotic gene. To test this hypothesis, expression of a few of the more prominent members of the bcl-2 family (bcl-2, bcl-x_L, bax, and the bcl-2 associated protein, bag-1) was examined by RT-PCR analysis as an initial screen.[8] In short, bcl-2 expression was not detected, while bcl-x_L, bax, and bag-1 expression was observed. Subsequent *in situ* RNA hybridization analyses revealed that both bag-1 and bax were similarly expressed in all stages of the tumor pathway whereas bcl-x_L was selectively upregulated in the tumor stage. The embryonic lethality of homozygous bcl-x_L KO mice precluded a loss-of-function test of this up-regulation, and so an overexpression study was undertaken.[8] Placing bcl-x_L under the control of the RIP promoter directed high level expression to the islets in all stages of the pathway. This expression resulted in an increase in the number and size of tumors in double transgenic RIP-Tag, RIP-bcl-x_L mice; interestingly the number of angiogenic precursors remained the same even though the apoptotic rate was suppressed in this stage. This result suggests that suppression of apoptosis is most critical in the progression from angiogenic precursor to end-stage tumor and that other events are rate limiting in angiogenic islets, e.g. forming new, functional blood vessels.

ANGIOGENESIS: PERSISTENT NEOVASCULARIZATION IS REQUIRED TO MAINTAIN A SUPPRESSED APOPTOTIC STATE

Over the last decade there has been an increased appreciation of the fact that all tumors require the growth of new blood vessels to expand, invade surrounding tissue, and potentially metastasize.[27,28] This requirement for neovascularization is a ripe target for anti-cancer therapeutic agents, the so-called angiogenesis inhibitors that intervene in this process.[29] Most of the work on angiogenic inhibitors has employed the subcutaneous injection of tumor cell lines and the subsequent treatment of mice with a significant tumor burden. The RIP-Tag model has been used to test the efficacy of angiogenesis inhibitors on endogenous, end-stage tumors, which develop in their natural microenviroment.[30,31] In addition, this transgenic mouse model affords an opportunity to test the potential of these inhibitors to block the progression of premalignant stages, something the transplant models do not permit.[31]

A common theme has emerged from the treatment of both transplanted and endogenous tumors with angiogenic inhibitors: with the compounds tested, reduction of tumor size correlated with an increase in the apoptotic rate (of the tumor cell component), showing no significant effect on tumor cell proliferation.[30–31] While the mechanism of this pro-apoptotic effect is not understood, two simple models come to mind: indirect killing via reduced O_2 content (hypoxia) or direct killing through the release of death-inducing cytokines such as TNF (see PERSPECTIVES).

LOH9: A PROPOSED TUMOR SUPPRESSOR GENE THAT REGULATES APOPTOSIS

Genetic alterations are central to the development of all cancers.[33] These genetic lesions fall into two classes: gain-of-function and loss-of-function. In fact, the discovery of the apoptotic regulator bcl-2 resulted from the study of a translocation that is common in B-cell lymphoma. This translocation juxtaposes the bcl-2 gene to the immunoglobulin heavy chain gene and de-regulates its expression, thereby suppressing apoptosis.[34,35] Many tumor suppressor genes (e.g. p53, bax) have central roles in regulating apoptosis.[10,36] Loss-of-function mutations in these genes result in an impaired ability of the cell to execute the apoptotic program in response to the aberrant proliferation inherent to all cancer cells.

Regions of loss of heterozygosity (LOH) are an indication of tumor suppressor genes (tsg) whose loss confers a growth advantage to the developing cancer cell. [33] In a genome wide search for regions of LOH in the RIP-Tag model, two statistically significant deletions were detected: one on chromosome 9 (LOH9; ~20% tumors) and another on chromosome 16 (LOH16; ~30% tumors).[37] In a subsequent study, it was shown that LOH16 is first lost in the transition from hyperplastic islet to angiogenic islet, while LOH9 is first detected in angiogenic islets and exhibits a higher rate of loss in end-stage tumors.[38] From this study we speculate that *Loh16* might be involved in the regulation of angiogenesis, while other lines of evidence implicate *Loh9* in the down-regulation of apoptosis. We will focus our discussion here on *Loh9*, although it is certainly possible that *Loh16* may function in multiple cellular processes, including apoptosis, as does the tumor suppressor gene p53.[10] Support for the notion that *Loh9* is involved in regulating apoptosis comes from analyzing the apoptotic potential of several clonal tumor cell lines that are either wild-type for chromosome 9 or harbor deletions on chromosome 9.[39] Interestingly, the majority of the cell population in lines that are wild-type for chromosome 9 (3/3) die within 24 hours of serum withdrawal while those that carry LOH9 (9/9) are relatively resistant to death in low serum.

PERSPECTIVES

A model emerges from our studies of the pattern and regulation of apoptosis during RIP-Tag tumorigenesis in which tumor cells utilize multiple independent pathways to escape apoptosis. This includes the activation of IGF-2, persistent neovascularization, up-regulation of bcl-x_L and genetic loss on chromosome 9 (FIG. 3). The fact that these tumors subvert four independent pathways is surprising in light of the fact that they are functionally p53 negative. It is commonly thought that loss of p53 in tumor cells results in complete abolition of the apoptotic program. However, early in the RIP-Tag progression there is a robust apoptotic response to the Tag driven proliferation whose suppression is necessary for tumor expansion. The potential for p53-independent apoptosis may be a function of the cell type in which the oncogene is expressed. In a model of Tag-induced tumors of the choroid plexus, apoptosis is largely p53-dependent.[40] Mice that express a mutant form of Tag (Tag_{121}) that does not bind to and inactivate p53 only develop hyperplastic nodules that exhibit a relatively high apoptotic rate (8%). The same transgenic mice develop choroid plexus tumors very rapidly when homozygous for a p53 null allele, likely as the result of a much reduced apoptotic rate (<1%). Clearly, these two different cell types, transformed by the same oncogene, have quite distinct apoptotic potential in the absence of p53 function. The results suggest that other cancers, even though p53 negative, may possess crucial stages in their development, analogous to the hyperplastic/angiogenic islet stages of RIP-Tag, where there is a robust p53-independent apoptotic response. As such, appreciation of p53-independent apoptotic regulatory mechanisms will undoubtedly be important in the treatment of human tumors.

The four distinct modulators of apoptosis described above have spawned four ongoing lines of inquiry. Our primary focus is to further understand their role and function in the context of RIP-Tag tumorigenesis. Moreover, it is our belief that studying this well-defined mouse model of islet cell carcinogenesis will also yield insight into the role and regulation of apoptosis in human cancers. In the next section we outline the important "second generation" questions and the approaches we are taking to address them.

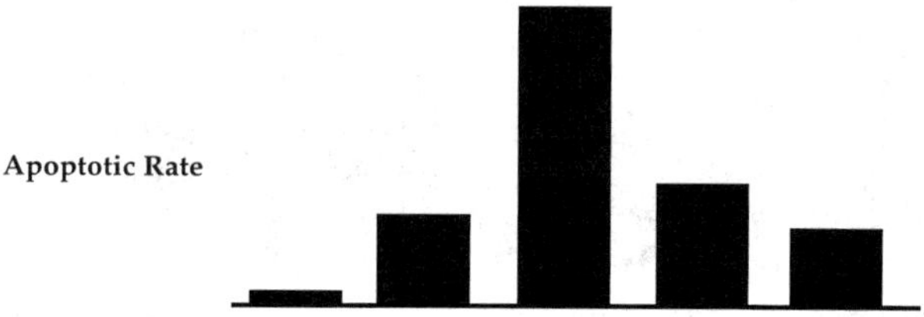

Islet stage	N	H	A	ST	LT
IGF-2	-	+/-	++	++	++
bcl-x_L	+	+	+	++	++
Angiogenesis	-	-	+	+	+
LOH9	-	-	+/-	?	++

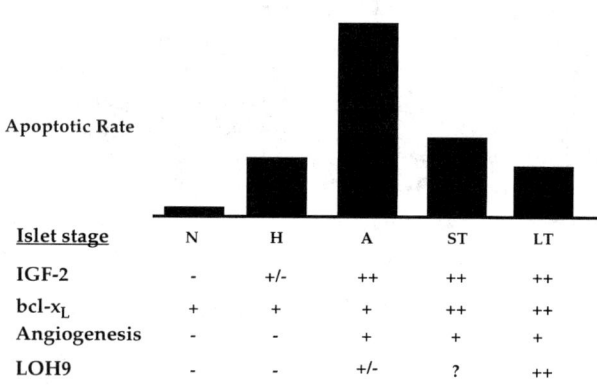

FIGURE 3. Tumor cells subvert multiple, independent pathways to escape apoptosis.[8] Alterations in four distinct regulators of apoptosis occur at different stages of tumorigenesis. Activation of IGF-2 and onset of angiogenesis occur as hyperplastic islets progress to angiogenic islets. Whereas bcl-x_L up-regulation and genetic loss on chromosome 9 (LOH9) are most pronounced in the tumor stage. Genetic or pharmacological modulation of any one of these regulators results in a significant alteration in the apoptotic rate.

IGF-II: DISSECTION OF THE SIGNALING PATHWAY *IN VIVO*

Activation of the IGF-1/IGF-2 signaling pathway is a feature of many epithelial cancers.[15] However, its functional significance in the developing tumor *in vivo* is unclear. Does IGF-1/IGF-2 signaling modulate proliferation, apoptosis, or some other process, such as invasion or metastasis? Taking advantage of the genetics of the mouse, we have definitively shown that in RIP-Tag carcinogenesis IGF-2 functions mainly to suppress apoptosis, although at some level it may also influence cellular malignancy. In this context, IGF-2 acts as a signal that confirms the decision to proliferate. In the absence of this confirmatory checkpoint the proliferating tumor cell dies. To further dissect the components of this pathway, in particular those that function downstream of IGF-2, we have shifted our focus towards the IGF-1R receptor, specifically addressing whether the IGF-1R is the signaling receptor for IGF-2. Indication that the IGF-1R does function in this capacity comes from cell culture studies in which a dominant negative IGF-1R sensitizes RIP-Tag tumor cell lines to treatment with death inducing agents such as etoposide.[41] We are taking a two-pronged approach to further examine this question *in vivo*: overexpression (via the RIP promoter) of either the wild-type (w.t.) IGF-1R or a dominant negative IGF-1R (IGF-1Rdn) known to abrogate receptor-signaling.[42] If IGF-1R is the primary signaling receptor for IGF-2, mice overexpressing the IGF-1Rdn should have a phenotype similar to that observed in IGF-2 KO experiments, namely reduction in tumor size and number. Likewise, an augmentation of the tumor phenotype is predicted in mice that overexpress the w.t. receptor.

BCL-2 FAMILY: ANALYSIS OF TUMOR STAGE PROTEIN PROFILES AND FUNCTIONAL TESTS

Alteration of the expression of bcl-2 family members in cancers is a well-documented phenomenon.[43,44] Both up-regulation of anti-apoptotic genes (e.g. bcl-2, bcl-x_L) and down-regulation of pro-apoptotic genes (e.g. bak, bax) have been reported. However, a strict functional link between bcl-2 family members and the development of human epithelial cancers had not been established until the recent identification of frameshift mutations in the bax gene in colon cancers that exhibit the microsatellite mutator phenotype (MMP^+).[36] Interestingly, colon cancers that follow a MMP^+ pathway are typically wild-type for p53, suggesting that selective pressure exists in these tumors for mutations in bax, a primary downstream target of p53 whose transactivation directly induces apoptosis (45). Our studies would predict that alterations in other p53 independent pathways might occur in MMP^+ colon cancers, in addition to mutations in bax, so as to suppress apoptosis.

Currently, we are examining the protein profile of bcl-2 family members such as bcl-x_L, bax, bak, and bad by Western blotting extracts derived from normal islets, angiogenic islets, and end-stage tumors. This approach should verify whether the RNA analyses are reflective of protein levels, in addition to determining whether protein modifications of bcl-2 family members, such as phosphorylation of bcl-x_L and bad, are likely to have a role in the down-modulation of apoptosis. Preliminary results indicate that the protein levels of both bax and bak are elevated in end-stage tumors relative to normal islets. This suggests that both bax and bak are up-regulated in response to oncogene induced proliferation in a p53-independent manner.

We are testing the functional contribution of both bax and bak by crossing KO alleles of these genes to RIP-Tag mice and subsequently determining tumor burden and the apoptotic and proliferative rates. If up-regulation of bax and bak drives the apoptosis in this pathway, then tumor growth should be accelerated in their absence due to a decrease in the rate of cell death. In addition, to determine if bcl-x_L function is required for the suppression of apoptosis, we intend to utilize the cre-loxP recombination system to "knockout" bcl-x_L in a β-cell specific manner in RIP-Tag mice.

ANGIOGENESIS: HOW DO ANGIOGENESIS INHIBITORS INDUCE TUMOR CELL APOPTOSIS?

Multiple studies indicate that disruption of the tumor vasculature through treatment with angiogenic inhibitors results in a significant increase in tumor cell apoptosis without an appreciable effect on proliferation.[30–32] The mechanism by which this apoptosis is induced is not clear; one possibility is that it results from reduced oxygen content (hypoxia) due to the perturbation of vessel function. Induction of apoptosis by hypoxia is a well-documented phenomenon.[46] Alternatively, the dying endothelium may release death-inducing factors, such as TNF, which act directly on the tumor cell. Support of each model comes from cell culture experiments in which cells cycling through S-phase as a result of c-myc overexpression exhibit an increased sensitivity to many apoptotic stimuli, including hypoxia and TNF.[47–50] In our system, localization of apoptotic bodies in regions of drug-treated tumors most removed from vessels would support the notion that hypoxia is the driving force behind tumor cell apoptosis.[46] This, however, is not the case. In general, there is a random distribution of apoptosis, and with certain compounds there is appreciable perivascular apoptosis.[31] To directly determine whether hypoxia is a likely inducer of tumor cell apoptosis, we are utilizing injectable compounds, such as pimonidazole hydrochloride, that bind to macromolecules under conditions of low oxygen.[51,52] Through the use of an anti-pimonidazole IgG, one can directly visualize tissue hypoxia *in situ*. If significant regions of hypoxia are observed in treated tumors in association with apoptotic bodies, it would be reasonable to conclude that it has a role in the observed tumor cell apoptosis. Likewise, if hypoxia is a driving force behind the tumor cell apoptosis, then common patterns of hypoxia-regulated gene expression should be observed in treated tumors. To address this possibility, we are using large-scale gene arrays to analyze patterns of gene expression in tumors that have been treated with a diverse set of angiogenic inhibitors. We are also developing protocols to isolate the endothelium (and subsequently mRNA) from both normal islets and tumors for gene discovery (e.g. cDNA library construction and sequencing) and gene expression analyses using array technology. The goal is to identify the genes underlying the morphological and functional differences that exist between normal and tumor endothelium.[53–55] Once in place, we hope to extend this approach to the analysis of endothelial cells isolated from treated and untreated tumors. This should unveil changes in gene expression that may function in the induction of tumor cell apoptosis, in addition to providing insight into the pathways that these novel compounds abrogate within the endothelium.

LOH9: IDENTIFICATION AND CHARACTERIZATION OF A TSG THAT MAY REGULATE APOPTOSIS

Correlative evidence suggests that a proposed tumor suppressor gene lost on chromosome 9 (*Loh9*) regulates apoptosis. We are using traditional positional cloning techniques along with a transfection based, functional "complementation" approach towards its identification. Recent fine structure mapping of LOH9 in a collection of ~ 450 F1 tumors (crosses: Cast × RIP-Tag, C3H and DBA × RIP-Tag, C57B6) has localized the critical region to ~ 1cm between the markers D9Mit12 and D9Mit115. We have begun to assemble a BAC contig across this genomic interval using these flanking markers and genes that map within the region as probes. Assuming, on average, that 1cM of genetic distance equals ~ 1Mb DNA, 10–20 BAC's (average BAC 120 kb) should cover the critical region. These BAC's will then be individually transfected into LOH9-containing cell lines and the apoptotic phenotype monitored in low serum. Any transfected BAC that restores apoptotic function to a LOH9-carrying cell (and presumably carries *Loh9*) will be sequenced and the identification of open reading frames will be carried out using standard sequence analysis programs. The final "proof" for any candidate gene will be the identification of mutations in the retained allele. Once the gene is identified we will focus on determining the function and mechanism of its action. What does the sequence suggest for gene function? Does *Loh9* function in a known apoptotic pathway or in a novel pathway? Is expression of *Loh9* altered in the ~ 80% of RIP-Tag tumors that do not exhibit LOH9? If *Loh9* is a "gatekeeper" whose loss of function is critical to the development of all RIP-Tag tumors, then one would predict epigenetic down-regulation in those tumors that are w.t. for chromosome 9. If this were not the case, one would infer that other points in the same pathway are disrupted, genetically or epigenetically. To determine whether *Loh9* may play a role in human disease, we will determine if it is mutated or its expression down-regulated in the major human epithelial cancers. To verify the role of *Loh9* in this tumor pathway, we will engineer both loss-of-function (KO) and gain-of-function alleles (overexpression via the RIP promoter). If loss of *Loh9* is truly a rate-limiting event in RIP-Tag tumorigenesis, then an exacerbation of the tumor phenotype is expected for mice that are homozygous for the KO. Conversely, a less severe tumor phenotype is expected in mice that constitively overexpress *Loh9*. In addition, these engineered alleles will undoubtedly shed light on the role of *Loh9* in the normal, non-pathological state.

SUMMARY

In the RIP-Tag mouse model of islet cell carcinoma, four different pathways (IGF-2 activation, neovascularization, up-regulation of bcl-x_L and *Loh9*) are utilized to suppress p53-independent apoptosis. These results contradict the notion that $p53^-$ tumor cells are not capable of executing the apoptotic program and, therefore, not amenable to therapy. We have shown that both the IGF-2 signaling pathway and the tumor vasculature are potential therapeutic targets, since a robust p53-independent apoptotic response is mounted when either is disrupted. Likewise, we predict that the appropriate modulation of bcl-x_L and *Loh9* function will result in increased apoptosis and therefore are potential targets for therapy. Indeed, a more complete understanding of all four of these modulators of p53-independent apoptosis should help guide the development of effective treatments for human cancer, half of which harbor mutations in p53.[56]

REFERENCES

1. Thompson, C.B. 1995. Apoptotis in the pathogenesis and treatment of disease. Science **267:** 1456–1462
2. Hanahan, D. 1985. Heritable formation of pancreatic β-cell; tumors in transgenic mice harboring recombinant insulin/simian virus 40 oncogenes. Nature **215:** 115–122
3. Ludlow, J.W. 1993. Interactions between SV40 large-tumor antigen and the growth suppressor pRb and p53. FASEB J. **7:** 866–871.
4. Alpert, S., D. Hanahan & G. Teitelman. 1988. Hybrid insulin genes reveal a developmental lineage for pancreatic endocrine cells and imply a relationship with neurons. Cell **53:** 295–308.
5. Tietelman, G., S. Alpert & D. Hanahan. 1988. Proliferation, senescence and neoplastic progression of beta cells in hyperplastic pancreatic islets. Cell **52:** 97–105.
6. Folkman, J., K. Watson, D. Ingber & D. Hanahan. 1989. Induction of angiogenesis during the transition from hyperplasia to neoplasia. Nature **339:** 58–61.
7. Perl, A-K., P. W. Wilgenbus, U. Dahl, H. Semb & G. Christofori. 1998. A causal role for E-cadherin in the transition from adenoma to carcinoma. Nature **392:** 190–193
8. Naik, P., J. Karrim & D. Hanahan. 1996. The rise and fall apoptosis during multistage tumorigenesis: down-modulation contributes to progression from angiogenic progenitors. Genes Dev. **10:** 2105–2116.
9. White, E. 1996. Life, death and the pursuit of apoptosis. Gene Dev. **10:** 1–15.
10. Levine, A.J. 1997. p53, the cellular gatekeeper for growth and division. Cell **88:** 323–331.
11. Efret, S., S. Baekkeskov, D. Lane & D. Hanahan. 1987. Coordinate expression of the endogenous p53 gene in beta cells of transgenic mice expressing hybrid insulin-SV40 T-antigen genes. EMBO J. **6:** 2299–2704.
12. Finegood, D.T., L. Scaglia & S. Bonner-Weir. 1995. Dynamics of β-cell mass in the growing rat pancreas. Diabetes **44:** 249–256
13. Christofori, G., P. Naik & D. Hanahan. 1994. A second signal supplied by insulin-like growth factor II in oncogene-induced tumorigenesis. Nature **369:** 414–418.
14. Naik, P., G. Christofori & D. Hanahan. 1994. Insulin-like growth factor is focally up-regulated and functionally involved as a second signal for oncogene-induced tumorigenesis. Cold Spring Harbor Symp. Quant. Biol. **LVIX:** 459–470.
15. Resnicoff, M. & R. Baserga. 1998. The role of the insulin-like growth factor I receptor in transformation and apoptosis. Ann. N.Y. Acad. Sci. **842:** 76–81
16. Schofield, P.N. 1992. The insulin-like growth factors: structure and biological function. Oxford University Press.
17. Rechler, M.M. 1991. Insulin-like growth factor II: Gene structure and expression into messenger RNA and protein. *In* Insulin-like Growth Factors: Molecular and Cellular Aspects. D. LeRoith, Ed.: 87. CRC Press, Boca Raton.
18. Rechler, M.M. & S.P. Nissley. 1990. Peptide growth factors and their receptors I. *In* Handbook of Experimental Pharmacology. M.B. Sporn & A.B. Roberts, Eds. **95:** 263. Springer-Verlag. Berlin, New York.

19. WERNER, H., M. WOLOSCHAK, B. STANNARD, Z. SHEN-ORR, C.T. ROBERTS, JR. & D. LEROITH. 1991. The insulin-like growth factor I receptor: molecular biology, heterogeneity and regulation. *In* Insulin-like Growth Factors: Molecular and Cellular Aspects. D. LeRoith, Ed.: 87. CRC Press, Boca Raton.
20. HAIG, D. & C. GRAHAM. 1991. Genomic imprinting and the strange case of the insulin-like growth factor II receptor. Cell **22:** 1045–1046.
21. FILSON, A.J., A. LOUVI, A. EFSTRATIADIS & E.J. ROBERTSON. 1993. Rescue of the T-associated maternal effect in mice carrying null mutations in Igf-2 and Igf2r, two reciprocally imprinted genes. Development **118:** 731–736.
22. KAUFFMANN-ZEH, A., P. RODRIGUEZ-VICIANA, E. ULRICH, C, GILBERT, P. COFFER, J. DOWNWARD & G. EVAN. 1997. Suppression of c-myc induced apoptosis by ras signaling through PI(3)K and PKB. Nature **385:** 554–558.
23. DATTA, S.R., H. DUDEK, X. TAO MASTERS, H.A. FU, Y. GOTOH & M.E. GREENBERG. 1997. Akt phosphorylation of Bad couples survival signals to the cell-intrinsic death machinery. Cell **91:** 231–241.
24. DELPESO, L., M. GONZALEZ-GARCIA, C. PAGE, R. HERRERA & G. NUNEZ. 1997. Interleukin-3-induced phosphorylation of Bad through the protein kinase Akt. Science **278:** 687–689.
25. CARDONE, M.H., N.ROY, H.R. STENNICKE, G.S. SALVESEN, T.F. FRANKE, E. STANBRIDGE, S. FRISCH & J.C. REED. 1998. Regulation of cell death protease caspase-9 by phosphorylation. Science **282:** 1318–1321
26. REED, J.C. 1997. Double identity for proteins of the Bcl-2 family. Nature **387:** 773–776.
27. FOLKMAN, J. 1995. Tumor angiogenesis. *In* The Molecular Basis of Cancer, J. Mendelsohn, P.M. Howley, M.A. Israel & L.A. Liotta, Eds.: 206–232. W.B. Saunders Co. Philadelphia.
28. HANAHAN, D. & J. FOLKMAN. 1996. Patterns and emerging mechanisms of the angiogenic switch during tumorigenesis. Cell: **86:** 353–364.
29. HANAHAN, D. 1998. A flanking attack on cancer. Nature Med. **4:** 13–14.
30. PARANGI, S., M. O'REILLY, G. CHRISTOFORI, L. HOLMGREN, J. GROSFELD, J. FOLKMAN & D. HANAHAN. 1996. Antiangiogenic therapy of transgenic mice impairs de novo tumor growth. Proc. Natl. Acad. Sci. USA **93:** 2002–2007.
31. BERGERS, G., K. JAVAHERIAN, K.M. LO, J. FOLKMAN & D. HANAHAN. 1999. Effects of angiogenesis on multistage carcinogenesis in mice. Science **284:** 808–812.
32. HOLMGREN, L., M.S. O'REILLY & J. FOLKMAN. 1995. Dormancy of micrometastases: balanced proliferation and apoptosis in the presence of angiogenesis suppression. Nat. Med. **1:** 149–153.
33. VOGELSTEIN, B. & K.W. KINZLER. 1993. The multistep nature of cancer. Trends Genet. **9:** 138–141.
34. BAKHSHI, A., J.P. JENSEN, P. GOLDMAN, J.J. WRIGHT, C.W. MCBRIDE, A.L. EPSTEIN & S.J. KORSMEYER. 1985. Cloning the chromosome breakpoint of t(14;18) human lymphomas clustering around JH on chromosome 14 near a transcriptional unit on 18. Cell **41:** 899–906.
35. CLEARY, M.L. & J. SKLAR. 1985. Nucleotide sequence of t(14;18) chromosomal breakpoint in follicular lymphoma and demonstration of a breakpoint cluster region near a transcriptionally active locus on chromosome 18. Proc. Natl. Acad. Sci. USA **82:** 7439–7443.
36. RAMPINO, N., H. YAMAMOTO, Y. IONOV, Y. LI, H. SAWAI, J.C. REED & M. PERUCHO. 1997. Somatic frameshift mutations in the BAX gene in colon cancers of the microsatellite mutator phenotype. Science **275:** 967–969.

37. DIETRICH W.F., E.H. RADANY, J.S. SMITH, J.M. BISHOP *et al.* 1994. Genome-wide search for loss of heterozygosity in transgenic mouse tumors reveals candidate tumor suppressor genes on chromosomes 9 and 16. Proc. Natl. Acad. Sci. USA **27:** 9451–9455.
38. PARANGI, S.W., W.F. DIETRICH, G. CHRISTOFORI, E.S. LANDER & D. HANAHAN. 1995. Tumor suppressor loci on mouse chromosomes 9 and 16 are lost at distinct stages of tumorigenesis in a transgenic model of islet cell carcinoma. Cancer Res. **55:** 6071–6076.
39. NAIK, P. S.-D. 1996. The regulation and role of cell survival and cell death in multistage tumorigenesis. *In* Biochemistry and Molecular Biology. University of California, San Francisco.
40. SYMONDS, H., L. KRALL, L. REMINGTON, M. AENZ-ROBLES, S. LOWE, T. JACKS & T. VAN DYKE. 1994. p53 dependent apoptosis suppress tumor growth and progression in vivo. Cell **78:** 703–711.
41. LAMM, G. & G. CHRISTOFORI. 1998. Impairment of survival factor function potentiates chemotherapy-induced apoptosis in tumor cells. Cancer Res. **58:** 801–807.
42. DUNN, S.E., M. EHRLICH, N.J. SHARP, K. REISS, *et al.* 1998. A dominant negative mutant of the insulin-like growth factor-I receptor inhibits the adhesion, invasion, and metastasis of breast cancer. Cancer Res. **58:** 3353–3361.
43. KRAJEWSKA, M., S.F. MOSS, S. KRAJEWSKI, K. SONG *et al.* 1996. Elevated expression of Bcl-X and reduced Bak in primary colorectal adenocarcinomas. Cancer Res. **15:** 2422–2427.
44. KRAJEWSKA, M., S. KRAJEWSKI, J.I. EPSTEIN, A. SHABAIK *et al.* 1996. Immunohistochemical analysis of bcl-2, bax, bcl-X, and mcl-1 expression in prostate cancers. Am. J. Pathol. **148:** 1567–1576.
45. 45. MIYASHITA, T. & J.C. REED. 1995. Tumor suppressor p53 is a direct transcriptional activator of the human bax gene. Cell **80:** 293–299.
46. GRAEBER, T.G., C. OSMANIAN, T. JACKS, D. E. HOUSMAN, C.J. KOCH, S.W. LOWE & A.J. GIACCIA. 1996. Hypoxia-mediated selection of cells with diminished apoptotic potential in solid tumors. Nature **379:** 88–91.
47. DONG, J., M. NAITO & T. TSURUO. 1997. c-Myc plays a role in cellular susceptibility to death receptor-mediated and chemotherapy-induced apoptosis in human monocytic leukemia U937 cells. Oncogene **7:** 639–647.
48. KLEFSTROM, J., E. ARIGHI, T. LITTLEWOOD, M. JAATTELA *et al.* 1997. Induction of TNF-sensitive cellular phenotype by c-Myc involves p53 and impaired NF-kappaB activation. EMBO J. **16:** 7382–7392.
49. JANICKE, R.U., F.H. LEE & A.G. PORTER. 1994. Nuclear c-Myc plays an important role in the cytotoxicity of tumor necrosis factor alpha in tumor cells. Mol. Cell Biol. **9:** 5661–5670.
50. ALARCON, R.M., B.A. RUPNOW, T.G. GRAEBER, S.J. KNOX *et al.* 1996. Modulation of c-Myc activity and apoptosis in vivo. Cancer Res. **56:** 4315–4319.
51. ARTEEL, GE, R.G.THURMAN, J.M. YATES & J.A. RALEIGH. 1995. Evidence that hypoxia markers detect oxygen gradients in liver: pimonidazole and retrograde perfusion of rat liver. Br. J. Cancer **72:** 889–895.
52. KIM, C.Y., M.H. TSAI, C. OSMANIAN, T.G. GRAEBER *et al.* 1997. Selection of human cervical epithelial cells that possess reduced apoptotic potential to low-oxygen conditions. Cancer Res. **57:** 4200–4204.
53. BROWN, J.M. & A.J. GIACCIA. 1998. The unique physiology of solid tumors: opportunities (and problems) for cancer therapy. Cancer Res. **58:** 1408–1416.
54. JAIN, R.K. 1988. Determinants of tumor blood flow: a review. Cancer Res. **48:** 2641–2658.

55. THURSTON, G., J.W. MCLEAN, M. RIZEN, P. BALUK *et al.* (1998). Cationic liposomes target angiogenic endothelial cells in tumors and chronic inflammation in mice. J. Clin. Invest. **101:** 1401–1413.
56. HAINAUT, P., T. HERNANDEZ, A. ROBINSON, P. RODRIGUEZ-TOME, T. FLORES, M. HOLLSTEIN, C.C. HARRIS & R. MONTESANO. 1998. IARC Database of p53 gene mutations in human tumors and cell lines: updated compilation, revised formats and new visualization tools. Nucleic Acids Res. **26:** 205–213.

Hyperoxia in Cell Culture

A Non-apoptotic Programmed Cell Death

JEFFREY A. KAZZAZ,[a,c] STUART HOROWITZ,[b] YUCHI LI,[a] AND LIN L. MANTELL[a]

[a]*CardioPulmonary Research Institute, Winthrop-University Hospital, SUNY Stony Brook School of Medicine, Mineola, New York 11501, USA*

[b]*The Heart and Lung Institute, Jewish Hospital, 217 East Chestnut St., Louisville, Kentucky 40202, USA.*

ABSTRACT: Here we discuss the morphological features and our current understanding of the pathways involved in non-apoptotic cell death from O_2 toxicity. Preliminary data on hyperoxic signaling indicate that NF-κB translocation (and presumptive activation) is not a result of the p42/p44 MAPK pathway, but a likely downstream consequence of activation of the JNK pathway. Our observations suggest the existence of multiple signal transduction pathways in hyperoxia-induced cell death: one involved in the stress response which appears to be NF-κB–dependent and another in cell death.

INTRODUCTION

Although apoptosis of epithelial cells is a prominent component of acute lung injury in vivo including hyperoxia,[1–3] direct exposure of cultured human alveolar epithelial (A549) cells to hyperoxia does not result in apoptosis.[4] These cells swell and die over a period of days of exposure to hyperoxia. We have made virtually identical observations on other transformed cell lines, including type II cell–derived lung epithelial cells isolated from mouse (MLE12) and rat (SV-40T2) as well HeLa cells (from cervical carcinoma). Exposure of two different primary cell types of human lung to hyperoxia in culture: vascular endothelial cells (HVEC-L) and bronchial epithelial cells (NHBE) resulted in a similar non-apoptotic cell death in response to hyperoxia. These data indicate that many epithelial cell lines (as well as endothelial HVEC-L cells) do not die via apoptosis when cultured in hyperoxia. We refer to cell death in hyperoxia as non-apoptotic (rather than necrotic) because, although cells eventually appear necrotic (morphologically), our data indicate that this death induces the activation of specific pathways, which are not typically associated with necrosis. In this report, we will present the morphological features and our current understanding of the pathways involved in non-apoptotic cell death from O_2 toxicity.

[c]Address for correspondence: CardioPulmonary Research Institute, Winthrop-University Hospital, 222 Station Plaza North, Suite 503-5, Mineola, NY 11501; 516-663-8899 (voice); 516-663-8873 (fax).

e-mail: jkazzaz@winthrop.org

CELL AND NUCLEAR MORPHOLOGY

Apoptosis is classically defined morphologically; cells and nuclei typically become shrunken. Therefore, we first addressed the nuclear morphology of A549 cells in hyperoxia at the electron and light microscope level.[4] As indicated above, these cells became swollen after 24 hours in 95% oxygen; their nuclei are enlarged with no apparent chromatin condensation (FIG. 1). This was in sharp contrast to the apoptotic morphology seen when apoptosis was induced with other oxidants. In addition, these cells were TUNEL-negative, and there was no the increase in fluorescent intensity or decrease in size when stained with the DNA-binding dye, DAPI. Rather, the nuclei were increased in size without increased fluorescent intensity. These data demonstrate that hyperoxic death is morphologically distinct from apoptosis.

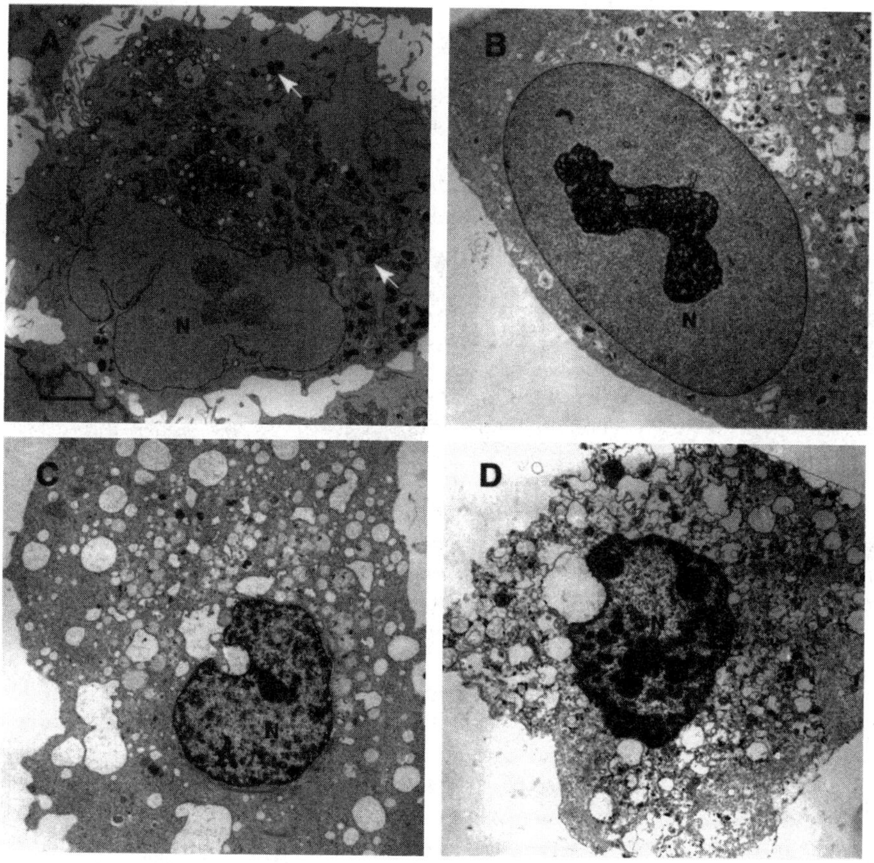

FIGURE 1. Electron micrographs of A549 cells exposed to room air, hyperoxia, superoxide, and hydrogen peroxide. **A:** room air control. **B:** cells exposed to 95% O_2 for 6 d. **C:** cells exposed to 5 mM H_2O_2 for 4 h. **D:** cells cultured in 5 mM paraquat for 18 h. (From Kazzaz *et al.*[4] Reprinted with permission from the *Journal of Biological Chemistry.*)

EXPRESSION AND REGULATION OF FOS AND JUN

We have begun to decipher the transcriptional regulatory events that occur in the early phases of oxidant-induced death. One early response to stress involves the transient expression of the c-Fos and c-Jun protooncogenes. The transcription complex known as AP-1 includes the Fos-Jun dimer and not only is reported to be redox sensitive,[5] but is also activated in lungs of hyperoxic rats.[6] To study AP-1 regulation during hyperoxia, mouse lung epithelial (MLE-12) cells were exposed to 95% O_2 for up to 24 hours. Cells were harvested at various time points and mRNA assayed by northern blots for Fos and Jun expression. Relative to control cells, both c-Fos and c-Jun transcript levels were elevated transiently (at 30 min) and then returned rapidly to baseline. Interestingly, both mRNAs were again increased in abundance at 16 to 24 h (Y. Li, unpublished observations).

The c-Fos promoter is sometimes regulated by the p42 and p44 Mitogen Activated Protein kinase, or MAP kinase cascade.[7] To investigate whether p42 and p44 MAP kinases are activated by hyperoxia, we used antibodies raised against the phosphorylated form of the proteins. No changes were detected in the levels of activated p42 or p44 during exposure to hyperoxia. In contrast, within 10 min of incubation in an apoptosis-inducing concentration of H_2O_2, there was a significant increase in phosphorylated p42 and p44, indicating that signaling events are different in these two modes of oxidant-induced lung epithelial cell death.

Increased abundance of c-Fos and c-Jun transcripts can be associated with activation of their corresponding proteins. Jun phosphorylation, necessary for activation,[8] was therefore assayed. Western blots were incubated with an antibody specific for phosphorylated c-Jun. A rise in the level of phosphorylated c-Jun was observed within 0.5 hours. The appearance of a second, slower-migrating band suggested the presence of additional phosphorylation probably ser-73. Similar to the mRNA, phosphorylated c-Jun levels decreased after this brief rise, but increased again at 16 to 24 hours. Jun phosphorylation can occur by Jun kinase (JNK).[9] To examine a possible role for JNK, we performed a JNK "pull-down" experiment. In this assay, cell lysates were incubated with affinity beads bound to GST-Jun fusion protein to "pull down" JNK. In this case, an increase in phosphorylated Jun was only evident at the late phase, at 16 to 24 h, suggesting that early and late signaling of Jun activation may be different (Y. Li, unpublished results).

NF-κB ACTIVATION DURING HYPEROXIA VERSUS HYDROGEN PEROXIDE EXPOSURE

Nonapoptotic cell death in hyperoxia occurs over several days—sufficient time for the elaboration of a program of gene expression. We therefore sought to understand transcription factor activity and signal transduction during hyperoxia, comparing it to cell death triggered by apoptotic stimuli. Apoptotic cell death can be prevented by the activation of NF-κB,[10–12] a multisubunit transcription factor that regulates genes involved in inflammation, infection, and stress.[13] Although cells might be protected from apoptosis, it is not known if NF-κB activation is part of a survival program used to escape other forms of cell death. We examined NF-κB in A549 cells exposed to concentrations of H_2O_2 that cause apoptosis versus non-apoptosis from hyperoxia. Despite the activation and induction of NF-κB by molecular O_2, the cells do not escape death.[14] These observations are summarized below.

Following release from the inhibitory binding protein I-κB, NF-κB translocates to the nucleus, regulating transcription.[15] NF-κB activation was studied during hyperoxia by immunolocalization of the p65 subunit of NF-κB. In A549 cells grown in room air, immunofluorescence was weak, and p65 was evident primarily in the cytoplasm. When cells were grown in 95% O_2, nuclear fluorescence was clearly evident by 30 min of hyperoxia, and it increased over the course of 1 day. The cells became slightly swollen by 24 h, and fluorescence became more intense not only in the nuclei, but also in the cytoplasm of many cells. Increased immunofluorescence suggested that the level of p65 protein also increased. RNA and protein blots show that indeed, NF-κB levels were increased during hyperoxic exposure. Western blots demonstrate increases as early as 30 min of exposure to 95% O_2, and peaked by 24 h. Northern blots showed that by 30 min of 95% O_2 exposure there was a slight increase in the steady state level of p65 mRNA. The message levels increased over the course of 1 d and remained elevated for two days. In contrast, cells undergoing H_2O_2-induced apoptosis showed no nuclear translocation of NF-κB. In addition, H_2O_2-induced apoptosis caused no increases in NF-κB protein or mRNA levels.The translocation of p65 to the nucleus is consistent with NF-κB activation in hyperoxia and suggests at least two hypotheses:

1. NF-κB is involved in initiating nonapoptotic cell death or
2. Although the cells are protected by NF-κB from apoptosis, another form of programmed cell death ensues.

A prediction from the first hypothesis is that by either preventing or knocking out NF-κB, the cells would undergo apoptosis in response to hyperoxia. The second hypothesis predicts that preventing NF-κB activation would either have no effect, or would increase cell death, because the protective pathways otherwise induced by NF-κB, would not be activated. We are currently trying to distinguish between these two hypotheses by testing the effect of overexpressing a dominant-negative form of I-κB on non-apoptotic cell death in epithelial cells.

EXPOSURE TO HYPEROXIA DELAYS APOPTOSIS

The observation that NF-κB is activated during hyperoxic exposure and that NF-κB activation can prevent apoptosis induced by various stimuli suggests that hyperoxic exposure is initiating a pathway that inhibits apoptosis. To test this notion we exposed A549 cells to various concentrations O_2 for 24 h then tested their sensitivity to undergo apoptosis with H_2O_2 exposure. On the face of it, this dual-oxidative-insult might be anticipated to kill the cells more rapidly than either insult alone. However, O_2 pre-exposure delayed the onset of apoptosis (L. Mantell, unpublished observations) suggesting that apoptotic pathways are inhibited.

Because molecular oxygen freely diffuses throughout the cell, one simple explanation for the delay is that hyperoxia is poisoning the cell, thereby blocking a key step in the apoptotic program. To test this, we performed similar experiments using a hyperoxia-resistant HeLa cell line called HeLa-80.[16] These cancer cells are resistant to—and continue to divide in—concentrations of oxygen as high as 80%,[16] and this resistance is stable. Pre-exposure of these cells to hyperoxia also delayed the onset of apoptosis induced by hydrogen peroxide (L. Mantell, unpublished observations). Because hyperoxia is not toxic to HeLa-80 cells, these results demonstrate that the delay we observe is not simply due to overall cell toxicity, but more likely to initiation of another program.

ATP IS NOT A LIMITING FACTOR

One of the factors determining whether a cell undergoes necrosis or apoptosis is the ATP levels in the cell.[17–20] Joenje and co-workers have demonstrated that hyperoxia results in respiratory failure and an increase in glycolysis in both HeLa and Chinese hamster ovary (CHO) cells[21,22] due, in part, to the inactivation of key mitochondrial enzymes. In addition, Allen and White have demonstrated that glucose can modulate cell death of A549 cells during hyperoxia under certain conditions.[20] Thus, one simple explanation of our data is that we are depleting the cells of ATP via glucose depletion. To test this hypothesis, we repeated our hyperoxic exposure experiments measuring the glucose levels daily. Our results indicate that cells are not deprived of glucose under our culture conditions (J.A. Kazzaz, unpublished results). Since these cells are using glycolysis as their source of energy, these data indicate that this cell death is not simply caused by ATP depletion.

SUMMARY AND DISCUSSION

These preliminary data on hyperoxic signaling indicate that NF-κB translocation (and presumptive activation) is not a result of the p42/p44 MAPK pathway, but a likely downstream consequence of activation of the JNK pathway. Moreover, c-Jun (and its phosphorylation) are increased in a two-phase fashion, the first occurring within 30 min, and the second after 16 hours. Reports in the literature that the Fos-Jun transcription complex AP-1 have demonstrated a role not only in the immediate-early response to stress but also in cell death (for references see review in ref. 23). Our observations suggest the existence of multiple signal transduction pathways in hyperoxia-induced cell death: one involved in the stress response which appears to be NF-κB–dependent and another in cell death (see FIG. 2).

We should note that there are reports of both epithelial- and non-epithelial cell lines dying via apoptosis in culture when exposed to hyperoxia.[24–26] This is not surprising considering that the apoptotic response to a particular stimulus is dependent on the cell type. For example, apoptosis is detected in neutrophils and neurons when exposed to hyperoxia. Taken together, these data suggest that the mode of cell death triggered by hyperoxia might differ among cell types.

Current thought on cell death is dichotomous: it is either apoptotic or necrotic. Yet a key issue is whether the non-apoptotic mode hyperoxic cell death represents a unique programmed cell death, or merely some point in the continuum between apoptosis and necrosis. The fact that none of the morphological criteria for apoptosis

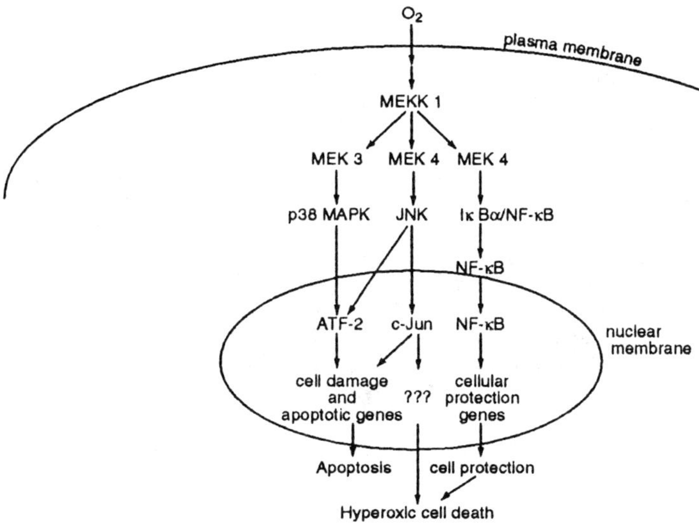

FIGURE 2. Proposed pathways leading to non-apoptotic cell death.

are evident in hyperoxia argues that hyperoxic cell death is distinct from apoptosis. Moreover, none of the mediators of apoptosis that we have studied to date (i.e., DNA fragmentation and caspases) are activated during hyperoxic cell death. Therefore, this is not apoptosis as currently defined either biochemically or morphologically. In addition, our data demonstrate that transcription factors are activated, which suggests that a "program" is initiated. Since necrosis is *never* described as cell death that initiates a program, this cell death is clearly not necrosis. Although we cannot formally dismiss the argument that this is some form of aborted apoptosis or "necrapoptosis"[27] until we have fully defined the biochemical pathways involved, our current thinking is that this represents another non-apoptotic cell death pathway.

REFERENCES

1. OTTERBEIN, L.E. *et al.* 1998. Pulmonary apoptosis in aged and oxygen-tolerant rats exposed to hyperoxia. Am. J. Physiol. (Lung Cell. Mol. Physiol.) **275:** L14–L20.
2. BARAZZONE, C. *et al.* 1998. Oxygen toxicity in mouse lung: pathways to cell death. Am. J. Respir. Cell Mol. Biol. **19:** 573–581.
3. MANTELL, L.L. *et al.* 1997. Unscheduled apoptosis during acute inflammatory lung injury. Cell Death and Diff. **4:** 604–607.
4. KAZZAZ, J.A. *et al.* 1996. Cellular oxygen toxicity—oxidant injury without apoptosis. J. Biol. Chem. **271**(25): 15182–15186.
5. XANTHOUDAKIS, S. *et al.* 1992. Redox activation of Fos-Jun DNA binding activity is mediated by a DNA repair enzyme. EMBO J. **11:** 3323–3335.
6. CHOI, A.M.K. *et al.* 1995. Molecular responses to hyperoxia in vivo: relationship to increased tolerance in aged rats. Am. J. Respir. Cell. Molec. Biol. **13**(1): 74–82.
7. HILL, C.S. *et al.* 1995. Transcriptional regulation by extracellular signals: mechanisms and specificity. Cell **80:** 199–211.

8. KYRIAKIS, J. *et al.* 1994. The stress-activated protein kinase subfamily of c-Jun kinases. Nature **369:** 159–160.
9. HIBI, M. *et al.* 1993. Identification of an oncoprotein- and UV-responsive protein kinase that binds and potentiates the c-Jun activation domain. Genes Dev. **7:** 2135–2148.
10. WANG, C.Y. *et al.* 1996. TNF- and cancer therapy-induced apoptosis: Potentiation by inhibition of NF-κB. Science **274:** 784–787.
11. BEG, A.A. *et al.* 1996. An essential role for NF-kappa B in preventing TNF-alpha–induced cell death. Science **274:** 782–784.
12. VAN ANTWERP, D. *et al.* 1996. Suppression of TNF-alpha-induced apoptosis by NF-kappaB. Science **274:** 787–789.
13. SCHRECK, R. *et al.* 1991. Reactive oxygen intermediates as apparently widely used messengers in the activation of the NF-kappa B transcription factor and HIV-1. EMBO J. **10:** 2247–2258.
14. LI, Y. *et al.* 1997. Nuclear factor-kappa B is activated by hyperoxia but does not protect from cell death. J. Biol. Chem. **272:** 20646–20649.
15. BALDWIN, A.J. 1996. The NF-kappa B and I kappa B proteins: new discoveries and insights. Annu. Rev. Immunol. **14:** 649–683.
16. JOENJE, H. *et al.* 1985. Some characteristics of hyperoxia-adapted HeLa cells. A tissue culture model for cellular oxygen tolerance. Lab. Invest. **52**(4): 420–428.
17. LELLI, J.L. *et al.* 1998. ATP converts necrosis to apoptosis in oxidant-injured endothelial cells. Free Radical Biol. Med. **25**(6): 694–702.
18. LEIST, M. *et al.* 1997. Intracellular adenosine triphosphate (ATP) concentration: a switch in the decision between apoptosis and necrosis. J. Exp. Med. **185**(8): 1481–1486.
19. RICHTER, C. *et al.* 1996. Control of apoptosis by the cellular ATP level. FEBS Lett. **378**(2): 107–110.
20. ALLEN, C. *et al.* 1998. Glucose modulates cell death due to normobaric hyperoxia by maintaining cellular ATP. Am. J. Physiol. (Lung Cell. Mol. Physiol.) **274:** L159–1174.
21. SCHOONEN, W.G. *et al.* 1990. Hyperoxia-induced clonogenic killing of HeLa cells associated with respiratory failure and selective inactivation of Krebs cycle enzymes. Mutation Res. **237**(3-4): 173–181.
22. SCHOONEN, W.G. *et al.* 1990. Respiratory failure and stimulation of glycolysis in Chinese hamster ovary cells exposed to normobaric hyperoxia. J. Biol. Chem. **265**(19): 1118–1124.
23. KARIN, M. *et al.* 1997. AP-1 function and regulation. Curr. Opin. Cell Biol. **9:** 240–246.
24. KATOH, S. *et al.* 1997. The rescuing effect of nerve growth factor is the result of up-regulation of bcl-2 in hyperoxia-induced apoptosis of a subclone of pheochromocytoma cells, PC12h. Neurosci. Lett. **232**(2): 71–74.
25. SHAIKH, A.Y. *et al.* 1997. Melatonin protects bovine cerebral endothelial cells from hyperoxia-induced DNA damage and death. Neuroscience Lett. **229**(3): 193–197.
26. JYONOUCHI, H. *et al.* 1998. The effects of hyperoxic injury and antioxidant vitamins on death and proliferation of human small airway epithelial cells. Am. J. Respir. Cell. Mo.l Biol. **19:** 426–436.
27. LEMASTERS, J.J.V. 1999. necrapoptosis and the mitochondrial permeability transition: shared pathways to necrosis and apoptosis. Am. J. Physiol. **276:** G1–G6.

Hyperoxia-induced Cell Death in the Lung—the Correlation of Apoptosis, Necrosis, and Inflammation

LIN L. MANTELL,[a,b,e] STUART HOROWITZ,[d] JONATHAN M. DAVIS,[c] AND JEFFREY A. KAZZAZ[a]

[a]*CardioPulmonary Research Institute,* [b]*Department of Thoracic Cardiovascular Surgery and* [c]*Department of Pediatrics, Winthrop-University Hospital, SUNY/Stony Brook School of Medicine, Mineola, New York 11501, USA*

[d]*The Heart and Lung Institute, Jewish Hospital, Louisville, Kentucky 40202, USA*

ABSTRACT: **Prolonged exposure to hyperoxia causes tissue damage in many organs and tissues. Since the entire surface area of lung epithelium is directly exposed to O_2 and other inhaled agents, hyperoxia leads to the development of both acute and chronic lung injuries.These pathologic changes in the lung can also be seen in acute lung injury (ALI) in response to other agents. Simple strategies to mitigate hyperoxia-induced ALI might not be effective by virtue of merely reducing or augmenting the extent of apoptosis of pulmonary cells. Identification of the specific cell types undergoing apoptosis and further understanding of the precise timing of the onset of apoptosis may be necessary in order to gain a greater understanding of the connection between apoptosis and tolerance to hyperoxia and ALI. Attention should also be focused on other forms of non-apoptotic programmed cell death.**

OXYGEN THERAPY AND ACUTE LUNG INJURY

Supraphysiological concentrations of O_2 and mechanical ventilation are used in the management of critically ill patients with a variety of clinical conditions. For example, in premature infants, impaired alveolar gas exchange requires the administration of elevated concentrations of inspired O_2 to maintain life. However, prolonged exposure to hyperoxia causes tissue damage in many organs and tissues.[1] Since the entire surface area of lung epithelium is directly exposed to O_2 and other inhaled agents, hyperoxia leads to the development of both acute and chronic lung injuries.[2,3] While prolonged exposure to hyperoxia in newborns can result in the development of bronchopulmonary dysplasia (BPD), in adults it can result in, or exacerbate complications of, the adult respiratory distress syndrome (ARDS).[4] Hyperoxic damage to lungs is characterized by destruction of the alveolar-capillary barrier leading to pulmonary edema, impaired gas exchange, and in the worst case, death. Severe pneumonitis is a prominent feature of hyperoxic lung injury, involving

[e]Address for correspondence: CardioPulmonary Research Institute, Winthrop-University Hospital, 222 Station Plaza North, Suite 505, Mineola, NY 11501; 516-663-8978 (voice); 516-663-8872 (fax).

e-mail: lmantell@winthrop.org

inflammatory infiltration of the interstitium and the alveolar lumen.[5] These pathologic changes in the lung can also be seen in acute lung injury (ALI) in response to other agents.

HYPEROXIA, REACTIVE OXYGEN INTERMEDIATES (ROI), AND ANTIOXIDANTS

Pulmonary O_2 toxicity is mediated in part by toxic and highly reactive oxygen radicals.[6,7] The terms oxygen radical or oxygen-derived free radical are generally used to describe reactive oxygen intermediates (ROI) that have an unpaired electron, or molecules that are formed by the univalent reduction of oxygen. The most commonly known ROI are superoxide ion ($\cdot O_2^-$), hydrogen peroxide (H_2O_2), hydroxyl radical ($\cdot OH$), hypochlorous acid (HOCl) and nitric oxide ($\cdot NO$). The superoxide anion ($\cdot O_2^-$) is extremely unstable and converts to H_2O_2 either spontaneously at 2×10^5 M/s (at pH 7.4) or catalytically, by superoxide dismutase (SOD) at 8×10^5 M/s.[8] Once formed, H_2O_2 can be converted by catalase to H_2O and O_2, or reduced to $\cdot OH$ in the presence of the transition metals Fe^{2+} or Cu^+. H_2O_2 can also be transformed by myeloperoxidase into hypochlorous acid (HOCl), in the presence of Cl^-.

Although ROI are formed as by-products of aerobic metabolism in cells, under normal physiologic conditions, over 90% of the O_2 consumed by cells is degraded to H_2O by the cytochrome oxidase system, thus avoiding the production of significant quantities of free radicals. Only 1 to 2% of O_2 is converted to superoxide by the electron transport chain in mitochondria.[9] The remaining O_2 is converted to ROI in other cellular compartments. The ROI generated under these conditions react either spontaneously with cellular macromolecules that are then scavenged by cellular antioxidants, or are scavenged directly. As a consequence, normal cellular steady-state levels of ROI are maintained at very low concentrations.

Hyperoxia introduces a large influx of O_2, which is converted to superoxide in many cell types. This conversion is carried out either by an electron transport chain in mitochondria or by microsomal respiratory chains. As a consequence, the elevated levels of superoxide derived from hyperoxia result in oxidation of macromolecules. Newborn infants requiring prolonged O_2 therapy show evidence of increased pulmonary protein oxidation (protein carbonyl formation) and lipid peroxidation.[10]

Several lines of evidence from both clinical and basic research indicate that elevated levels of ROI are associated with a number of human diseases, especially when inflammatory injury is involved. Under these pathological conditions, large amounts of ROI, generated locally or systemically, react with cellular or extracellular targets, such as proteins, lipids, and nucleic acids, to induce cellular and tissue damage that leads to organ injury.[11–13]

The overproduction of ROI during hyperoxia has raised the question of how endogenous antioxidant systems respond to hyperoxic exposure. Much attention has been focused on the classic antioxidant enzymes (AOE), including superoxide dismutases (SOD), catalase, and glutathione peroxidases (GPx), which could play a crucial role in the response to oxidant stress by promoting the metabolism of ROI.[4,5,14] In adult animals, the activities of these enzymes respond differently to hyperoxia in various species, and there is no simple O_2 dose response. For example,

SOD, catalase, and GPx enzyme activities do not increase at all in response to hyperoxia in adult rabbits.[15] In contrast, adult rats exposed to 80 to 85% O_2 exhibit modest increases in the activities of SOD[16,17] and GPx,[18] although exposure to 100% O_2 results in a decrease in catalase activity in the lung.[19] Newborns are different; unlike the response in adult rabbits, the activities of SOD, catalase and cellular GPx are induced by hyperoxia (100% O_2) in normal newborn (term) rabbits.[20] These observations provide a possible explanation for the observations that neonates of some species (including rabbits), are relatively resistant to hyperoxia when compared to adults of the same species. Perhaps the capacity to induce antioxidant enzymes plays an adaptive role, rendering the animal tolerant.[22] Data supporting this notion come from experiments in rats where tolerance to O_2 toxicity is induced by pretreatment with hyperoxia, ozone,[23] tumor necrosis factor,[24] IL-1,[24,25] diethyldithiocarbamate[27] or intravenous endotoxin. In some species, these agents can induce AOE activities prior to subsequent exposure to hyperoxia. However, the pleiotropic effects of these treatments preclude straightforward interpretation of the results.

The correlation of the higher tolerance to hyperoxia-induced injury with induced levels of AOE activities provides the basis for the consideration of therapeutic use of antioxidant enzymes. Formal proof that the constitutive level of at least one AOE is important to survival in hyperoxia comes from a study of EC (extracellular)-SOD-null mice that were exposed to 100% O_2. These knockout mice were found to be sensitive to hyperoxia, relative to control diploids.[28] Therefore, even if endogenously-increased AOE activities are not themselves protective, antioxidant augmentation could still be of great therapeutic value. For example, transgenic mice engineered so that overexpression of Mn-SOD is targeted to alveolar type II epithelial cells survive substantially longer in 100% O_2 than do nontransgenic controls.[29] Therefore, cell-specific overexpression of AOE might be used therapeutically, to augment endogenous enzymes that are insufficient to deal with the increased ROI-associated acute lung injury.[30]

ROI AND APOPTOSIS

The highly elevated levels of ROI following hyperoxia exposure overwhelm the antioxidant systems in many pulmonary cell types, which therefore become targets of ROI-induced damage. Investigators have focused on the cell-specific effects of a variety of oxidative insults, and have found that no cell type is completely resistant to injury. Although endothelial and epithelial cells are sources of ROI generation, they can also be critical targets and their injury can lead to lung edema. When repair mechanisms fail or become overwhelmed, the cells die.

It is thought that cells die via one of the two modes, either apoptosis or necrosis.[31–33] Necrosis is the result of unscheduled, acute injury. Necrosis is unregulated and in living animals is accompanied by an inflammatory response, which can be local or systemic. On the other hand, apoptosis is physiologically-regulated, and occurs during normal cell turnover and development.[34,35] When apoptosis is triggered, cells actively participate in their own demise by following a specific program of events. Apoptosis is generally considered to be the predominant mode of cell death from oxidative insults.[36,37] However, nonapoptotic, programmed cell death can also result from such insults (see Kazzaz *et al.,* this volume). ROI can not only trigger apoptosis, but can also result from apoptosis stimulated by other, non-oxidative events. For example, apoptosis caused by steroid treatment or serum deprivation involves an obligatory membrane lipid peroxidation step.[38,39] Both apoptosis and associated membrane lipid peroxidation can be prevented by overexpression of some antioxidants.[40] Thus, tissue injury resulting from ROI might be expected to involve apoptosis.

In typical nucleated mammalian cells undergoing apoptosis, chromosomal DNA is hydrolyzed by endogenous endonucleases, yielding oligonucleosome-size fragments, chromatin condensation and nuclear shrinkage.[31–33] Overall cell shrinkage also occurs, probably as a result of increased intracellular protease activity.[41–44] Apoptotic cells are easily distinguished from normal cells in culture, because of their small size and condensed chromatin, which latter can be visualized using dyes (such as DAPI and Hoechst) that fluoresce when bound to DNA.[45–47] Genomic DNA isolated from apoptotic cells, but not normal healthy cells, usually contains a low molecular weight fraction of DNA that can be visualized by agarose gel electrophoresis as 200 bp nucleosome ladders.[46,48,49] Moreover, the 3′-OH ends created by activated endonucleases are substrates for the enzyme T4 terminal deoxynucleotidyl transferase, which has been used for *in situ* assays of apoptosis. Used together with labeled dUTP, the TUNEL (Terminal dUTP Nick End Labeling) assay can be used to identify apoptotic cells in fixed cell cultures or tissue sections, since only the apoptotic nuclei will incorporate dUTP at detectable levels.[50,51]

HYPEROXIC LUNG INJURY AND CELL DEATH

In addition to the increased levels of ROI, a marked inflammatory response in the lung takes place following exposure to hyperoxia. Necrosis is therefore expected to be the only mode of cell death following hyperoxia, according to the current paradigm of inflammation occurring secondary to necrosis.[52] However, there have been few studies testing whether apoptosis occurs during acute tissue injury in complex organs, including the lung, which is estimated to contain as many as 60 different cell types.[53]

Lung injury is known to be multifaceted, involving cascades of events that can be initiated by a wide variety of primary insults.[54] Because of the complex nature of acute lung injury, we hypothesized that cell death might be multimodal, involving not only necrosis, but also apoptosis.

We examined apoptosis in a mouse model of acute lung injury induced by hyperoxia. The TUNEL assay has provided a powerful tool to assess apoptosis *in situ.* In a mouse model of hyperoxic lung injury, TUNEL-positive, apoptotic nuclei were clearly evident at a time when lung injury was severe, relative to control unexposed mice.[55] To independently confirm that the TUNEL assay was an accurate marker of apoptosis in the lung, isolated genomic DNA from exposed and control lungs was studied. Hyperoxic lungs exhibited nucleosomal DNA ladders that are typical of apoptosis, and these ladders were not detected in control lungs. Moreover, this response was not generalized to other, uninjured, organs, since ladders were not evident in the liver DNA of the same animals.[55]

Similar observations were made in other labs and in other hyperoxic animals, indicating that apoptosis is a prominent feature of hyperoxic lung injury, and is specific to the lung.[55–58] However, except for the obvious apoptosis in airway epithelium, it is difficult to definitively identify the cell types involved, especially in the gas exchange regions. Such identification requires electron microscopy or dual labeling methods.[59]

The observation that apoptosis is a feature of hyperoxic acute lung injury is somewhat surprising in light of the current dogma that apoptosis is thought not to initiate inflammation.[52] Taken together with previous morphological descriptions of tissue necrosis during acute lung injury, these observations suggest a dual mode of cell death. Moreover, the notion that apoptosis occurs in the absence of an inflammatory response should be re-evaluated. Because inflammation is a prominent component in several models of lung injury,[55] it is difficult to determine whether apoptosis results from the increased ROI burden, or is secondary to mediators that are elaborated by inflammatory cells. In either case, apoptotic cells themselves are likely to add to the overall oxygen radical burden, because membrane lipid peroxidation appears to be a necessary component of the apoptotic cascade.[60,61]

EXPRESSION OF APOPTOSIS-RELATED GENES

Relatively few studies have examined the expression of apoptosis-related genes in hyperoxic lung injury. Activation of caspase family members is one of the essential events leading to apoptosis. Caspase-1, also known as IL-1β–converting enzyme (ICE), has been shown to play important roles in both apoptosis and inflammation. To determine if ICE activation occurs in hyperoxia-induced ALI, we examined the expression of ICE in untreated and hyperoxic mouse lungs and found that ICE expression was induced relative to controls. In particular, the activated form of ICE (p20) was present exclusively in nuclei of injured lung cells.[55]

Recently, several other genes involved in the regulation of apoptosis have been examined in mouse lungs.[57,62,63] Individual members of the Bcl-2 gene family can either inhibit or promote apoptosis, depending on the relative abundance of the individual gene product at particular time. To begin to understand the role of the Bcl-2 family in hyperoxic cell death, northern blots were probed for Bcl-x and Bax, two members whose increased expression is associated with promotion of cell death. Prolonged hyperoxic exposure was associated with an 8-fold increase in the relative abundance of bax mRNA, and a 12-fold increase in bcl-x.[57] These data provide the rationale for further investigation into the Bcl-2 family in hyperoxic ALI.

It is also known that hyperoxia induces DNA damage. The protooncogene p53 has been shown to play an essential role in the cellular response to DNA damage by regulating apoptosis and cellular proliferation. Perhaps alteration of expression of protooncogene p53 at both mRNA and protein levels is an upstream event of pulmonary apoptosis. Interestingly, hyperoxia induces p53 expression in mouse lungs.[57,62] Immunohistochemistry studies revealed that this upregulation of p53 may occur primarily in airway epithelial cells.[62] In addition, Fas, which is associated with the onset of apoptosis, was upregulated in the hyperoxic mouse lungs. p21—a cyclin-dependent kinase inhibitor involved in inhibition of cell proliferation and regulation of apoptosis—increases at both mRNA and protein levels following hyperoxic exposure. Moreover, during recovery from hyperoxia, a decrease in p21 expression was observed.[63]

Alteration in the expression of these apoptosis regulation-related genes in pulmonary cells does not necessarily indicate that they are important for survival from hyperoxia. Neither Fas- nor p53-null mice showed any changes in susceptibility to O_2 toxicity (as measured by wet: dry lung weight ratios, or pulmonary edema), suggesting that these pathways are not of central importance in hyperoxic cell death in the lung.[57] Therefore, while they might participate in programmed pulmonary cell death, they appear to be dispensable, perhaps reflecting redundant pathways.

THE ROLE OF APOPTOSIS IN THE TOLERANCE TO HYPEROXIA

The presence of apoptosis in the acutely injured lung could be part of a protective mechanism to limit the extent of injury or inflammation, as has been proposed during resolution of ARDS.[64,65] If apoptosis is also protective during acute lung injury, animals that are relatively resistant to such injury might be predicted to exhibit apoptosis sooner than sensitive animals. To examine this possibility, we studied two different animal models with differential sensitivity to hyperoxia: inbred mice[66] and adult vs. newborn rabbits.[67] Quantitative TUNEL assays indicate that extent of apoptosis correlates with the severity of hyperoxia induced ALI, and inversely correlates with resistance.[55] In these models, there was no association of apoptosis and resistance. Alternatively, apoptosis is simply a measure of the severity of lung injury. Overexpression of interleukin-11 (IL-11) has been shown to induce tolerance to hyperoxia and mitigate the development of ALI.[58] Interestingly, the extent of apoptosis assayed either by TUNEL or DNA fragmentation was higher in the lungs of the wild-type mice compared to their transgenic littermates overexpressing IL-11 specifically in the lung.[58] Likewise, in a study of cell death in hyperoxic adult rats,[56] it was observed that rats show more apoptosis in their lungs with increasing exposure to hyperoxia. However, the extent of pulmonary apoptosis was significantly diminished in lungs showing marked sensitivity to hyperoxia, suggesting that the tolerance to hyperoxic lung injury may be mediated through increasing the extent of apoptosis in these animals.[56] In summary, the percent of cells undergoing apoptosis is highly variable in models of acquired (vs. inbred) resistance to hyperoxia in rats and mice, suggesting that there is no simple quantitative relationship between the extent of apoptosis and resistance or sensitivity to hyperoxia.[56]

Apoptosis is also involved in the remodeling at the air-lung interface to restore lung epithelium during the repair phase of ALI. An important event is the removal of excess, hyperplastic type II cells. Apoptosis of type II cells is prominent in patients with resolving acute lung injury, while minimal apoptosis is observed in patients with chronic pneumonia.[68] In addition, bronchoalveolar lavage from patients during the repair phase of ALI is able to induce apoptosis in cultured pulmonary cells. In contrast, lavage fluid from either normal individuals or patients during the injury phase does not possess this capability.[64]

It is therefore important to point out that simple minded strategies to mitigate hyperoxia-induced ALI might not be effective by virtue of merely reducing or augmenting the extent of apoptosis of pulmonary cells. Identification of the specific cell types undergoing apoptosis and further understanding of the precise timing of the onset of apoptosis may be necessary in order to gain a greater understanding of the connection between apoptosis and tolerance to hyperoxia and ALI. Furthermore, more attention should focus on other forms of non-apoptotic programmed cell death by hyperoxia.[69,70]

REFERENCES

1. WISPE, J. & R. ROBERTS. 1987. Molecular basis of pulmonary oxygen toxicity. Clin. Perinatol. **14:** 651–666.
2. NORTHWAY, W.J. *et al.* 1967. Pulmonary disease following respiratory therapy of hyaline membrane disease: Bronchopulmonary dysplasia. New Engl. J. Med. **276:** 357–368.
3. HOROWITZ, S. & J.M. DAVIS. 1997. Lung injury when development is interrupted by premature birth. *In* Lung Growth and Development. J.A. McDonald, Ed.: 577–610. Marcel Dekker, New York.
4. PAPPAS, C. *et al.* 1983. Effect of prolonged exposure to 80% oxygen on the lung of the newborn mouse. Lab. Invest. **48:** 735–748.
5. FREEMAN, B. *et al.* 1986. Antioxidant enzyme activity in alveolar type II cells after exposure of rats to hyperoxia. Exp. Lung Res. **10:** 203–222.
6. FREEMAN, B. *et al.* 1982. Hyperoxia increases oxygen radical production in rat lung homogenates. Arch. Biochem. Biophys. **216:** 477–484.
7. FREEMAN, B.C. & J.D. CRAPO. 1981. Hyperoxia increases oxygen radical production in rat lungs and lung mitochondria. J. Biol. Chem. **256:** 10986–10992.
8. MOSLEN, M. 1994. Reactive oxygen species in normal physiology, cell injury and phagocytosis. Adv. Exp. Med. Biol. **366:** 17–27.
9. GRISHAM, M. 1992. Reactive metabolites of oxygen and nitrogen in biology and medicine. RG Lands, Austin.
10. VARSILA, E. & P.E. ANDERSSON. 1995. Early protein oxidation in the neonatal lung is related to development of chronic lung disease. Acta Paediatr. **84:** 1296–1299.
11. GHIO, A. & D. QUIGLEY. 1994. Complexation of iron by humic-like substances in lung tissue: role in coal workers' pneumoconiosis. Am. J. Physiol. **267:** L173–L179.
12. GRONECK, P. & C. SPEER. 1995. Inflammatory mediators and bronchopulmonary dysplasia. Arch. Dis. Child Fetal Neonatal Ed. **73:** F1–F3.
13. GROEN, H. *et al.* 1994. Clinical evaluation of lymphocyte sub-populations and oxygen radical production in sarcoidosis and idiopathic pulmonary fibrosis. Respir. Med. **88:** 55–64.
14. MCCORD, J. 1993. Human disease, free radicals, and the oxidant/antioxidant balance. Clin. Biochem. **26:** 351–357.
15. BAKER, R. *et al.* 1989. Development of O2 tolerance in rabbits with no increase in antioxidant enzymes. J. Appl. Physiol. **66:** 1679–1684.
16. CRAPO, J. & J. MCCORD. 1976. Oxygen-induced changes in pulmonary superoxide dismutase assayed by antibody titrations. Am. J. Physiol. **231:** 1196–1203.
17. VINCENT, R. *et al.* 1994. Quantitative immunocytochemical analysis of Mn SOD in alveolar type II cells of the hyperoxic rat. Am. J. Physiol.-Lung Cell Mol. Pharmacol. **11:** L475–L481.
18. KENNEDY, C. *et al.* 1992. Application of the EPR spin-trapping technique to the detection of radicals produced in vivo during inhalation exposure of rats to ozone. Toxicol. Appl. Pharmacol. **114:** 41–46.

19. AHOTUPA, M. *et al.* 1992. Pro-oxidant effects of normobaric hyperoxia in rat tissues. Acta Physiol. Scand. **145:** 151–157.
20. FRANK, L. & I. SOSENKO. 1991. Failure of premature rabbits to increase antioxidant enzymes during hyperoxic exposure: increased susceptibility to pulmonary oxygen toxicity compared with term rabbits. Pediatr. Res. **29:** 292–296.
21. FINI, M. *et al.* 1987. Cloning of a complementary DNA for rabbit proactivator. A metalloproteinase that activates synovial cell collagenase, shares homology with stromelysin and transin, and is coordinately regulated with collagenase. Arthritis Rheum. **30:** 1254–1264.
22. YAM, J. *et al.* 1978. Oxygen toxicity: Comparison of lung biochemical responses in neonatal and adult rats. Pediatr. Res. **12:** 115–119.
23. JACKSON, R. & L. FRANK. 1984. Ozone-induced tolerance to hyperoxia in rats. Am. Rev. Respir. Dis. **129:** 425–429.
24. JENSEN, J. *et al.* 1992. Role of tumor necrosis factor in oxygen toxicity. J. Appl. Physiol. **72:** 1902–1907.
25. LEWIS-MOLOCK, Y. *et al.* 1994. Lung manganese superoxide dismutase increases during cytokine-mediated protection against pulmonary oxygen toxicity in rats. Am. J. Respir. Cell. Mol. Biol. **10:** 133–141.
26. TSAN, M. *et al.* 1991. Interleukin 1 protects rats against oxygen toxicity. J. Appl. Physiol. **71:** 688–697.
27. MANSOUR, H. *et al.* 1986. Diethyldithiocarbamate provides partial protection against pulmonary and lymphoid oxygen toxicity. J. Pharmacol. Exp. Ther. **236:** 476–480.
28. CARLSSON, L.M. et al. 1995. Mice lacking extracellular superoxide dismutase are more sensitive to hyperoxia. Proc. Natl. Acad. Sci. USA **92:** 6264–6268.
29. WISPE, J. *et al.* 1992. Human Mn-superoxide dismutase in pulmonary epithelial cells of transgenic mice confers protection from oxygen injury. J. Biol. Chem. **267:** 23937–23941.
30. TURRENS, J. 1991. The potential of antioxidant enzymes as pharmacological agents in vivo. Xenobiotica **21:** 1033–1040.
31. FARBER, E. 1994. Programmed cell death: Necrosis versus apoptosis. Modern Pathol. **7:** 605–609.
32. MAJNO, G. & I. JORIS. 1995. Apoptosis, oncosis, and necrosis: an overview of cell death. Am. J. Pathol. **146:** 3–15.
33. STELLER, H. 1995. Mechanisms and genes of cellular suicide. Science 267: 1445-1449.
34. WYLLIE, A.H. 1995. The genetic regulation of apoptosis. Curr. Opin. Genet. Develop. **5:** 97–104.
35. WHITE, E. 1996. Life, death, and the pursuit of apoptosis. Gene Develop **10:** 1–15.
36. PAYNE, C.M. *et al.* 1995. Apoptosis overview emphasizing the role of oxidative stress, DNA damage and signal-transduction pathways. Leuk Lymphoma **19:** 43–93.
37. SLATER, A.F.G. *et al.* 1995. The role of intracellular oxidants in apoptosis. Biochim. Biophys. Acta **1271:** 59–62.
38. FAVIER, A. *et al.* 1994. Antioxidant status and lipid peroxidation in patients infected with HIV. Chemico-Biol. Interact. **91:** 165–180.
39. ZAMZAMI, N. *et al.* 1995. Sequential reduction of mitochondrial transmembrane potential and generation of reactive oxygen species in early programmed cell death. J. Exp. Med. **182:** 367–377.
40. HOCKENBERY, D.M. *et al.* 1993. Bcl-2 functions in an antioxidant pathway to prevent apoptosis. Cell **75:** 241–251.
41. HALE, A.J., *et al.* 1996. Apoptosis: molecular regulation of cell death. Eur. J. Biochem. **236:** 1-26.
42. SLATER, A.F.G. *et al.* 1996. Intracellular redox changes during apoptosis. Cell Death Differentiation **3:** 57-62.

43. NICHOLSON, D.W. 1996. ICE/CED3-like proteases as therapeutic targets for the control of inappropriate apoptosis. Nat. Biotechnol. **14:** 297-301.
44. PATEL, T. *et al.* 1996. The role of proteases during apoptosis. FASEB J 10: 587-597.
45. YANAGISAWA-SHIOTA, F. *et al.* 1995. Endonuclease activity and induction of DNA fragmentation in human myelogenous leukemic cell lines. Anticancer Res. **15:** 259–265.
46. BORTNER, C.D. *et al.* 1995. The role of DNA fragmentation in apoptosis. Trends Cell Biol. **5:** 21–26.
47. BELLOC, F. *et al.* 1994. A flow cytometric method using Hoechst 33342 and propidium iodide for simultaneous cell cycle analysis and apoptosis determination in unfixed cells. Cytometry **17:** 59–65.
48. WONG, P. *et al.* 1994. The use of C0t-1 probe DNA for the detection of low levels of DNA fragmentation. Biochem. Cell Biol. **72:** 649–653.
49. LEIST, M. *et al.* 1994. Murine hepatocyte apoptosis induced in vitro and in vivo by TNF-α requires transcriptional arrest. J. Immunol. **153:** 1778–1788.
50. GAVRIELI, Y. *et al.* 1992. Identification of programmed cell death in situ via specific labeling of nuclear DNA fragmentation. J. Cell Biol. **119:** 493–501.
51. TORNUSCIOLO, D.R.Z. *et al.* 1995. Simultaneous detection of TdT-mediated dUTP-Biotin nick end labeling (TUNEL)-positive cells and multiple immunohistochemical markeres in single tissue sections. Biotechniques **19:** 800–805.
52. THOMPSON, C.B. 1995. Apoptosis in the pathogenesis and treatment of disease. Science **267:** 1456–1462.
53. STONE, K.C. *et al.* 1992. Allometric relationships of cell numbers and size in the mammalian lung. Am. J. Respir. Cell. Mol. Biol. **6:** 235–243.
54. SPRAGG, R.G. & R.M. SMITH. 1991. Biology of lung injury. *In* The Lung: Scientific Foundations. R.G. Crystal & J.B. West, Eds. Raven Press, Ltd. New York.
55. MANTELL, L.L. *et al.* 1997. Unscheduled apoptosis during acute inflammatory lung injury. Cell Death Diff. **4:** 604–607.
56. OTTERBEIN, L.E. *et al.* 1998. Pulmonary apoptosis in aged and oxygen-tolerant rats exposed to hyperoxia. Am. J. Physiol. (Lung Cell. Mol. Physiol.) **275:** L14–L20.
57. BARAZZONE, C. *et al.* 1998. Oxygen toxicity in mouse lung: Pathways to cell death. Am. J. Respir. Cell Mol. Biol. **19:** 573–581.
58. WAXMAN, A. *et al.* 1998. Targeted lung expression of interleukin-11 enhances murine tolerance of 100% oxygen and diminishes hyperoxia-induced DNA fragmentation. J. Clin. Invest. **101:** 1970–1982.
59. PIEDBOEUF, B. *et al.* 1996. In vivo expression of intercellular adhesion molecule 1 in type II pneumocytes during hyperoxia. Am. J. Respir. Cell Mol. Biol. **15:** 71–77.
60. GIROTTI, A. 1998. Lipid hydroperoxide generation, turnover, and effector action in biological systems. Lipid Res. **39:** 1529–1542.
61. KOWALTOWSKI, A. & A. VERCESI. 1999. Mitochondrial damage induced by conditions of oxidative stress. Free Radic. Biol. Med. **26:** 463–471
62. O'REILLY, M. *et al.* 1998. Exposure to hyperoxia induces p53 expression in mouse lung epithelium. Am. J. Respir. Cell. Mol. Biol. **18:** 43–50.
63. O'REILLY, M. *et a*l. 1998. Accumulation of p21(Cip1/WAF1) during hyperoxic lung injury in mice. Am. J. Respir. Cell. Mol. Biol. **19:** 777–785.
64. POLUNOVSKY, V.A. et al. 1993. Role of mesenchymal cell death in lung remodeling after injury. J. Clin. Inves. **92:** 388–397.
65. UHAL, B.D. *et al.* 1995. Fibroblasts isolated after fibrotic lung injury induce apoptosis of alveolar epithelial cells in vitro. Am. J. Physiol.-Lung Cell. Mol. Physiol. **13:** L819–L828.
66. HUDAK, B. *et al.* 1993. Inter-strain variation in susceptibility to hyperoxic injury of murine airways. Pharmacogenetics 3: 135–143.

67. VENESS-MEEHAN, K.A. *et al.* 1991. Cell-specific alteration in expression of hyperoxia-induced messenger RNAs of lung. Am. J. Respir. Cell Mol. Biol. **5:** 516–521.
68. BARDALES, R. *et al.* 1996. Apoptosis is a major pathway responsible for the resolution of type II pneumocytes in acute lung injury. Am. J. Pathol. **149:** 845–852.
69. KAZZAZ, J.A. *et al.* 1996. Cellular oxygen toxicity—oxidant injury without apoptosis. J. Biol. Chem. **271:** 15182–15186.
70. LI, Y. *et al.* 1997. Nuclear factor-kappaB is activated by hyperoxia but does not protect from cell death. J. Biol. Chem. **272:** 20646–20649.

Apoptosis in Coxsackievirus B3–induced Myocarditis and Dilated Cardiomyopathy

S.A. HUBER,[a,c] R.C. BUDD,[b] K. ROSSNER,[b] AND M.K. NEWELL[b]

[a]*Department of Pathology, University of Vermont, Burlington, Vermont 05405, USA*

[b]*Department of Medicine, University of Vermont, Burlington, Vermont 05405, USA*

ABSTRACT: Group B coxsackieviruses (CVB), which infect the myocardium, cause myocarditis and dilated cardiomyopathy. However, not all infections of the myocardium result in disease. In the mouse model, CVB infection stimulates autoimmune T cell response to cardiac antigens, and these autoimmune effectors cause myocyte necrosis and cardiomyopathy. Induction of pathogenic autoimmunity depends upon CD4$^+$ Thl (interferon-γ positive) cells while Th2 (IL-4 positive) cell responses promote disease resistance. T lymphocytes expressing the γ-δ T cell receptor (γδ$^+$) constitute up to 12% of the inflammatory cells in the heart and are crucial to maintaining a dominant Th1 response phenotype. γδ$^+$ lymphocytes modulate T cell responses by selectively lysing CD4$^+$ Th2 cells. Thl cells are not killed by γδ$^+$ cells. Lysis requires direct cell:cell interaction between the γδ$^+$ cell and CD4$^+$ Th2 target and is most likely mediated through Fas:FasL interaction. These studies demonstrate a novel mechanism for immune modulation of cytokine responses *in vivo*.

Antigenic stimulation can induce naïve CD4$^+$ T cells to differentiate into populations with defined cytokine profiles.[1–3] The best described of these are Thl cells, which make interferon-gamma (IFNγ), interleukin-2 (IL-2), and tumor necrosis factor-beta (TNFβ), and Th2 cells, which make IL-4, IL-5, IL-6, and IL-10. Often, susceptibility or resistance to specific diseases corresponds to a preferential Thl or Th2 cell response in the individual.[4–8] Thl cells usually promote cellular immunity while Th2 cells interact with B lymphocytes in humoral immunity. Allergic asthma depends upon Th2 cells because the IL-4 produced by these lymphocytes facilitates immunoglobulin class switching to the IgE isotype.[9] Similarly, in both systemic lupus erythematosis and rheumatoid arthritis, Th2 bias is found.[5] Diseases dependent on cellular immune responses often show a Thl cell dominance. These include experimental allergic and Theiler's virus–induced encephalopathies,[6,10] multiple sclerosis,[6] diabetes,[7] inflammatory arthritides,[8,11] and coxsackievirus B3–induced myocarditis.[12,13]

Presently, there are three major mechanisms for imposing immune bias towards a particular Th cell phenotype. These are:

1. factors influencing differentiation of the naïve or precursor Th0 cell into a Thl or Th2 cell[1,14–17];

[c]Address correspondence to: Sally Huber, Ph.D., Department of Pathology, University of Vermont, Burlington, VT 05405; 802-656-8944 (voice); 802-656-8965 (fax).
e-mail: shuber(a)salus.uvm.edu

2. factors promoting survival of differentiated Thl or Th2 cells at the termination of an immune response[18,19]; and
3. factors causing selective elimination of differentiated Thl or Th2 cell.[12,13]

The best investigated of these mechanisms is the first, i.e., factors involved in differentiation of Th cells. Activation of Th0 cells in the presence of IL-12, IFNγ, or low antigen concentrations causes preferential Thl cell bias. Similarly, presence of IL-4 favors a Th2 cell response. Certain antigen-presenting cells, such as dendritic cells or B lymphocytes, are also more likely to stimulate differentiation to Thl or Th2 cells, respectively.[20,21] The type of antigen-presenting cell may influence Th0 cell differentiation either by variable expression of specific accessory molecules, or by altered antigen processing. The accessory molecule B7-1 reportedly favors Thl cell responses while B7-2 promotesTh2 cell differentiation.[14] Antigen-presenting cells differing in relative B7-l/B7-2 expression could significantly affect the dominant Th cell response. Similarly, different antigen presenting cells can use separate proteases with distinct peptide cleavage sites in antigen processing.[22] This can result in dissimilar sets of peptides produced from the same antigen. Individual T cell epitopes may themselves promote Th cell differentiation to a specific subtype through avidity of binding to MHC class II molecules with tightly bound epitopes inducing predominant Thl cell responses.[23]

Once differentiated, Thl cells are more sensitive to activation-induced cell death (AICD) than Th2 cells.[18,19] Thl cells express both Fas and FasL, resulting in autocrine cell death after growth factor withdrawal. Th2 cells either lack or express lower levels of FasL, which promotes their survival after an immune response. AICD should produce a long-term bias of antigen-specific T cells toward the Th2 phenotype. To counterbalance AICD, we believe that specific lymphocytes expressing the γδ T cell receptor (TcR) are capable of selectively killing Th2 cells while sparing Thl cells. Several studies have shown that $\gamma\delta^+$ T cells modulate $CD4^+$ Th cell responses either toward a Thl or Th2 cell phenotype.[9,24] Generally, most studies suggest that $\gamma\delta^+$ lymphocytes modulate Th cell responses through the release of cytokines (IFNγ or IL-4) early in the $CD4^+$ T cell response, thus driving Th0→Thl or Th0→Th2 differentiation. While this certainly may represent *one* mechanism for $\gamma\delta^+$ cell involvement, we believe that $\gamma\delta^+$ cells can also directly kill Th2 cells, most likely through Fas-dependent mechanisms.

Two lineages of T lymphocytes are distinguished by their TcR. The majority population in peripheral blood and lymphoid organs express α and β polypeptide chains, the $\alpha\beta^+$ T cells. These are the classical antigen-specific T lymphocytes, which react to processed antigen bound to major histocompatibility complex (MHC) molecules.[25] A separate population, which primarily localizes to the epithelial tissues, expresses γ and δ polypeptide chains ($\gamma\delta^+$ T cells).[26] Unlike the $\alpha\beta^+$ T cells, lymphocytes expressing the γδ TcR recognize antigen directly without requiring antigen processing. Frequently, $\gamma\delta^+$ cell recognition is either independent of MHC molecules or occurs in the absence of peptides binding to MHC antigen.[27,28] $\gamma\delta^+$ T cells are often categorized as a form of "innate" immunity.[29] Innate immunity refers to broadly reactive immune factors that are capable of impeding infections during the time required for antigen-specific immunity to develop. The major components of the innate immune response are macrophages, natural killer cells, and $\gamma\delta^+$ T cells. While some $\gamma\delta^+$ cells apparently react to the infecting organism,[30] others recognize either heat shock proteins,[26] nucleotide triphosphates[31] or prenyl pyrophosphates,[32] which are common to multiple organisms and stressed or damaged host tissues.

γδ⁺ T cells often accumulate at sites of inflammation in both autoimmune and infectious diseases. In some cases, the γδ⁺ T cells may prove beneficial to the individual and dampen the immune-mediated tissue injury. Experimental murine models of arthritis and lupus-like disease are aggravated by depletion of γδ⁺ T cells.[33,34] In contrast, γδ⁺ T cells promote pathogenicity in coxsackievirus B3 (CVB3)–induced myocarditis in mice.[12,13] Two variants of CVB3 have been selected that both infect and replicate in the heart. The variant, designated H3 induces severe myocardial inflammation (FIG. 1) and high animal mortality in BALB/c mice. The second variant, designated H310A1, produces minimal myocarditis and no animal deaths, despite high concentrations of virus in the heart. Precursor frequency analysis of the virus-reactive CD4⁺ T cell population indicates that pathogenicity correlates with a preferential Thl (IFNγ⁺) cell response while resistance correlates with a Th2 (IL-4⁺) response (FIG. 2). γδ⁺ T cells constitute between 5 and 12% of the inflammatory cell infiltrate in the hearts of H3 virus–infected mice. These γδ⁺ T cells were isolated from the myocarditic animals by positive selection on anti-γδ TcR antibody-coated plates. Adoptive transfer of 5000 γδ⁺ T cells by intravenous injection into H310A1 virus–infected animals was sufficient to restore myocarditis susceptibility in the

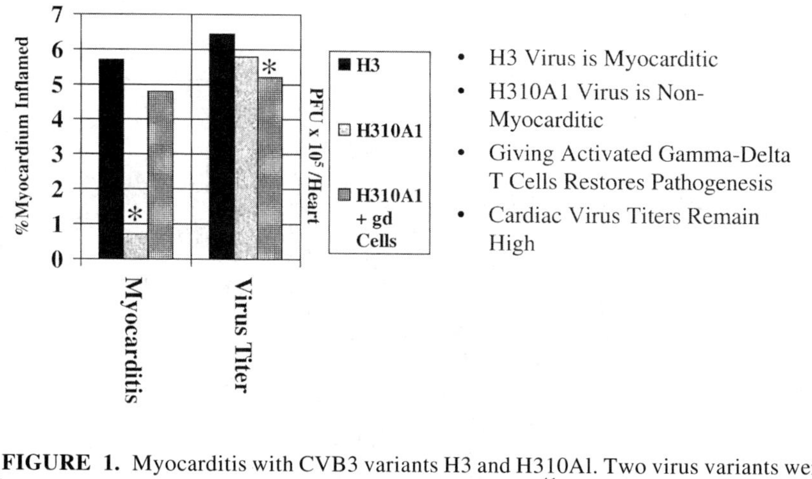

FIGURE 1. Myocarditis with CVB3 variants H3 and H310Al. Two virus variants were isolated and characterized as described in Van Houten *et al.*[41] BALB/c mice were infected intraperitoneally with 10^4 plaque-forming units (PFU) virus. Seven days later, hearts were removed and divided in half. Half of each heart was formalin-fixed, paraffin-embedded, sectioned, stained with hematoxylin and eosin, and evaluated by image analysis[42] for percent of the myocardium undergoing inflammation. The remaining heart tissue was homogenized and titered for virus according to established protocol.[41] Cells expressing the γδ TcR (gd) were isolated from the hearts of H3 virus-infected BALB/c donor mice 7 days after infection by positive selection on anti-γδ TcR antibody-coated plates as described in Huber *et al.*[12] 5000 purified γδ⁺ T cells were injected intravenously through the tail vein into BALB/c recipients on the same day as the animals were infected with H310A1 virus. Results represent mean values of 5 mice/group. *Group significantly different from H3 virus–infected animals at $p < 0.05$.

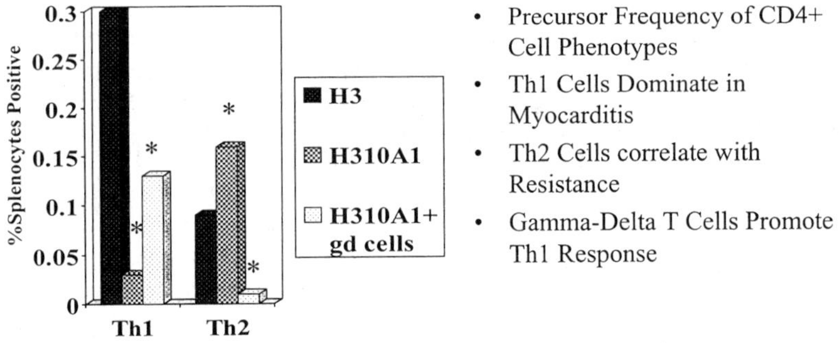

FIGURE 2. γδ cells promote myocarditis. Precursor frequency analysis of virus-specific CD4$^+$ Thl and Th2 cell populations. Spleens from the animals described in FIGURE 1 were pressed through fine mesh screens to form single cell suspensions. The cells were centrifuged on ficoll-hypaque to remove red blood cells and CD4$^+$ T cells were purified as described in Huber *et al.*[12] Between 0 and 1000 CD4$^+$ T cells were co-cultured with 5×10^4 irradiated (2000R) BALB/c monocytes, 20 U/ml rIL-2 and 10 pg/ml CVB3 for 14 days. The remaining viable cells were washed and stimulated with 5 μg/ml concanavalin A. Supernatants were retrieved and assayed for IFNγ and IL-4 by capture ELISA. Estimates of precursor frequency was obtained by the maximum-likelihood method of Good *et al.*[43] Results represent mean values of 5 mice/group. *Group significantly different from H3 infected animals at $p < 0.05$.

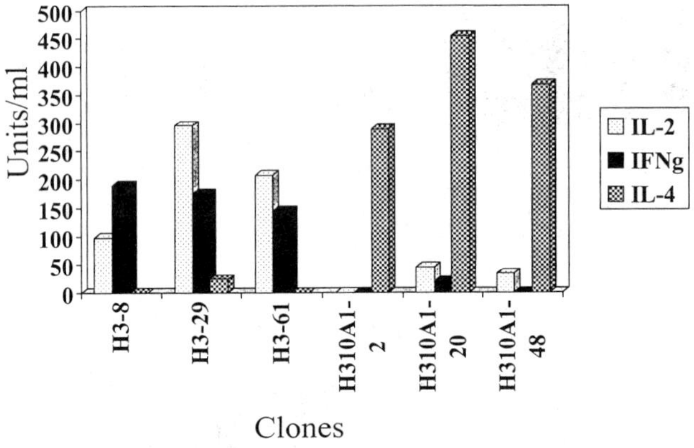

FIGURE 3. Cytokine profiles of CD4$^+$ T cell clones. BALB/c mice were infected i.p. with 10^4 PFU H3 or H310A1 virus. Seven days later, CD4$^+$ T cells were isolated and co-cultured with irradiated splenocytes and 1 μg/ml virus. Wells showing growth after 14 days were subcloned at 0.3 cells/well and maintained by restimulation at 14 day intervals with fresh irradiated splenocytes and virus. For cytokine analysis, individual clones were incubated with 50 ng/ml PMA and 500 ng/ml ionomycin for 24 hours. Supernatants were retrieved and assayed for IFNγ, IL-2 and IL-4 by capture ELISA. Results represent calculated cytokine concentration/ml of original supernatant.

recipients and shift the dominant $CD4^+$ cell response from a Th2 to a Thl phenotype (FIGS. 1 and 2). This effect required activated $\gamma\delta^+$ T cells since cells isolated from the spleens of uninfected mice were incapable of altering H310A1 virus–induced resistance. The change was not due to transfer of the pathogenic H3 virus with the $\gamma\delta^+$ T cells. When the $\gamma\delta^+$ T cells were homogenized and titered on HeLa cell monolayers, no virus was observed. Secondly, adoptive transfer of the activated $\gamma\delta^+$ T cell population into uninfected BALB/c recipients did not result in detectable infection or myocarditis. From these results, we hypothesized that $\gamma\delta^+$ T cells modulated myocarditis-susceptibility through their effect on Th subset responses.

To determine how $\gamma\delta^+$ T cells modulate Th subset responses, studies were performed in vitro using virus-specific $CD4^+$ T cell clones derived from H3 (Thl) and H310A1 (Th2) virus–infected animals.[13] Briefly, BALB/c mice were infected intraperitoneally with 10^4 plaque forming units (PFU) of either H3 or H310A1 virus. Seven days later, $CD4^+$ T cells were obtained and cultured with irradiated syngeneic splenocytes and 1 μg/ml of the respective virus. Wells showing positive cell growth 14 days later were subcloned 0.3 cells/well and maintained by restimulation at 14 day intervals with fresh virus and irradiated splenocytes. $CD4^+$ T cell clones were phenotyped for cytokine production by culturing 105 cells from each clone with 50 ng/ml PMA and 500 ng/ml ionomycin for 24 hours. Supernatants were retrieved and assayed for cytokine by capture ELISAS. Results of selected clones are given in FIGURE 3. Three Thl clones (H3-8, H3-29, and H3-61) were identified by production of IFNγ and IL-2, but not IL-4. Three Th2 clones (H310A1-H2, H310A1-H20 and H310A1-H48) produced IL-4 but little or no IFNγ.

Next, the $CD4^+$ T cell clones were obtained 10 days after the last antigen stimulation. These lymphocytes were co-cultured with isolated $\gamma\delta^+$ T cells at a ratio of 1:10 for six hours. The $\gamma\delta^+$ T cells were removed by adsorption to anti-$\gamma\delta^+$ TcR antibody–coated plates and the non-adherent $CD4^+$ cells were stained with propidium iodide. Additional $CD4^+$ cells were incubated with 100 ng/ml hamster anti-Fas (Pharmingen) then incubated on mouse anti-hamster IgG-coated plates to cross-link Fas on the T cell surface. CD4+ cells were evaluated for DNA content by flow cytometry (FIG. 4).Cross-linking Fas resulted in apoptosis of 29% of Thl and 20% of Th2 cells (DNA less than 2n). Co-culture with $\gamma\delta^+$ T cells resulted in apoptosis of 36% of the Th2 cell clone but only 6% of the Thl cell clone. Thus, Fas cross-linking kills both Thl and Th2 cells with a bias in killing towards the Thl target. $\gamma\delta^+$ T cells preferentially kill Th2 cells *in vitro.*

The next question is how do $\gamma\delta^+$ T cells kill $CD4^+$ T cells? $\gamma\delta^+$ T cells both produce cytokines, such as TNFα, and express high levels of FasL. Since antibody cross-linking of Fas results in lysis of both Thl and Th2 cells, one might hypothesize that soluble mediators are the more likely mechanism of Th2 cell death. To confirm or refute this hypothesis, $\gamma\delta^+$ T cells and ^{51}Cr-labeled H310Al-H48 Th2 cells were cultured either together or separated by membrane inserts (0.2 μm pore size, Nunc). When $\gamma\delta^+$ T cells are cultured on the same side of the membrane as the labeled Th2 cell targets (E+T*/0), excellent cytotoxicity of the target cells was observed (FIG. 5). Minimal killing occurred when the effectors and labeled Th2 cells were cultured on opposite sides of the insert (E/T*). Since soluble factors should readily diffuse through the membrane, these results strongly indicate that direct cell:cell interac-

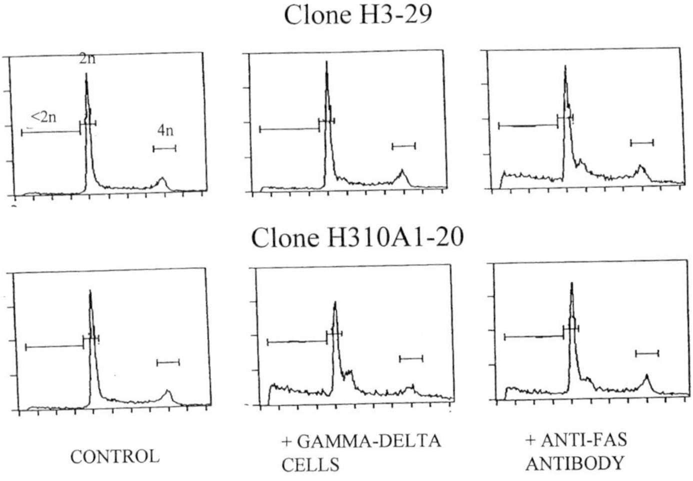

Percent Apoptotic Cells

Clone*	Control	+ Gamma	+ Anti-Fas
Thl	3.8 ± 1.5	6.4 ± 2.5	29.8 ± 3.3
Th2	4.2 ± 2.0	36.4 ± 5.7	20.1 ± 2.7

* mean ± SEM of 3 clones

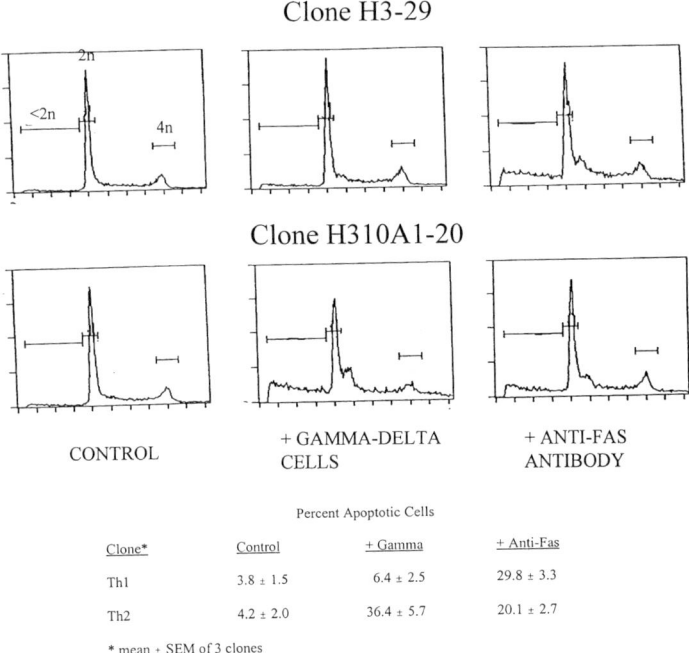

FIGURE 4. Apoptosis of CD4$^+$ T cell clones. Thl (H3-29) and Th2 (H310Al-20) clones were obtained 10 days after the last antigen stimulation. The cells were co-cultured with 1×10^5 $\gamma\delta^+$ T cells (effector:target cell ratio of 10: 1). The $\gamma\delta^+$ T cells were isolated from the hearts of day 7 H3 virus-infected BALB/c mice as described in Huber *et al.*[12] Additional CD4$^+$ T cells from each clone were incubated with 100 ng/ml hamster anti-Fas (Jo2) antibody, washed, then incubated on anti-hamster IgG coated plates to cross-link the anti-Fas antibody. After incubation for 6 h, CD4$^+$ T cells were retrieved, stained with propidium iodide and evaluated by flow cytometry[44] for apoptotic (less than 2n DNA) cells.

tions are necessary for $\gamma\delta^+$ T cell–mediated killing. One possibility might be that $\gamma\delta^+$ T cells must recognize target cells in order to release soluble factors. To evaluate this possibility, $\gamma\delta^+$ T cells were cultured with unlabeled Th2 cells on one side of the membrane and labeled Th2 cells were added to the other side (E+T/T*). Again, little cytotoxicity was observed. Finally, anti-TNFα antibody added to the culture had no effect on killing of the Th2 cell targets. In experiments not shown here, cytotoxicity of Th2 cell targets was blocked by Fas-Fc, but not by concanamycin A, a specific inhibitor of perforin-mediated cell death.[35] These results indicate that soluble factors are not responsible for $\gamma\delta^+$ T cell–mediated killing of Th2 cell targets.

Antibody cross-linking of Fas lyses both Thl and Th2 cells (FIG. 4). Yet, $\gamma\delta^+$ T cells selectively kill Th2 cells through direct cell:cell contact and this cytotoxicity is blocked by FasFc, which suggests involvement of FasL. These data appear contradictory. One possible explanation is that $\gamma\delta^+$ T cells kill through FasL-Fas interac-

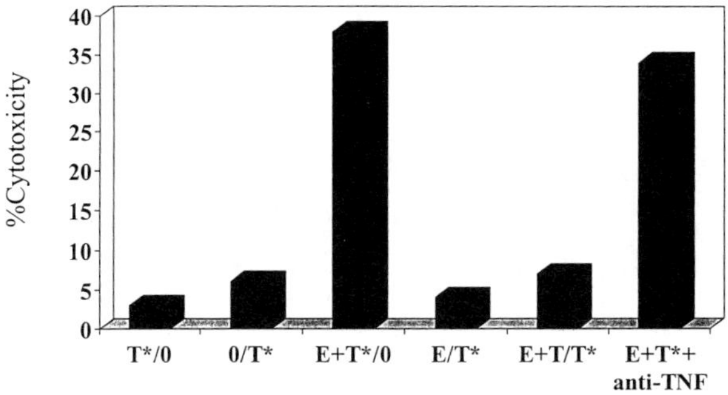

FIGURE 5. $\gamma\delta^+$ T cell killing of Th2 cells requires direct contact. 0.2 micron membrane filter inserts (NUNC) for tissue culture wells were used. Cells were placed either outside (outside/–) or inside (–/inside) of the inserts. 3×10^5 $\gamma\delta^+$ T cells (E) were cultured for 6 h with 1×10^3 H310A-48 (Th2) cells which were either ^{51}Cr-labeled (T*) or unlabeled (T). Additional wells cultured $\gamma\delta^+$ T cells and labeled Th2 cell targets together and added 10 µg/ml anti-TNFα (Pharmingen). Cultures were done in quadruplicate. *Cytotoxicity is significantly greater than zero at $p < 0.05$.

tion, but simultaneously provide secondary signals either to abort apoptosis in Thl cells (target cell rescue), or enforce the death signal in Th2 cells. Fas signaling will stimulate either proliferation or death pathways depending upon various factors, including the stage of the cell cycle when Fas engagement occurs. Signaling through CD28 or replenishment of T cell growth factors (IL-2, IFNγ) may also abort apoptosis.[36] Should $\gamma\delta^+$ T cells provide secondary rescue signals exclusively for Thl cells, this would lead to differential killing only of the Th2 cell targets. Alternatively, at least certain subpopulations of $\gamma\delta^+$ T cells recognize class II MHC IE molecules.[37,38] In studies reported by this group (Huber *et al.*, manuscript submitted), activation of murine $CD4^+$ T cells will upregulate expression of IE. Class II MHC molecule cross-linking will lead to cell death.[39,40] Thus, $\gamma\delta^+$ T cells may kill Th2 cells through cross-linking both Fas and MHC class II IE molecules if Th2 cells upregulate IE and Thl cells do not.

The significance of this work is that it provides an important counterbalance to differentiation and AICD in regulating Th cell responses. Immune responses *in vivo* are likely to be complex. Factors favoring both Thl and Th2 cell differentiation may co-exist in peripheral lymphoid organs. Use of redundant regulatory mechanisms could ensure that specific types of immunity needed by the host to combat infections are available.

ACKNOWLEDGMENTS

Supported by AHA grant-in-aid 9750081N (SAH); and Grants HL58583 (SAH); IA33470 (MKN); and AR43520 (RCB) from the National Institutes of Health

REFERENCES

1. Fitch, F. *et al.* 1993. Differential regulation of T lymphocyte subsets. Annu. Rev. Immunol. **11:** 29–48.
2. Mosmann, T. & R. Coffinan. 1989. Thl and Th2 cells: different patterns of lymphokine secretion lead to different functional properties. Annu. Rev. Immunol. **7:** 145–173
3. Paul, W.E. & R.A. Seder. 1994. Lymphocyte response and cytokines. Cell **76:** 241–251.
4. Coffinan, R.L. *et al.* 1989. Antibody to interleukin-5 inhibits helminth-induced eosinophilia in mice. Science **245:** 308–310.
5. Fuss, I.J. *et al.* 1997. Characteristic T helper 2 T cell cytokine abnormalities in autoimmune lymphoproliferative syndrome, a syndrome marked by defective apoptosis and humoral autoimmunity. J. Immunol. **158:** 1912–1918.
6. Ruddle, N. *et al.* 1990. An antibody to lymphotoxin and tumor necrosis factor prevents transfer of experimental allergic encephalomyelitis. J. Exp. Med. **166**: 991–1001.
7. Heath, W.R. *et al.* Autoimmune diabetes as a consequence of locally produced interleukin-2. Nature **359:** 547–549.
8. Saxne, T. *et al.* 1988. Detection of tumor necrosis factor alpha but not tumor necrosis factor beta in rhematoid arthritis synovial fluid and serum. Arthritis Rheum. **31:** 1041–1051.
9. McMenamin, C. *et al.* 1994. Regulation of IgE responses to inhaled antigen in mice by antigen-specific $\gamma 8^+$ T cells. Science **265:** 1869–1873.
10. Peterson, J. *et al.* 1993. Split tolerance of Thl and Th2 cells in tolerance to Theiler's murine encephalomyelitis virus. Eur. J. Immunol. **23:** 46–52.
11. Yssel, H. *et al.* 1991. *Borrelia burgdorferi* activates T helper type 1-like T cell subset in Lyme arthritis. J. Exp. Med. **174:** 593.
12. Huber, S. *et al.* 1996. Modulation of cytokine expression by $CD4^+$ T cells during coxsackievirus B3 infections of BALB/c mice initiated by cells expressing the $\gamma 8^+$ T cell receptor. J. Virol. **70:** 3039–3045.
13. Huber, S. *et al.* 1994. Augmentation of pathogenesis of coxsackievirus B3 infections in mice by exogenous administration of interleukin-1 and interleukin-2. J. Virol. **68:** 195–206.
14. Kuchroo, V.K. *et al.* 1995. B7-1 and B7-2 costimulatory molecules activate differentially the Thl/Th2 developmental pathways: application to autoimmune disease therapy. Cell **80:** 707–718.
15. Hseih, C.S. *et al.* 1993. Development of Thl $CD4^+$ T cells through IL-12 produced by Listeria-induced macrophages. Science **260:** 547–549.
16. Le Gros, G. *et al.* 1990. Generation of interleukin-4 (IL-4)-producing cells in vivo and in IL-2 and IL-4 are required for in vitro generation of IL-4-producing $CD4^+$ T cells. J. Exp. Med. **172:** 921–929.
17. Rincon, M. *et al.* 1997. Interleukin (IL)-6 directs the differentiation of IL-4 producing $CD4^+$ T cells. J. Exp. Med. **185:** 461–469.
18. Zhang, X. *et al.* 1997. Unequal death in T helper cell (Th)l and Th2 effectors: Thl, but not Th2, effectors undergo rapid Fas/FasL-mediated apoptosis. J. Exp. Med. **185:** 1837–1849.
19. Ramsdell, F. *et al.* 1994. Differential ability of Thl and Th2 T cells to express Fas-ligand and to undergo activation-induced cell death (AICD). Int. Immunol. **6:** 1545.
20. Everson, M.P. *et al.* 1998. Dendritic cells from Peyer's patch and spleen induce different T helper cell responses. J. Interferon Cytokine Res. **18:** 103–115.
21. Flynn, S. *et al.* 1998. CD4 T cell cytokine differentiation: The B cell activation molecules OX40 ligand, instructs CD4 T cells to express interleukin 4 and upregulates expression of the chemokine receptor, Blr-1. J. Exp. Med. **188:** 297–304.

22. ZISMAN, E. & E. MOZES. 1994. Processing requirements of two acetylcholine receptor derived peptides for binding to antigen presenting cells and stimulation of murine T cell lines. Int. Immunol. **6:** 683–691.
23. CARBALLIDO, J. *et al.* 1997. The intensity of T cell receptor engagement determines the cytokine pattern of human allergen-specific T helper cells. Eur. J. Immunol. **27:** 515–523.
24. HSIEH, J. *et al.* 1996. In vitro cytokine production in murine listerosis. Evidence for immunoregulation by gamma delta$^+$ T cells. J. Immunol. **156:** 232–237.
25. GARBOCZI, D.N. *et al.* 1996. Structure of the complex between human T-cell receptor, viral peptide and HLA-A2. Nature **384:** 134–141.
26. BOM, W. *et al.* 1990. Recognition of heat shock proteins and gamma delta cell function. Immunol. Today **11:** 40–43.
27. SCHILD, H. *et al.* 1994. The nature of major histocompatibility complex recognition by gamma-delta T cells. Cell **76:** 29–37.
28. WEINTRAUB, B.C. *et al.* 1994. Gamma delta T cells can recognize nonclassical MHC in the absence of conventional antigenic peptides. J. Immunol. **153:** 3051–3058.
29. HUBER, S.A. *et al.* 1998. Enteroviruses and myocarditis: viral pathogenesis through replication, cytokine induction and immunopathogenicity. Adv. Virus Res. **51:** 35–80.
30. SCIAMMAS, R. *et al.* 1994. Unique antigen recognition by a herpesvirus-specific gamma delta cell. J. Immunol. **152:** 5392–5397.
31. CONSTANT, P. *et al.* 1994. Stimulation of human gamma delta T cells by nonpeptide mycobacterial ligands. Science **264:** 267–270.
32. TANAKA, Y. *et al.* 1995. Natural and synthetic non-peptide antigens recognized by human gamma delta T cells. Nature **375:** 155–158.
33. PETERMAN, G.M. *et al.* 1993. Role of gamma delta T cells in murine collagen-induced arthritis. J. Immunol. **151:** 6546–6558.
34. PELEGRI, C., *et al.* 1996. Depletion of gamma delta T cells does not prevent or ameliorate, but rather aggravates rat adjuvant arthritis. Arthritis Rheum. **38:** 204–215.
35. KATAOKA, T. *et al.* 1996. Concanamycin A, a powerful tool for characterization and estimation of contribution of perforin and Fas based lytic pathways in cell mediated cytotoxicity. J. Immunol. **156:** 3678.
36. SHI, Y. *et al.* 1995. CD28-mediated signaling in vivo prevents activation-induced apoptosis in the thymus and alters peripheral lymphocyte homeostasis. J. Immunol. **155:** 1829.
37. LEFRANCOIS, L. *et al.* 1990. Extrathymic selection of TCR gamma delta$^+$ T cells by class II major histocompatibility complex molecules. Cell **63:** 333–340.
38. MATIS, L.A. *et al.* 1987. Major histocompatibility complex-linked specificity of gamma delta receptor-bearing T lymphocytes. Nature **330:** 262–264.
39. TRUMAN, J.-P. *et al.* 1994. Lymphocyte programmed cell death is mediated via HLA class II DR. Int. Immunol. **6:** 887–892.
40. TRUMAN, J.-P. *et al.* 1997. HLA class II-mediated death is induced via Fas/FasL interactions in human splenic B lymphocytes. Blood **89:** 1996–2001.
41. VAN HOUTEN, N. *et al.* 1991. Selection of an attenuated coxsackievirus B3 variant using a monoclonal antibody reactive to myocyte antigen. J. Virol. **65:** 1286–1290.
42. KNOWLTON, K. *et al.* 1996. A mutation in the puff region of VP2 attenuates the myocarditic penotype of an infectious CDNA of the Woodruff variant of coxsackievirus B3. J. Virol. **70:** 7811–7818.
43. GOOD, M.D. *et al.* 1983. Analysis of true anti-hapten cytotoxic clones in limited dilution microcultures after correction for "anti-self" activity: precursor frequencies, Ly-2 and Thy-I phenotype, specificity and statistical method. J. Immunol. **130:** 2046–2055.

44. HUBER, S. 1997. Coxsackievirus-induced myocarditis is dependent on distinct immunopathogenic responses in different strains of mice. Lab. Invest. **76:** 691–701.

The Immune Response to Apoptotic Cells

DROR MEVORACH[a]

The Laboratory for Cellular and Molecular Immunology, Division of Medicine, Tel-Aviv Sourasky Medical Center, Sackler Faculty of Medicine, Tel-Aviv, Israel

ABSTRACT: **Programmed cell death (PCD) can be divided into two distinct but linked sequential processes, killing of the cells and removal of the dead cells, which may be a neighboring cell or a professional phagocyte. Following internalization of the apoptotic cell, the phagocyte typically triggers neither the development of a pro-inflammatory response nor the production of autoantibodies directed against apoptotic self antigens. Since apoptotic cells are characterized by translocation of autoantigens such as nucleosomes to the surface of the cell, we tested the hypothesis that excess or abnormally processed apoptotic cells can generate autoantibodies. We have found that syngeneic apoptotic load can induce transient hypergammaglobulinemia, anti-DNA, anticardiolipin, and glomerular depositions in normal mice. Furthermore, we also found that one of the important mechanisms of uptake of apoptotic cells involves opsonization by the complement system, suggesting that deficient states could lead to aberrant handling of apoptotic cells. Therefore, conditions in which apoptotic cells become immunogenic may explain antigen selection in inflammatory and autoimmune conditions, such as in systemic lupus erythematosus (SLE).**

INTRODUCTION

Although cell death by apoptosis is measured in hours, the removal of apoptotic fragments is normally so rapid that apoptotic cells are rarely seen, even in tissues such as the thymus where up to 95% of cells undergo apoptosis.[1] It is thought that uptake of these cells by specific receptors in phagocytes results in the disposal of cellular contents without the induction of inflammation.[2]

The immune system is an example of a system in which PCD is a major mechanism of regulation. In the lymph nodes, PCD is a basic mechanism that allows the termination of the normal immune response and the prevention of the formation of activated autoimmune B and T cells.[3] The development of autoimmune diseases is seen when autoreactive immune cells fail to undergo PCD, as illustrated in mice and humans deficient in fas (CD95).[4]

The mechanisms whereby apoptotic cells are efficiently identified, removed, and degraded by phagocytes in mammalian cells are not well understood. Studies published from 1990 through 1998 showed several receptors to be important in the uptake of apoptotic cells. Several integrins, the vitronectin receptor, $\alpha_v\beta_3$,[5] $\alpha_v\beta_5$,[6] and the β_2 integrins, CD11b/CD18 and CD11c/CD18,[7] were all shown to mediate uptake

[a]Address for correspondence: The Laboratory for Cellular and Molecular Immunology Division of Medicine, Tel-Aviv Sourasky Medical Center, 6 Weizmann Street, Tel-Aviv 64239, Israel; 972-3-697-4923 (voice); 972-3-692-5757 (lab); 972-2-643-3935 (fax).
e-mail: mevdm@netvision.net.il

of apoptotic cells. Class A scavenger receptors,[8] class B scavenger receptors,[9,10] the CD68 receptor, which is the receptor for oxidized LDL,[11] and CD14,[12] were also shown to have a role in uptake of apoptotic cells. Phosphatidylserine (PS) appears on the outer leaflet upon the initiation of PCD. Several studies proposed the existence of the so-called PS receptor,[13] that has not yet been characterized. Finally, inhibition of the ABC1 cassette transporter, a transmembrane protein that transports substrates (ions, lipids) across the membrane, decreases uptake of apoptotic cells.[14]

Since self-apoptotic cells are a major source of self-antigens, it is essential that their removal will not be accompanied by the development of an autoreactive immune response. Indeed, it has been suggested by *in vitro* studies that removal of apoptotic cells by macrophages may be associated with inhibition of the release of pro-inflammatory cytokines.[15,16] *In vitro* phagocytosis of apoptotic cells suppresses the release of GM-CSF, IL-8, TNF-α, and thromboxane B2, but not TGF-β and PGE2.[16] IL-10 was shown to have anti-inflammatory properties manifested by suppression of the release of pro-inflammatory cytokines such as IL-8 and TNF-α.[17] However, it is not clear whether IL-10 secretion following phagocytosis of apoptotic cells is increased.[15,16]

To understand the mechanism of development of autoimmune disease, we tried to define the immune response, *in vivo,* to an i.v. load of syngeneic apoptotic thymocytes, and to evaluate the role of the complement system in uptake of apoptotic cells. Our preliminary results showed transient elevations in anti-ssDNA, anticardiolipin, and total IgG in normal mice,[18] and acceleration of the autoimmune disease in pro-autoimmune mice.[19] *In vitro* studies showed that the complement system is involved in removal of apoptotic cells.[7]

METHODS

Immunization Protocol

C3H SnJ (C3H), BALB/c, and C57BL/6 (B6) mice were obtained from the Jackson Laboratories, Bar Harbor, ME. Thymocytes and splenocytes were prepared from 6–8 week old mice, and syngeneic cells were injected i.v. as described.[18]

Apoptosis of Murine Thymocytes

Apoptosis was induced by irradiation (600-cGy from a ^{137}cesium source). Apoptosis was defined by Annexin-V staining of the cells and by staining of DNA by propidium iodide.[18] Apoptotic cells were stained with FITC-conjugated annexin V (Nexins), propidium iodide, and PE-conjugated IgG2b mouse monoclonal anti-human-iC3b (Quidel, CA). Flow cytometry analysis was performed as described before.[7,18]

Immune Response

Serum samples were obtained immediately prior to immunization and at biweekly intervals following immunization for up to 30 weeks. The immune response was evaluated by quantifying serum immunoglobulin concentrations and autoantibody production.

Clinical and Pathological Evaluation

Mice were examined bimonthly for clinical signs of disease and for hematuria or proteinuria using N Multistix SG (Bayer, IN). Histological evaluation of kidneys was performed as described elsewhere.[18]

Phagocytosis Assays

Interaction between human macrophages and apoptotic cells was performed as described elsewhere.[7]

RESULTS

Transient Elevations of ANA, anticardiolipin, anti-ssDNA, and Total IgG Were Seen in Normal Strains of Mice Immunized with Syngeneic Apoptotic Thymocytes

When 10^7 apoptotic thymocytes in four weekly intervals were injected to C3H mice, up to two- to threefold transient elevations in total IgG, anti-ssDNA and anticardiolipin were observed. The elevations were mainly IgM, although IgG elevations were seen as well.[18] In FIGURE 1A, positive ANA is shown in serum of immunized C3H mice. As shown in FIGURE 2, pre-incubation with ssDNA or cardiolipin mi-

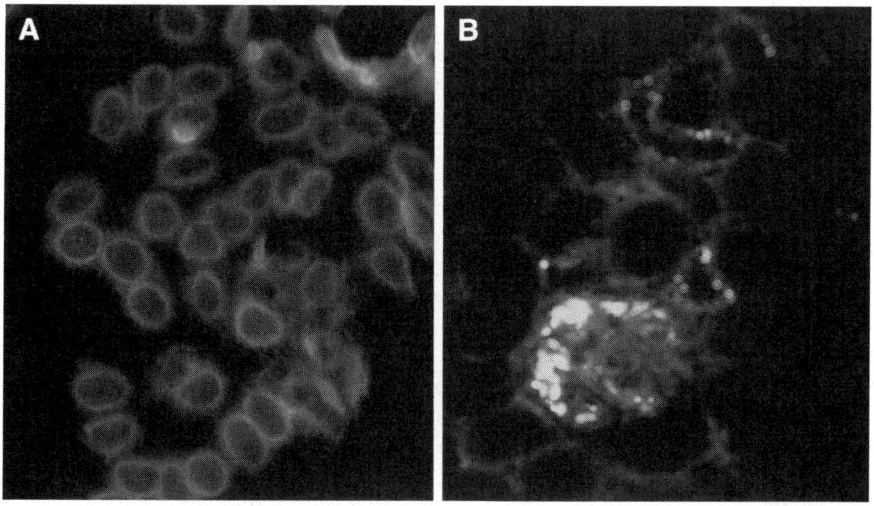

FIGURE 1. C3H mice, immunized with syngeneic apoptotic thymocytes, develop anti-nuclear antibody and IgG mesangial depositions. C3H mice were immunized with 10^7 syngeneic apoptotic cells in four weekly intervals. The mice developed anti-nuclear antibodies between 2–6 weeks after the first immunization. In most mice the higher titer declined to normal levels within four weeks. **A:** Serum of immunized mice was examined by indirect immunofluoresence staining of Hep-2 cells. It shows anti-nuclear staining in a peripheral and nucleolar pattern. **B:** Kidney sections of the immunized mice show mesangial IgG depositions in the glomeruli. Some staining of the tubuli is seen also (×600).

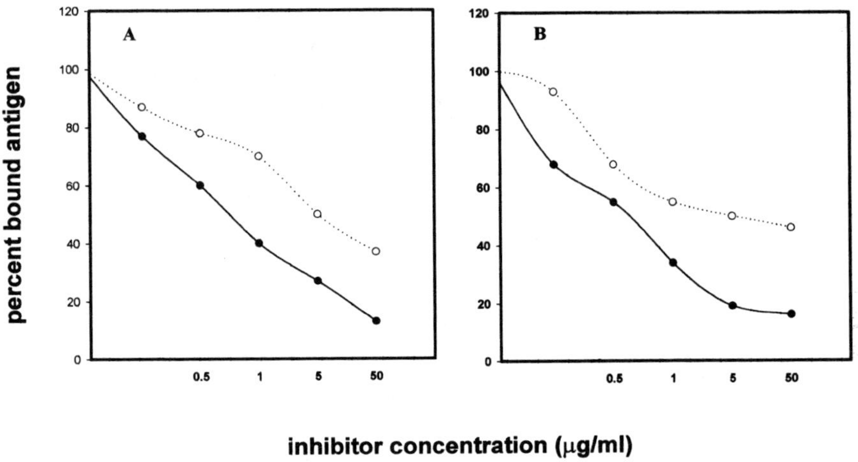

FIGURE 2. The anti-ssDNA and anticardiolipin autoantibodies, induced by immunization with apoptotic cells, show partial cross reactivity. To examine whether the autoantibodies detected in immunized C3H mice were specific and/or cross-reactive, sera were incubated with either 0–50 μmol of ssDNA (*closed circles*) or 0–50 μmol cardiolipin micelles (*open circles*) as described.[18] The degree of inhibition of anti-ssDNA by ssDNA and cardiolipin micelles is shown in **A** (IgM), and **B** (IgG). Both autoantibodies were specific as shows the inhibition curve. As some sub-populations of anti-DNA and anticardiolipin may show cross-reactivity, we examined it by cross-inhibition using cardiolipin micelles for anti-ssDNA, and ssDNA for anticardiolipin. Cross-reactivity is seen in ~5-fold concentration of the inhibitor.

celles inhibited anti-ssDNA and anticardiolipin, respectively. At higher concentrations of inhibitors, cross-reactivity between anti-DNA and anticardiolipin was seen, as ssDNA and cardiolipin micelles inhibited anticardiolipin and anti-ssDNA, respectively. Polyclonal activation was seen also as increase in total IgG was documented; however, elevated titers of other autoantibodies like anti-dsDNA, and anti-Sm were not seen.[18] These results indicate that autoantibodies like anti-ssDNA and anticardiolipin may represent "natural autoantibodies" as a part of clearance mechanism, via opsonization with antibodies, in response to intra-vascular load of apoptotic cells.

Mesangial Depositions of IgG in Mice Immunized with Apoptotic Cells

When kidney sections of immunized mice were examined, mesangial IgG depositions were seen in mice immunized with apoptotic cells (FIG. 1B), but not in mice immunized with viable syngeneic splenocytes.[18] Glomerulonephritis was not observed in most mice when kidneys were examined 4–6 months following the immunization, indicating that full-blown autoimmunity did not develop in immunized mice. These results may suggest that an intra-vascular load of apoptotic cells may be deposited in the kidney either directly or via antibody.

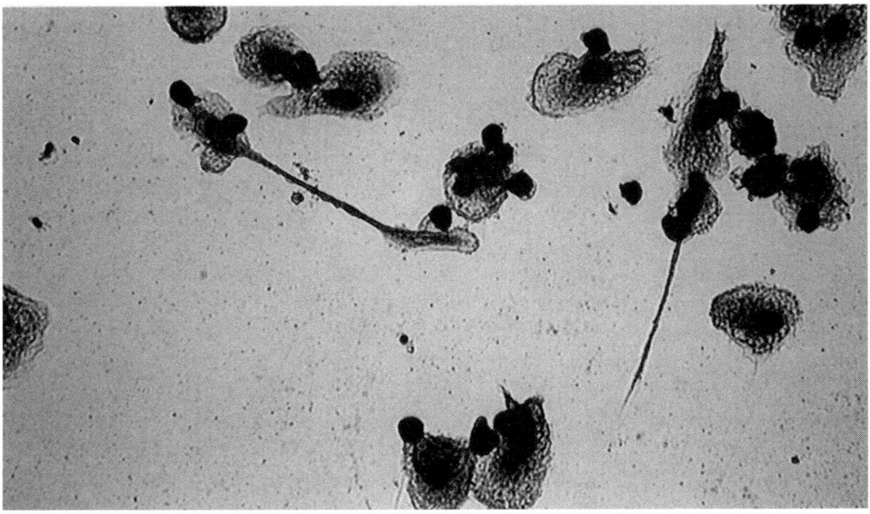

FIGURE 3. Uptake of apoptotic cells by human macrophages. *In vitro* clearance of apoptotic Jurkat cells by human monocytes-derived macrophages. 1–3 apoptotic cells are seen attached to or within each macrophage. The 1 hour–phagocytosis assay was done in the presence of 15% serum. The presence of serum increased uptake 3- to 5-fold, until a phagocytosis index of 150–190 was reached.[7]

Serum Complement Is Important in Uptake of Apoptotic Cells by Human Macrophages

In vitro studies examining the role of serum proteins in uptake of apoptotic cells showed an important role for the complement cascade. In the presence of 15% serum, both homologous (the same human donor or two human donors of apoptotic cells and macrophages) and heterologous (mice apoptotic cells and human macrophages) phagocytosis interactions were increased in 2.5- and 10-fold, respectively.[7] Serum depletion of classical or alternative components of the complement or heat inactivation resulted in the loss of 80–100% of this increase. With some variability, this observation was seen in uptake of T, B, lymphocytes, neutrophils, human cell lines, and mice thymocytes. The uptake of apoptotic cells by human macrophages is shown in FIGURE 3. Ongoing studies examine the role of the complement, *in vivo*, in response to a load of apoptotic cells.

DISCUSSION

The clearance of apoptotic cells may be a significant contributor to several vascular, neurological and autoimmune diseases. It has long been appreciated that DNA and histones are major autoantigens in SLE. More recently has evidence been provided that the DNA histone complex, i.e., nucleosomes, are the preferred targets of autoantibodies in SLE.[20] The question thus arises as to how nucleosomes and several other intracellular antigen targets can be immunogenic in SLE. Apoptotic cells form cytoplasmic blebs, some of which are shed as apoptotic bodies, and exposure of keratinocytes to UV light leads to the translocation of the nuclear autoantigens Ro (SSA) and La (SSB) to the surface of the cell.[21,22] Casciola Rosen *et al.*[23] demonstrated that when apoptosis in induced in keratinocytes following UV light exposure, the cell surface expression not only of Ro and La but also of nucleosomes and ribosomes can be explained by translocation of certain intracellular particles to the apoptotic surface blebs. Another translocation that occurs during apoptosis is that of PS, an acidic phospholipid that normally resides on the inside of the cell, but flips to the outside of the cell membrane when the cell undergoes apoptosis.[24,25] PS, like cardiolipin, is a major autoantigen for anti-phospholipid antibodies in SLE. Taken together with the immunization studies in mice, these findings provide a unifying hypothesis to explain antigen selection in SLE, viz. that SLE patients respond to exposure of intracellular proteins of cells undergoing apoptosis. Indeed, patients with SLE show also an immune response to post-translational protein modifications during apoptosis (for review see ref. 26).

Why do SLE patients form an immune response to apoptotic material? SLE patients have reduced uptake of apoptotic cells by macrophages *in vitro.*[27,28] The presence of autoantibodies, such as anticardiolipin, may lead to pro-inflammatory uptake.[29] In addition, as shown in our studies, apoptotic load even in normal strains of mice induced autoantibodies and glomerular depositions. Our *in vitro* observation that the complement system is crucial for uptake of apoptotic cells by human macrophages[7] may present an additional factor since most patients deficient in early components of the complement system, develop SLE, and it was shown that C1q-deficient mice develop autoantibodies and glomerulonephritis characterized by apoptotic bodies depositions in the glomeruli.[30] As shown in our studies in normal mice strains, apoptotic load by itself is probably not sufficient for the establishment of autoimmunity. In this regard, the role of dendritic cells in the uptake of apoptotic cells, may be found to be important, not only in cross-priming, but also in cross-tolerance.[31] As shown by our studies with normal and pro-autoimmune mice strains, additional conditions, such as activation of the phagocytes and the existence of autoreactive B and T cells, seem to be necessary for the development of autoimmune response.

ACKNOWLEDEGMENTS

The immunization and clearance studies were performed by D.M at the laboratory of Dr. K.B. Elkon, Hospital for Special Surgery, New York. Dr. K.B. Elkon and Dr. R. Lockshin kindly reviewed the manuscript.

REFERENCES

1. SURH, C.D. & J. SPRENT. 1994. T cell apoptosis detected in situ during positive and negative selection in the thymus. Nature **372:** 100–103.

2. SAVILL, J., V. FADOK, P. HENSON & C. HASLETT. 1993. Phagocyte recognition of cells undergoing apoptosis. Immunol. Today **14:** 131–136.
3. VAN PARIJS, L & A.K. ABBAS. 1998. Homeostasis and self-tolerance in the immune system: turninglymphocytes off. Science **280:** 243–248.
4. VAISHNAW, A.K, J.D. MCNALLY & K.B. ELKON. 1997. Apoptosis in rheumatic diseases. Arthritis Rheum. **40:** 1917–1927
5. SAVILL, J., N. DRANSFIELD HOGG & C. HASLETT. 1990. Vitronectin receptor-mediated phagocytosis of cells undergoing apoptosis. Nature **343:** 170–173.
6. ALBERT, M.L, *et al.* 1998. Immature dendtritic cells phagocytose apoptotic cells via $\alpha_v\beta_5$ and CD36, and cross-present antigens to cytotoxic T lymphocytes. J. Exp. Med. **188:** 1359–1368.
7. MEVORACH, D., J. MASCARENHAS, D. GERSOV & K.B. ELKON. 1998. Complement-dependent clearance of apoptotic cells by human macrophages. J. Exp. Med. ***188:*** 2313–2320.
8. REN, Y., R.L. SILVERSTEIN, J. ALLEN & J. SAVILL. 1995. CD36 gene transfer confers capacity for phagocytosis of cells undergoing phagocytosis. J. Exp. Med. **181:** 1857–1862.
9. PLATT, N., H. SUZUKI, Y. KURIHARA, T. KODAMA & S. GORDON. 1996. Role for the class A macrophage scavenger receptor in the phagocytosis of apoptotic thymocytes in vitro. Proc. Natl. Acad. Sci. USA **93:** 12456–12460.
10. FUKASAWA, M., H. ADACHI, K. HIROTA, M. TSUJIMOTO, H. ARAI & K. INOUE. 1996. SRB1, a class B scavenger receptor, recognizes both negatively charged liposomes and apoptotic cells. Exp. Cell Res. **222:** 246–250.
11. SAMBRANO, G.R. & D. STEINBERG. 1995. Recognition of oxidatively damaged and apoptotic cells by an oxidized low density lipoprotein receptor on mouse peritoneal macrophages: role of membrane phosphatidylserine. Proc. Natl. Acad. Sci. USA **92:** 1396–1400.
12. DEVITT, A., O.D. MOFFATT, C. RAYKUNDALIA, J.D. CAPRA, D.L. SIMMONS & C.D. GREGORY. 1998. Human CD14 mediates recognition and phagocytosis of apoptotic cells. Nature **392:** 505–509.
13. FADOK, V.A., D.R. VOELKER, P.A. CAMPBELL, J.J. COHEN, D.L. BRATTON & P.M. HENSON. 1992. Exposure of phosphatidyl serine on the surface of apoptotic lymphocytes triggers specific recognition and removal by macrophages. J. Immunol. **148:** 2207–2216.
14. LUCIANI, M.-F. & G. CHIMINI. 1996. The ATP binding cassette transporter, ABC1, is required for the engulfment of corpses generated by apoptotic cell death. EMBO J. **15:** 226–235.
15. VOLL, R.E., M. HERRMANN, E.A. ROTH, C. STACH, J.R. KALDEN & I. GIRKONTAITE. 1997. Immunosuppresive effects of apoptotic cells. Nature **390:** 350–351.
16. FADOK, V.A., D.L. BRATTON, A. KONOWAL, P.W. FREED, J.Y. WESTCOTT & P.M. HENSON. 1998. Macrophages that have ingested apoptotic cells in vitro inhibit proinflammatory cytokine production through autocrine/paracrine mechanisms involving TGF-beta, PGE2, and PAF. J. Clin. Invest. **101:** 890–898.
17. FIORENTINO, D.F. *et al.* 1991. IL-10 inhibits cytokine production by activated macrophages. J. Immunol. **1147:** 3815–3822.
18. MEVORACH, D., J.-L. ZHOU, X. SONG & K.B. ELKON. 1998. Systemic exposure to irradiated apoptotic cells induces autoantibody production. J. Exp. Med. **188:** 387– 392.
19. MEVORACH, D., L. ZHOU, M. MADAIO, A. MARSHAK-ROTHSTEIN & K.B. ELKON. 1998. Apoptotic cells are targeted antigens in lupus-like disease developing in MRL/lpr and MRL/$^+$/gld mice. . Arthritis Rheum. **41:** (9) suppl.
20. MOHAN, C., S. ADAMS, V. STANIK, S.K. DATTA. 1993. Nucleosome: A major immunogen for pathogenic autoantibody-inducing T cells of lupus. J. Exp. Med. **177:** 1367–1381.

21. LeFeber, W.P. D.A. Norris, S.R. Ryan, *et al.* 1984. UV light induces binding of antibodies to selected nuclear antigens on cultured human keratinocytes. J. Clin. InvesT. **74:** 1545–1551.
22. Golan, T.D., K.B. Elkon, A.E. Gharavi & J.G. Krueger. 1990. Enhanced membrane binding of autoantibodies to cultured keratinocytes of SLE patients after UVB/UVA irradiation. J. Clin. Dermatol. **45**: 245–251.
23. Casciola-Rosen, L.A., G. Anhalt & A. Rosen. 1994. Autoantigens targeted in systemic lupus erythematosus are clustered in two populations of surface structures on apoptotic keratinocytes. J. Exp. Med. **179:** 1317–1330.
24. Martin, S.J., C.P.M. Reutelingsperger, A.J. McGahon, J.A. Rader, R. van Schie, D.M. LaFace & D.R. Green. 1995. Early redistribution of plasma membrane phosphatidylserine is a general feature of apoptosis regardless of the initiating stimulus: Inhibition by overexpression of Bcl-2 and Ab1. J. Exp. Med. **182:** 1545–1556.
25. Verhoven, B., R.A. Schlegel & P. Williamson. 1995. Mechanisms of phosphatidylserine exposure, a phagocyte recognition signal, on apoptotic T lymphocytes. J. Exp. Med. **182:** 1597–1601.
26. Utz, P.J. & P. Anderson. 1998. Posttranslational protein modifications, apoptosis, and the bypass of tolerance to autoantigens. Arthritis Rheum. **41:** 1152–1160.
27. Mevorach, D., D. Gershov, S. Ng, J. Salmon & K.B. Elkon. 1997. Decrease in uptake of apoptotic cells in patients with systemic lupus erythematosus. Arthritis Rheum. **40:** (9) suppl.
28. Hermann, M. *et al.* 1998. Impaired phagocytosis of apoptotic material by monocytes-derived macrophages from patients with systemic lupus erythematosus. Arhritis Rheum. **41:** 1241–1250.
29. Manfredi, A.A. *et al.* 1998. Apoptotic cell clearance in systemic lupus erythematosus. I. Opsonization by antiphospholipid antibodies. Arthritis Rheum. **41:** 205–214
30. Botto, M., C. Dell'Agnola, A.E. Bygrave, E.M. Thompson, T. Cook, F. Petry, M. Loos, P.P. Pandolfi & M.J. Walport. 1998. Homozygous C1q deficiency causes glomerulonephritis associated with multiple apoptotic bodies. Nature Genet. **19:** 56–59.
31. Inaba, K. *et al.* 1998. Efficient presentation of phagocytosed cellular fragments on the major histocompatability complex class II products of dendtritic cells. J. Exp. Med. **188:** 2163–2173.

Programmed Cell Death as a Mechanism of CD4 and CD8 T Cell Deletion in AIDS

Molecular Control and Effect of Highly Active Anti-retroviral Therapy

MARIE-LISE GOUGEON[a] AND LUC MONTAGNIER

Unité d'Oncologie Virale and CNRS ERS 572, Département SIDA et Rétrovirus, Institut Pasteur, 28 rue du Dr. Roux, 75724 Paris Cedex 15, France

ABSTRACT: Infection with human immunodeficiency virus (HIV) results in the progressive destruction of CD4 T lymphocytes, generally associated with progression of the disease.The progressive disappearance of CD4 T lymphocytes leads to the lack of control of HIV replication and to the development of severe immune deficiency responsible for the occurrence of opportunistic infections associated with AIDS. In this review we discuss premature lymphocyte apoptosis in the context of HIV infection as the consequence of the continuous production of viral proteins, leading to an unbalanced immune activation and to the triggering of apoptotic programs. The chronic immune activation induces the continuous expression of death factors which could turn lymphocytes, including CD4 T cells, CD8 CTL or APC, into effectors of apoptosis, leading to the destruction of healthy activated non-infected cells. Thus, programmed cell death would significantly contribute to peripheral T cell depletion in AIDS, particularly if the Th cell renewal is impaired. Under potent anti-retroviral therapies, a complete normalization of lymphocyte apoptosis is observed, concomitant with a partial restoration of the number and the functions of the immune system.

INTRODUCTION

Human immunodeficiency virus (HIV) infection results in the progressive destruction of CD4 T lymphocytes, generally associated with disease progression. The first identified receptor for HIV was the CD4 molecule and recent studies to delineate the molecular basis of cellular tropism led to the identification of coreceptors for HIV. The receptor CXCR4 was identified as the co-receptor responsible for the efficient entry of T-tropic strains of HIV-1 into target cells, and the β-chemokine receptor CCR5 was identified as the co-receptor for M-tropic HIV-1.[1] CD4 T lymphocytes are the orchestrator of the immune system. First, through the production of cytokines, they help the effectors of innate immunity, such as natural killer cells (NK), γδ T lymphocytes or monocytes, in eliminating virus-infected cells. In addition, they are essential to the specific activation and maturation of B lymphocytes into antibody-secreting plasmocytes, they are required for the differentiation of

[a]Address for correspondence: 33 1 45 68 8907 (voice); 33 1 45 68 8909 (fax). e-mail: mlgougeo@pasteur.fr

$CD8^+$ T cells into virus-specific cytotoxic T lymphocytes (CTL), and they are a source of chemokines, suppressor factors of HIV replication. Therefore, the progressive disappearance of CD4 T lymphocytes leads to the lack of control of HIV replication and to the development of severe immune deficiency responsible for the occurrence of opportunistic infections associated with AIDS.[1] Despite years of investigation, the mechanisms responsible for the deletion of this lymphocyte subset are still not elucidated.

CD4 T cell destruction can be mediated directly by HIV as a consequence of viral gene expression, or indirectly through priming for apoptosis of uninfected cells when triggered by different agents. In addition to these pathways, a complementary cytopathic effect is probably provided by the immune system, since infected cells may be killed by HIV-specific CTL or antibody-dependent cell-mediated cytotoxicity (ADCC). The current understanding of CD4 T cell homeostasis in the course of HIV infection is that the progressive depletion of CD4 T lymphocytes is the consequence of both their destruction by several mechanisms dependent on the virus, and the lack of compensation by the production of new CD4 T cells, because of a possible blockade of the CD4 T cell renewal machinery at the level of the bone marrow or of the thymus. In this article, we discuss the possible contribution of programmed cell death (PCD) by apoptosis on the destruction of T lymphocytes, the mechanisms involved in this process, the consequences of excessive apoptosis on the effectors of the immune system and the influence of highly active anti-retroviral therapies (HAART).

INFLUENCE OF HIV-1 GENES ON THE INDUCTION OF APOPTOSIS

Several HIV-1 gene products can influence directly the survival of the infected cell or of bystander cells. Infection of CD4 T cell cultures with HIV is associated with a cytopathic effect of the virus, leading to the death by apoptosis of both infected and non-infected cells. Apoptosis is triggered by the viral envelope glycoprotein gp160, expressed on the surface of infected cells, which binds to accessible CD4 receptors on the surface of neighboring cells.[2,3] Both gp120 and gp41 are required for triggering apoptosis and no other gene besides the envelope is involved. Thus, chronically HIV-infected cells can serve as effector cells to induce apoptosis in uninfected target CD4 T cells. During this process, which involves syncytia formation and cell-to-cell spread of HIV infection, the anti-retroviral drug AZT blocks the spread of HIV infection without any apparent effect on apoptosis.[4] Further studies on the apoptotic pathway involved in gp120-dependent apoptosis of uninfected CD4 T cells showed that it involved caspases although it was not mediated by the CD95 or TNF-RI molecules.[5] Tat, a viral transcription factor, was found to up-regulate Bcl–2 expression, protecting cells from apoptosis.[6] In contrast, establishment of stable Tat-expressing cell lines or addition of exogenous Tat has been reported to sensitize cells to CD95-, T-cell receptor (TCR)- or CD4-induced apoptosis.[7,8] In these studies, Tat alone was insufficient to induce apoptosis, but it appeared to sensitize cells to apoptosis triggered by a second signal, such as CD95 or TCR signaling. A recent study has suggested that Bcl-2 is a critical cellular determinant in the tendency toward an acute or a persistent infection. Indeed, HIV replication in susceptible CD4 T or monocytic cell lines first results in a decrease of Bcl-2, permitting an initial boost of replication, and then the replication is negatively controlled by Bcl-2 to reach a balance characterized by low virus production and a level of Bcl-2 compatible with cell survival.[9] HIV*vpr* gene, which is required for productive infection of nondividing cells, was found to induce arrest of cells in the G2/M phase of the cell cycle and then to trigger apoptosis.[10] This was observed in human T cells and in fibroblasts.[11] Another HIV gene, *vpu*, increases the susceptibility of infected peripheral T cells and Jurkat T cells to CD95-induced apoptosis.[12] Thus it appears from these *in vitro* studies that HIV can influence the cell death machinery and trigger apoptosis, not only in infected cells, but also in bystander non-infected cells. These observations raise an important question in HIV pathogenesis: is virus killing limited to infected T cells *in vivo*?[13,14]

PREMATURE PRIMING FOR APOPTOSIS OF NON-INFECTED PERIPHERAL T LYMPHOCYTES FROM HIV-INFECTED SUBJECTS: RELATION TO IMMUNE ACTIVATION AND DISEASE EVOLUTION

Premature Priming for Apoptosis of CD4 and CD8 T Cells from HIV⁺ Patients

It was reported several years ago that peripheral blood T cells from HIV-infected persons are highly prone to apoptosis induced *in vitro*.[15–18] Indeed, the incubation in medium alone of freshly isolated peripheral blood mononuclear cells (PBMCs) from HIV-infected individuals induces a rapid spontaneous apoptosis,[15,16,18] which is observed in 15–50% of total lymphocytes whereas only 2–5% of lymphocytes from control subjects die under the same culture conditions. Moreover, activation-induced apoptosis is also observed in patients' lymphocytes, following stimulation with mitogens, superantigens or anti-TCR antibodies, whereas these stimuli are not proapoptotic in lymphocytes from control donors.[15–18] Although it was first reported that the increased priming for apoptosis in HIV infection exclusively concerned the CD4 subset,[17] it rapidly became clear that the CD8 subset was similarly primed for apoptosis.[16,18,20] In fact, a phenotypic study of apoptotic cells on a large cohort of HIV-positive patients revealed that not only T cells but all blood mononuclear cells, including B cells, T cells, NK cells, granulocytes and monocytes, had an increased fragility upon short-term culture.[19] These observations were confirmed *in vivo*, in lymph nodes of HIV-infected patients, in which apoptosis was detected not only in CD4 but also in CD8 T cells, B cells and dendritic cells,[21,22] and also in tonsillar tissue from HIV-infected donors, which showed increased apoptosis in both CD4 and CD8 T cells compared to uninfected donors.[23]

Owing to the low fraction of infected CD4 T cells and to the premature priming for apoptosis of non-infected lymphocytes in HIV^+ patients, it was suggested that bystander inappropriate cell death was responsible for CD4 T cell depletion[13,14] and CD8 T cell destruction[24] in AIDS. This was supported by the observation that apoptotic T cells in lymph node sections of HIV-infected children and SIV-infected macaques were dominant in uninfected bystander cells whereas infected cells were not found to be apoptotic.[25] This has been confirmed by studies highlighting the excessive number of apoptotic cells over infected cells in lymph nodes of HIV-infected adults[21] and showing the great frequency in patients' lymph nodes of T lymphocytes expressing the tissue transglutaminase (tTG), a Ca^{2+}-dependent enzyme that cross-links intracellular proteins during the apoptotic process and whose expression underlines a pre-apoptotic stage.[22]

Relation between Apoptosis and HIV-driven Immune Activation

Apoptosis plays a crucial role in the homeostatic control of cell numbers following antigenic stimulation, ensuring the clearance of primed lymphocytes in order to terminate an immune response and to avoid autoimmune reactions.[26] Nevertheless, this normal process of elimination of activated cells might be detrimental for the immune system in the case of a chronic infection such as that induced by HIV. Indeed, a general state of immune activation is observed in the asymptomatic phase of HIV-infection both in lymphoid tissue and peripheral blood lymphocytes, and persists throughout the entire course of HIV infection. This is reflected by follicular hyperplasia in lymphoid tissue and the expression of activation markers such as HLA-DR, CD45R0 and CD38 in CD4 and CD8 T cells.[21,27] Although HIV replication is dramatically down-regulated under the influence of the specific immune response, HIV is never eliminated and its persistence associated with the unceasing expression of HIV antigens is probably the primary mechanism for this chronic immune stimulation. In addition, exogenous factors, such as opportunistic pathogens, stimulate the production of proinflammatory cytokines, including TNFα, IL1β and IL-6, which drive cellular activation and viral replication.[28] This unbalanced immune activation might be the primary mechanism responsible for the premature cell death in AIDS. This is suggested by the following observations:

1. apoptotic cells in patients' lymphoid tissues and in blood exhibit an activated phenotype[19,21];
2. there is a statistically significant correlation between the intensity of spontaneous or TCR-triggered apoptosis in both CD4 and CD8 subsets and their *in vivo* activation state[19];
3. recent studies performed in West Africa comparing patients infected with HIV-1 or HIV-2 showed that the low pathogenicity of HIV-2 infection is associated with a lower level of immune activation and less T cell apoptosis[29]; and
4. the lack of chronic immune activation in the non-pathogenic HIV-1 infection in chimpanzees is associated with a very low level of T cell apoptosis.[16,30,31]

Contribution of PCD to Disease Evolution and AIDS Pathogenesis

A series of observations reported in HIV-infected persons and in simian models of lentiviral infection argue for a correlation between the intensity of T cell apoptosis and the pathogenicity of the infection. First, the proportion of CD4 and CD8 T lymphocytes undergoing apoptosis spontaneously or after ligation of the TCR or the CD95 receptor is increasing with disease evolution, evaluated by the *in vivo* reduction of the CD4 T cell number.[32–34] Second, there is a correlation between the intensity of lymphocyte apoptosis and resistance (in long-term non-progressors) or susceptibility (in rapid progressors) to AIDS development (ref. 35; M-L. Gougeon and H. Lecoeur, unpublished observations). Third, comparative studies in pathogenic models of lentiviral infection, including macaques infected with SIV[16,36] or cats infected with FIV[37] vs. non-pathogenic models, including SIV-infected African green monkeys[36] or chimpanzees infected with HIV or SIVcpz[16,30,31] revealed that increased lymphocyte apoptosis was observed only in pathogenic lentiviral infections. Interestingly, a recent study reported the case of two female chimpanzees that showed a progressive loss of CD4 T cells associated with high viral burdens, hyperimmune activation and increased levels of CD4 T cell apoptosis following inoculation of HIV-1 isolated from a chimpanzee infected with the virus for 8 years. By contrast, no apoptosis and no activation was observed in animals without loss of CD4 T cells.[38] These observations provide additional evidence that a correlation exists between immune activation, T cell loss, and apoptosis, and that apoptosis can significantly contribute to AIDS pathogenesis. As detailed below, it could be the mechanism responsible for the clearance of activated but healthy T cells and consequently contribute to the impoverishment of the pool of effectors (T helper, Th; and cytotoxic T lymphocytes, CTL).

MOLECULAR CONTROL OF HIV-DEPENDENT APOPTOSIS: CONTRIBUTION TO THE DESTRUCTION OF THE EFFECTORS OF THE IMMUNE SYSTEM

Negative Regulation of Bcl-2 Expression in CD8 T Cells: Consequences for the Anti-Viral Cytotoxic Function

Bcl-2 and its homologous proteins play a key role in the control of cell death of T and B cell lineages during lymphoid development, ensuring their appropriate selection. In differentiated mature T lymphocytes, regulation of Bcl-2 expression might be crucial for the development and persistence of a memory T cell response following an immune activation.[26] In order to determine whether the priming for apoptosis of lymphocytes from HIV-infected donors was associated with a differential expression of Bcl-2, freshly isolated PBMCs from HIV-infected donors at different stages of the disease were analyzed by *f*luorescence-*a*ctivated *c*ell *s*orting (FACS) for intracellular Bcl-2 expression.[39] A decreased Bcl-2 expression was consistently detected *ex vivo* in a fraction of CD8 T lymphocytes from HIV-positive donors, whereas it was never observed in lymphocytes from control donors. Interestingly, *in vivo* low Bcl-2 expression in CD8 T lymphocytes was associated with their priming for spontaneous apoptosis after a short- term culture, and a significant correlation was observed between the *in vivo* level of Bcl-2 expression and the propensity of corresponding cells to undergo apoptosis, either spontaneously or following CD95 ligation.[39] *Ex vivo* phenotypic characteristics of the low Bcl-2 CD8 T cells suggested that they were cytotoxic, because of their activated state and the expression the cytotoxic granules TIA-1 and perforin.[39] This subset, characterized as $CD8^+$ $CD45R0^+$ $TIA\text{-}1^+$ and Bcl-2 low, that we found in the blood, is also highly expressed (up to 60% of CD8 T cells) *in vivo* in lymph nodes of HIV-infected patients.[40]

A strong HIV-specific cytotoxic response is generated rapidly after HIV infection and persists during the chronic phase of the infection. However, this cytotoxic response is markedly lost on the onset of AIDS symptoms, and the molecular mechanisms involved are still unknown. One of the hallmarks of HIV infection is the defective production of IL-2, which is linked to the progressive depletion of CD4 T cells, the major source of IL-2.[41,42] IL-2 can upregulate *in vitro* Bcl-2 expression in lymphocytes from chronically HIV-infected patients.[43] Thus, the *in vivo* deficiency in IL-2 production would prevent the up-regulation of Bcl-2 molecule on cytotoxic T lymphocytes, which thus could not be rescued from apoptosis. Therefore, the loss of anti-viral cytotoxic activity in the course of HIV infection might be related to an abnormal priming for apoptosis of CTL, consequently to both a persistent virus-driven immune stimulation and the gradual loss of survival factors.

Positive Regulation of the CD95 System: Consequences for CD4 T Cell Depletion

The CD95 molecule belongs to the tumor necrosis factor receptor (TNFR) superfamily that includes various molecules involved in immune regulation, such as the TNF receptors I and II, CD27, CD30, and CD40.[44] It is characterized by three extracellular cysteine-rich domains found in all family members and an intracellular death domain shown to transduce signals for apoptosis in the TNFR and the CD95 molecules.[44,45] T cell receptor triggering in activated peripheral T cells may induce apoptosis that involves autocrine suicide or paracrine death mediated via CD95 receptor/ligand interaction. The CD95 ligand (CD95L) is a type II transmembrane protein produced by activated T cells and constitutively expressed in a variety of tissues. While the expression of CD95 is likely to be ubiquitous on activated immune cells, that of CD95L is more restricted to activated professional killer cells such as $CD8^+$ and $CD4^+$ cytotoxic T cells, NK cells, and APC.[46–48]

The *in vivo* involvement of the CD95 pathway in T cell apoptosis during HIV infection is supported by a series of observations. An increased expression of CD95 is detected in both CD4 and CD8 T lymphocytes from patients, and at the AIDS stage up to 80–90% of T cells are $CD95^+$.[31,39,49,50] This is associated in patients with the appearance of cells susceptible to CD95-induced apoptosis, whose proportion increases with disease progression.[31,34,50] CD95L is also upregulated in both CD4 and CD8 T lymphocytes from patients, which thus become possible effectors of apoptosis.[34] Finally, a significant increase in macrophage-associated CD95L is detected in lymphoid tissue from HIV^+ subjects, which is correlated with the degree of tissue apoptosis.[51] All these observations suggest that significant dysregulation of both CD95 and CD95L expression occurs in HIV infection. Experiments performed in HIV-infected chimpanzees showed that their resistance to AIDS, correlated with the lack of CD4 T cell depletion, is associated with the lack of susceptibility of their T lymphocytes to CD95-induced apoptosis, arguing for an involvement of the CD95 system in CD4 T cell depletion.[31]

Which are the possible effectors of CD95-induced apoptosis in AIDS? The CD95-based cytotoxic activity could be mediated by both activated CD4 and CD8 T cells and also by HIV-infected antigen-presenting cells (APC). The up-regulation of CD95L on CD4 T cells induced either by direct *in vitro* infection with HIV[52] or through the effect of viral proteins such as gp120, Tat or Nef[8,48] make them possible effectors in killing CD95 expressing cells. This is corroborated by the demonstration that activated CD4 T lymphocytes, expressing CD95L, can kill CD95-expressing CD8 T lymphocytes.[53] The cytotoxic function of macrophages was suggested by the observation that CD95L expression was induced on APC either as a consequence of *in vitro* HIV infection[48] or following incubation with HIV proteins, gp120 and Tat.[8,48] $CD8^+$ LAK (lymphokine-activated killer) cells were also identified as killers of HIV-infected CD4 T cells *in vitro*. However, the involvement of the CD95 system in this cytotoxicity was not investigated. In a recent study we have asked whether professional CTL, specific for HIV peptides, were potential effectors of the destruction of CD95-expressing activated lymphocytes. Indeed, an anti-Nef HLA class I restricted CTL clone, derived from an HIV-infected subject, was able to mediate both perforin- and Fas-mediated dependent cytotoxic activities on Nef-presenting target cells and Fas-expressing compliant cells, respectively.[54] Thus, in addition to being protective through the elimination of HIV-infected cells, anti-viral CTL could be deleterious through the destruction of CD95-expressing cells, abundant in HIV-infected patients because of the persistent stimulation of the immune system.

Other Cell Death Genes Involved in HIV-induced Apoptosis: Contribution of the CXCR4 Coreceptor

Several examples of CD95-independent apoptosis were recently reported in HIV infection. In an *in vitro* system of direct infection by HIV of PBMC or T cell lines, the majority of HIV-induced T cell death involved direct loss of infected cells rather than death of uninfected bystander cells, and the CD95 pathway was not involved.[55] However, because it was reported that necrosis is the major mechanism involved in the direct killing by cytopathic HIV of CD4 T cells whereas apoptosis is involved in immune cell-mediated killing,[56] it cannot be excluded that in the study by Gandhi *et al.*[55] CD4 T cells were mostly dying by necrosis, which would explain the non-involvement of an apoptosis death factor such as CD95. Another example of CD95-independent apoptosis was reported in experiments where primary uninfected CD4 T cells died of apoptosis when they were in contact with HIV-infected or HIV gp120-expressing cells. Apoptosis was blocked by inhibitors of caspases but not by CD95 or TNF-R1 molecules.[5] In fact, several recent studies suggested that, in addition to CD95L, other members of the TNF family are involved in HIV-induced apoptotic cell death. TRAIL (TNF-related-apoptosis-inducing-ligand) was identified as an apoptotic inducing factor in T cells from HIV-infected patients, but not in normal T cells even after prolonged activation *in vitro*.[57,58] Apoptosis in CD8 T cells was reported to involve the TNF/TNF-R system. Indeed, binding of HIV gp120 or SDF-1 (stromal-derived factor 1), to the chemokine receptor CXCR4, induces the upregulation of membrane TNF on macrophages and TNF-RII on peripheral CD8 T cells, leading to apoptosis of CD8 T cells.[59] The SDF-1 receceptor CXCR4, when triggered by HIV gp120, was also found to induce in normal CD4 T lymphocytes a rapid cell death, independent of known caspases and lacking oligonucleosomal DNA fragmentation, but showing several features of apoptosis. Apoptosis triggered via CXCR4 was exclusively observed in CD4 but not in CD8 T cells, was independent of CD95, and was inhibited by SDF-1.[60] The induction of apoptosis through CXCR4 by gp120 or SDF-1 was also reported on human neuronal cells, in the absence of CD4 molecule. Therefore, receptors for HIV including CD4 or CXCR4 can trigger apoptotic programs in normal T lymphocytes upon binding of HIV or its envelope.

The Dysregulation of Cytokine Synthesis in HIV Infection Is Related to the Priming of T Helper Cells for Apoptosis

Alterations in cytokine production were reported to occur in the course of HIV infection and to be associated with disease progression.[41] The synthesis of two functionally distinct families of cytokines was analyzed: type 1 cytokines (IL-2, IFNγ, IL-12, TNFα) mainly involved in cell-mediated immunity and in the destruction of intracellular parasites, and type 2 cytokines (IL-4, IL-5, IL-6, IL-10, IL-13) controlling the activation and differentiation of B cell and immunity against extracellular pathogens. In HIV-infected patients, a progressive change in the balance of type 1 vs. type 2 cytokines was proposed to contribute to susceptibility to AIDS.[41] However, depending on the methods used for cytokine detection and on the lymphoid organs analyzed, the Th1→Th2 shift has not been systematically observed. A new method of single cell analysis by flow cytometry allows the enumeration of Th1/Th2 subsets derived from peripheral T cells stimulated in short-term cultures and the obtainment of information on the number and the phenotype of cells that are potentially capable of producing a given cytokine. This approach revealed that HIV infection is associated with an alteration in the Th1 profile, characterized by a significant decrease in the number of IL-2 or TNFα-producing T cells, while the number of IFNγ producers remained normal.[42] The disappearance of IL-2-producing T cells, a good indicator of disease progression, was associated with the progressive shrinkage of the naïve $CD45RA^+CD4^+$ T cell compartment, while the persistence of IFNγ producers was associated with the expansion of the CD8 T cell compartment.[42] With that experimental approach, no increase in the proportion of Th2 cells was detected in blood T cells, although some HIV^+ patients with hyper IgE syndrome showed an increased number of $IL\text{-}13^+$ T cells. However, an increased production by patients' monocytes of type 2 cytokines such as IL-6 and IL-10 is not excluded.

Because the rate of T cell apoptosis is increased early in HIV infection, we have asked whether alterations in the representation of some Th1 subsets was the consequence of a differential susceptibility to activation-induced apoptosis. This was performed by a multiparametric flow cytometric approach,[42] combining at the single cell level the detection of surface antigens, intracellular cytokines and apoptosis. Exogeneous cytokines can modulate the susceptibility of lymphocytes to apoptosis[26] and we have found that the intrinsic capacity of lymphocytes to produce a given cytokine upon activation can also influence their survival. Indeed, T lymphocytes committed to IFNγ or TNFα production were more sensitive to activation-induced apoptosis than lymphocytes committed to IL-2-production.[42] This gradient of susceptibility to activation-induced apoptosis (IL-2 < IFNγ < TNFα) was detected in both CD4 and CD8 subsets, as well as in control donors and HIV-infected patients. The differential intrinsic apoptosis susceptibility of Th1 effectors was found to be tightly regulated by Bcl-2 expression. For example, the increased susceptibility to apoptosis of patients' IL-2 producers was found related to a decreased Bcl-2 expression. In addition, significant correlations were observed between the progressive drop in the proportion of IL-2– or TNFα-synthesizing T cells and their susceptibility to activation-induced apoptosis.[42] These observations suggest that exacerbation of PCD in HIV-infection contributes to the disappearance of Th1 effectors, and consequently to the alteration of HIV-specific cytotoxic function.

Regulation by Cytokines of HIV-dependent Apoptosis

Because cytokines can regulate the survival of activated cells, the influence of type 1 vs. type 2 cytokines was tested on T cells of HIV-infected patients. The addition of type 1 cytokine IL-2 was found to block *in vitro* spontaneous apoptosis[16] and activation-induced apoptosis[61,62] of patients' T cells. In contrast, the type 2 cytokines IL-4 and IL-10 had no effect or enhanced apoptosis. However, activation-induced apoptosis and CD95-mediated apoptosis could be blocked by antibodies against IL-4 and IL-10 and enhanced by anti-IL-12 antibodies.[61,62] Because IL-15 shares many biological properties with IL-2, we examined the effects of exogenous IL-15 on lymphocytes of HIV-infected individuals and found that it could act as a potent survival factor in the prevention of spontaneous apoptosis. This was associated with an ability to up-regulate Bcl-2 expression.[43] These observations indicate that Th1 cytokines, such as IL-2 and IL-15, are able to prevent HIV-dependent apoptosis. IL-2 probably plays a pivotal role in anti-HIV immunity through its involvement in Th and CTL functions, its ability to prevent PCD and to promote T cell activation. Its requirement for efficient control of HIV infection is suggested by studies performed in HIV-infected chimpanzees, indicating that this non-pathogenic infection is correlated with the maintenance of IL-2 producing T cells.[31] The essential role of IL-2 in the control of HIV infection is confirmed by the benefits of *in vivo* IL-2 infusions in HIV^+ patients, resulting in both a clinical and an immunological improvement, characterized by an important and stable rise in CD4 T cells.[63]

RESTORATION OF THE IMMUNE SYSTEM UNDER ANTI-RETROVIRAL THERAPY: IMPACT OF CONTROLLED PCD

The recent availability of anti-retroviral therapies that reduce viral load to undetectable levels and concomitantly increase CD4 counts will help to determine whether, in the absence of detectable virus in the blood, the immune system can regenerate. In fact, the increase in CD4 T cells is not observed in all the patients, and the functional alterations in addition to the skewed TCR repertoire of the CD4 Th subset are only partially corrected under anti-retroviral therapy.[63] The mechanisms that account for the rise in blood CD4 T cells following HAART are currently not completely understood. The initial rise of CD4 T cells would be due to the migration of memory T cells from the lymphoreticular tissues, where they are not trapped anymore by the virus, to the blood[64] and after several weeks of HAART the sustained rise in CD4 T cells would be the result of their peripheral proliferation, due both to the removal of HIV-induced suppression[65] and to the regulation of apoptosis.[33] Indeed, rapidly after initiation of HAART, an important drop in spontaneous, activation-induced and CD95-triggered apoptosis is observed in both CD4 and CD8 T cells from all treated patients[33] (FIG. 1). This occurs before the decrease in immune activation, and the resistance to CD95-induced apoptosis precedes the down-regulation of CD95 expression. Thus, suppression of the plasmatic viral load is associated with the regulation of apoptosis, which reaches the normal values detected in T cells from healthy donors. The renewal capacities of the immune system in the context of HIV infection are not known. It was proposed by Heeney[66] that the ability of chimpanzees to maintain immunological integrity in the face of persistent HIV infection was

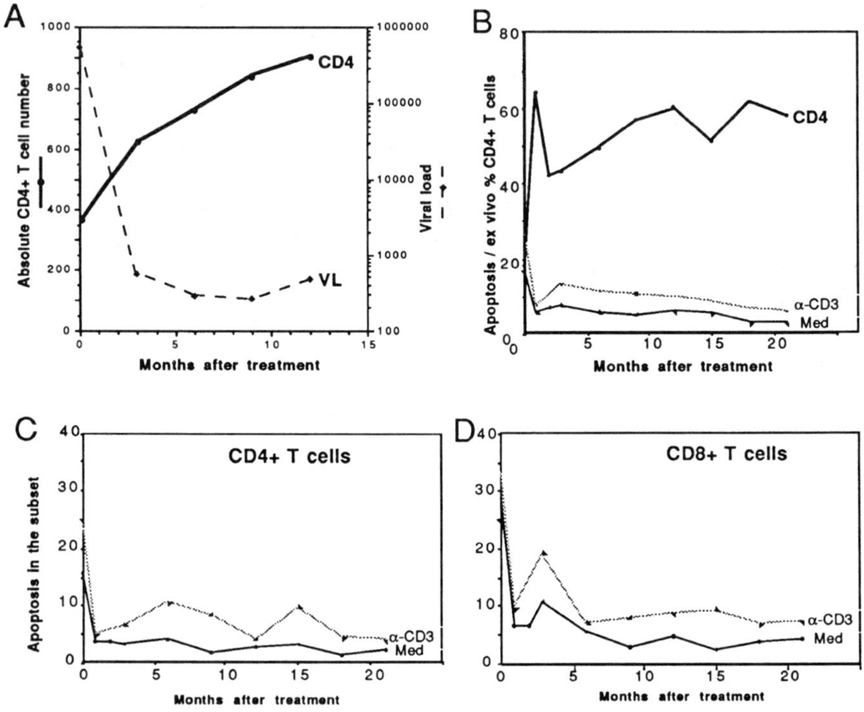

FIGURE 1. Influence of HAART on apoptosis of T lymphocytes from a representative chronic HIV-infected subject. T cell apoptosis was measured following a short-term culture of PBMC in medium or in the presence of anti-CD3 mAbs, and quantification of apoptotic cells within the T cell subsets was performed following a co-staining with 7-AAD and anti-CD4 or anti-CD8 mAbs, as described.[19] **A** shows that, following potent anti-retroviral therapy, a stable increase in the absolute number of blood CD4 T cells and a decrease in the plasmatic viral load are observed. **B** shows that the increase in CD4 T cells is concomitant with the decrease in both spontaneous and activation-induced lymphocyte apoptosis. **C** and **D** show that the negative control of apoptosis under HAART concerns both CD4 and CD8 T cell subsets. *Med*, medium only (control).

associated with the maintenance of the primary and secondary lymphoid environments important for T cell renewal, considering that they are partly destroyed in infected humans. Recent analyses of the *in vivo* ability of thymus from HIV-positive patients to compensate for HIV-dependent CD4 T cell destruction by producing new CD4 T cells argue for alterations in thymic activity in some patients, which are partly cured following HAART. More studies are needed to evaluate the renewal capacities of the immune system when HIV is neutralized.

CONCLUDING REMARKS

Homeostasis is maintained by an extremely complex set of regulatory processes that differ markedly in quiescent and activated cells, and the main role of apoptosis is to limit the clonal expansion of lymphocytes during an immune response. As discussed in this review, premature lymphocyte apoptosis in the context of HIV infection is the consequence of the continuous production of viral proteins, leading to an unbalanced immune activation and to the triggering of apoptotic programs. The chronic immune activation induces the continuous expression of death factors which could turn lymphocytes, including CD4 T cells, CD8 CTL or APC, into effectors of apoptosis, leading to the destruction of healthy activated non-infected cells, which upregulate death factors and downregulate survival factors. Thus, PCD would significantly contribute to peripheral T cell depletion in AIDS, particularly if the Th cell renewal is impaired. Under potent anti-retroviral therapies, a complete normalization of lymphocyte apoptosis is observed, concomitant with a partial restoration of the number and the functions of the immune system.

ACKNOWLEDGMENTS

The authors thank the members of the group for their contribution, particularly F. Boudet, S. Garcia, H. Lecoeur, and E. Ledru. This work was supported by grants from the Agence Nationale de Recherche sur le SIDA (ANRS), the Fondation pour la Recherche Médicale (FRM) (Sidaction), the Centre National de la Recherche Scientifique (CNRS), the Pasteur Institute and the EU (contracts ERB-IC15-CT97-0901 and BMH4-CT97-2055).

REFERENCES

1. Levy, J.A. 1998. HIV and the pathogenesis of AIDS, 2nd edit. American Society of Microbiology. Washington, DC.
2. Terai, C., R.S. Kornbluth, C.D. Pauza, D.D. Richman & D.A. Carson. 1991. Apoptosis as a mechanism of cell death in cultured T lymphoblasts acutely infected with HIV-1. J. Clin. Invest. **87:** 1710–1715.
3. Laurent-Crawford, A.G., B. Krust, S. Muller, Y. Rivière, M.A. Rey-Cuillé J.-M. Béchet, L. Montagnier & A. Hovanessian. 1991. The cytopathic effect of HIV is associated with apoptosis. Virology **185:** 829–839.
4. Gougeon, M-L., A.G. Laurent-Crawford, A.G. Hovanessian & L. Montagnier. 1993. Direct and indirect mechanisms mediating apoptosis during HIV infection: contribution to in vivo CD4 T cell depletion. Sem. Immunol. **5:**187–194.
5. Ohnimus, H., M. Heinkelein & C. Jassoy. 1997. Apoptotic cell death upon contact of $CD4^+$ T lymphocytes with HIV glycoprotein-expressing cells is mediated by caspases but bypasses CD95 (Fas/Apo-1) and TNF receptor 1. J. Immunol. **159:** 546–5252.
6. Zauli, G., D. Gibellini, A. Caputo, A. Bassini, M. Negrini, M. Monne, M. Mazzoni & S. Capitani. 1995. The HIV-1 tat protein upregulates bcl-2 gene expression in Jurkat T-cell lines and primary peripheral blood mononuclear cells. Blood **86:** 3823–3828.
7. Li, C.J., D.J. Friedman, C. Wang, V. Metelev & A.B.Pardee. 1995. Induction of apoptosis in uninfected lymphocytes by HIV-1 Tat protein. Science **268:** 429–432.

8. WESTENDORP, M.O. *et al.* 1995. Sensitization of T cells to CD95-mediated apoptosis by HIV-1 Tat and gp120. Nature **375:** 497–501.
9. AILLET, F. *et al.* 1998. HIV induces a dual regulation of Bcl-2, resulting in persistent infection of $CD4^+$ T or monocytic lines. J. Virol. **72:** 9698–9705.
10. STEWART, S.A, B. POON, J.B.M. JOWETT & I.S.Y. CHEN. 1997. HIV-1 Vpr induces apoptosis following cell cycle arrest. J. Virol. **71:** 5579.
11. YAO, X *et al.* 1998. Vpr stimulates viral expression and induces cell killing in HIV-1 infected dividing Jurkat T cells. J. Virol. **72:** 4686–4693.
12. CASELLA, C.R., RAPAPORT E.L. & T.H. FINKEL. 1999. Vpu increases susceptibility of HIV-1 infected cells to Fas killing. J. Virol. **73:** 92–100.
13. AMEISEN, J.C. & A. CAPRON. 1991. Cell dysfunction and depletion in AIDS: the program cell death hypothesis. Immunol. Today **12:** 102–105.
14. GOUGEON, M-L. & L. MONTAGNIER. 1993. Apoptosis in AIDS. Science **260:** 1269–1270.
15. GOUGEON, M-L. *et al.* 1991. Evidence for an engagement process towards apoptosis in lymphocytes of HIV infected patients. C. R. Acad. Sci. **312:** 529–537.
16. GOUGEON, M-L. *et al.*1993. Programmed cell death of T lymphocytes in AIDS related HIV and SIV infections. AIDS Res. Hum. Retrov. **9:** 553–563.
17. GROUX, H. *et al.* 1992. Activation-induced death by apoptosis in lymphocytes from human immunodeficiency virus-infected asymptomatic individuals. J. Exp. Med. **175:** 331–340.
18. MEYAARD, L. *et al.* 1992. Programmed death of T cells in HIV-1 infection. Science **257:** 217–219.
19. GOUGEON, M-L. *et al.* 1996. Programmed cell death in peripheral lymphocytes from HIV-infected persons: the increased susceptibility to apoptosis of CD4 and CD8 T cells correlates with lymphocyte activation and with disease progression. J. Immunol. **156:** 3509–.
20. LEWIS, D.E., D.S. TANG, A. ADU-OPPONG, W. SCHOBER & J.R. RODGERS 1994. Anergy and apoptosis in $CD8^+$ T cells from HIV-infected persons. J. Immunol. **153:** 412.
21. MURO-CACHO, C.A., G. PANTALEO & A. FAUCI. 1995. Analysis of apoptosis in lymph nodes of HIV-infected persons. Intensity of apoptosis correlates with the general state of activation of the lymphoid tissue and not with the stage of disease or viral burden. J. Immunol. **154:** 5555.
22. AMENDOLA, A., M-L. GOUGEON, F. POCCIA, A. BONDURAND, L. FESUS & M. PIACENTINI. 1996. Induction of tissue transglutaminase in HIV pathogenesis. Evidence for a high rate of apoptosis of CD4 T lymphocytes and accessory cells in lymphoid tissues. Proc. Natl. Acad. Sci. USA **93:** 11057.
23. ROSOK, J.E. *et al.* 1998. Correlates of apoptosis of $CD4^+$ and $CD8^+$ T cells in tonsillar tissue in HIV type 1 infection. AIDS Res. Hum. Retrov. **14:** 1635–1643.
24. GOUGEON, M-L. 1995. Does apoptosis contribute to CD4 T cell depletion in HIV infection? Cell Death Differ. **2:** 1–6.
25. FINKEL, T.H. *et al.* 1995. Apoptosis occurs predominantly in bystander cells and not in productive cells of HIV- and SIV-infected lymph nodes. Nature Med. **1:** 129.
26. AKBAR, A.N. & M. SALMON. 1997. Cellular environments and apoptosis: tissue microenvironments control activated T-cell death. Immunol. Today **18:** 72.
27. GIORGI, J.V., Z. LIU, L.E. HULTIN, W.G. CUMBERLAND, K. HENNESSEY & R. DETELS 1993. Elevated level of $CD38^+CD8^+$ cells in HIV infection add to the prognostic value of low $CD4^+$ T cell level: results of 6 years follow-up. J. Acquired Immune Deficiency Syndrome **6:** 904.
28. BLANCHARD, A., L. MONTAGNIER & M-L. GOUGEON. 1997. Influence of microbial infections on the progression of HIV disease. Trends Microbiol. **5:** 326–331.

29. MICHEL, P., A. TOURE-BALDE, C. ROUSSILHON, G. ARIBOT, J.L. SARTHOU & M-L. GOUGEON. 1999. Reduced immune activation and T cell apoptosis in HIV-2 compared to HIV-1 infected West African patients. Correlation of T cell apoptosis with seric β2 microglobulin and with disease evolution. J. Infect. Dis. In press.
30. HEENEY, J. *et al.* 1993. The resistance of HIV-infected chimpanzees to progression to AIDS correlates with absence of HIV-related dysfunction. J. Med. Primatol. **22:** 194.
31. GOUGEON, M-L. *et al.* 1997. Lack of chonic immune activation in HIV-infected chimpanzees correlates with the resistance of T cells to Fas/Apo-1 (CD95)-induced apoptosis and preservation of a Th1 phenotype. J. Immunol. **158:** 2964.
32. BÖHLER, T., C. BÄUMLER, I. HERR, A. GROLL, E. KURZ & K-M. DEBATIN. 1997. Activation of the CD95 system increases with disease progression in HIV1-infected children and adolescents. Pediatr. Infect. Dis. J. **16:** 754.
33. GOUGEON, M-L., H. LECOEUR & Y. SASAKI. 1999. The CD95 system in HIV infection. Impact of HAART. Immunol. Lett. **66:** 97–103.
34. SLOAND, E.M. *et al.* 1997 Role of Fas ligand and receptor in the mechanism of T-cell depletion in AIDS: effect on $CD4^+$ lymphocyte depletion and HIV replication. Blood **89:** 135.
35. LIEGLER, T.J. *et al.* 1998. Diminished spontaneous apoptosis in lymphocytes from HIV-infected long term non progressors. J. Infect. Dis. **178:** 669–679.
36. ESTAQUIER, J.T. *et al.* 1994. Programmed cell death and AIDS: significance of T cell apoptosis in pathogenic and nonpathogenic primate lentiviral infections. Proc. Natl. Acad. Sci. USA **91:** 9431.
37. HOLZNAGEL, E. *et al.* 1998. The role of in vitro–induced lymphocyte apoptosis in feline immunodeficiency virus infection: correlation with different markers of disease progression. J. Virol. **72:** 9025–9033.
38. DAVIS, I.C., M. GIRARD & P. FULTZ. 1998. Loss of $CD4^+$ T cells in HIV-1 infected chimpanzees is associated with increased lymphocyte apoptosis. J. Virol. **72:** 4623–4632.
39. BOUDET, F., H. LECOEUR & M-L. GOUGEON. 1996. Apoptosis associated with ex vivo down-regulation of bcl-2 and up-regulation of Fas in potential cytotoxic $CD8^+$ T lymphocytes during HIV infection. J. Immunol. **156:** 2282.
40. BOFILL, M. *et al.* 1995. Presence of $CD3^+CD8^+$Bcl-2low lymphocytes undergoing apoptosis and activated macrophages in lymph nodes of HIV-1^+ patients. Am. J. Pathol. **146:** 1542.
41. CLERICI, M. & G.M. SHEARER. 1994. The Th1/Th2 hypothesis of HIV infection: new insights. Immunol. Today **15:** 575–580.
42. LEDRU, E., H. LECOEUR, S. GARCIA, T. DEBORD & M-L. GOUGEON. 1998. Differential susceptibility to activation-induced apoptosis among peripheral Th1 subsets: correlation with Bcl-2 expression and consequences for AIDS pathogenesis. J. Immunol. **160:** 3194–3206.
43. NAORA, H. & M-L. GOUGEON. 1999. IL-5 is a potent survival factor in the prevention of spontaneous but not CD95-induced apoptosis in CD4 and CD8 T lymphocytes of HIV-infected individuals. Correlation with its ability to increase Bc1-2 expression. Cell Death Diff. In press.
44. NAGATA, S. 1997. Apoptosis by death factor. Cell **88:** 355.
45. PETER, M.E. & P.H. KRAMMER. 1998. Mechanisms of CD95 (APO-1/Fas)-mediated apoptosis. Curr. Opin. Immunol. **10:** 545–551.
46. OSHIMI, Y., S. ODA, Y. HONDA, S. NAGATA & S. MIYAZAKI. 1996. Involvement of Fas ligand and Fas-mediated pathway in the cytotoxicity of human natural killer cells. J. Immunol. **157:** 2909.
47. SUDA, T. *et al.* 1995. Expression of the Fas ligand in cells of T cell lineage. J. Immunol. **154:** 3806.
48. BADLEY, A.D. *et al.* 1996. Upregulation of Fas ligand expression by human immunodeficiency virus in human macrophages mediates apoptosis of uninfected T lymphocytes. J. Virol. **70:** 199.

49. DEBATIN, K-M. *et al.* 1994. High expression of APO-1 (CD95) on T lymphocytes from HIV-infected children. Blood **83:** 3101–.
50. KATSIKIS, P.D. *et al.* 1995. Fas antigen stimulation induces marked apoptosis of T lymphocytes in human immunodeficiency virus-infected individuals. J. Exp. Med. **181:** 2029–.
51. DOCKRELL, D.H. *et al.* 1998. The expression of Fas ligand by macrophages and its upregulation by HIV infection. J. Clin. Invest. **101:** 2394–2405.
52. MITRA, D. *et al.* 1996. HIV-1 upregulates Fas ligand expression in $CD4^+$ T cells in vitro and in vivo: association with Fas-mediated apoptosis and modulation by aurintricarboxylic acid. Immunology **87:** 581–.
53. PIAZZA, C. *et al.* 1997. $CD4^+$ T cells kill $CD8^+$ T cells via Fas/Fas ligand-mediated apoptosis. J. Immunol. **158:** 1503.
54. GARCIA, S., M. FEVRIER, G. DADAGLIO, H. LECOEUR, Y. RIVIERE & M-L. GOUGEON. 1997. Potential deleterious effect of anti-viral cytotoxic lymphocytes through the CD95 (Fas/APO-1)-mediated pathway during chronic HIV infection. Immunol. Lett. **57:** 53.
55. GANDHI, R.T., B.K. CHEN, S.E. STRAUS, J.K. DALE, M.J. LENARDO & D. BALTIMORE. 1998. HIV-1 directly kills $CD4^+$ T cells by a Fas-independent mechanism. J. Exp. Med. **187:** 1113–1122.
56. WANG, L.Q. *et al.* 1998. Apoptotic killing of $CD4^+$ T lymphocytes in HIV-1 infected PHA-stimulated PBL cultures is mediated by $CD8^+$ LAK cells. Virology **241:** 169–180.
57. KATSIKIS, P.D. *et al.* 1997. Interleukin-1-beta converting enzyme-like protease involvement in Fas-induced and activation-induced peripheral blood T cell apoptosis in HIV infection-TNF-related apoptosis-inducing ligand can mediate activation-induced T cell death in HIV infection. J. Exp. Med. **186:** 1365–1372.
58. JEREMIAS, I., I. HERR, T. BOEHLER & K-M. DEBATIN 1998. TRAIL/Apo2-ligand-induced apoptosis in human T cells. Eur. J. Immunol. **28:** 143–152.
59. HERBEIN, G. *et al.* 1998. Apoptosis of $CD8^+$ T cells is mediated by macrophages through interaction of HIV gp120 with chemokine receptor CXCR4. Nature **395:** 189–194.
60. BERNDT, C., B. MOPPS, S. ANGERMULLER, P. GIERSCHIK & P.H. KRAMMER. 1998. CXCR4 and CD4 mediate a rapid CD95-independent cell death in CD4 T cells. Proc. Natl. Acad. Sci. USA **95:** 12556–12561.
61. CLERICI, M. *et al.* 1994. Type 1/type 2 cytokine modulation of T cells programmed cell death as a model for HIV pathogenesis. Proc. Natl. Acad. Sci. USA **91:** 11811–11817.
62. ESTAQUIER, J. *et al.* 1995. Th1/Th2 cytokines and T cell death: preventive effect of IL-12 on activation-induced and CD95 (Fas/APO-1)-mediated apoptosis of $CD4^+$ T cells from HIV-infected persons. J. Exp. Med. **182:** 1759–1765.
63. CONNORS, M. *et al.* 1997. HIV infection induces changes in $CD4^+$ T cell phenotype and depletions within the $CD4^+$ T cell repertoire that are not immediately restored by antiviral or immune based therapies. Nature Med. **5:** 533.
64. PAKKER, *et al.* 1998 Biphasic kinetics of peripheral blood T cells after triple combination therapy in HIV-1 infection: a composite redistribution and proliferation. Nature Med. **4:** 208–214.
65. HELLERSTEIN, M. *et al.* 1999. Directly measured kinetics of circulating T lymphocytes in normal and HIV-infected humans. Nature Med. **5:** 83–89.
66. HEENEY, J.L. 1995. AIDS: a disease of impaired Th-cell renewal? Immunol. Today **16:** 515.

Index of Contributors